Listeria monocytogenes:
Pathogenesis and Host Response

Listeria monocytogenes:
Pathogenesis and Host Response

Edited by

Howard Goldfine
University of Pennsylvania
Philadelphia, PA, USA

and

Hao Shen
University of Pennsylvania
Philadelphia, PA, USA

 Springer

Howard Goldfine
University of Pennsylvania
School of Medicine
Department of Microbiology
Philadelphia, PA 19104-6076
USA

Hao Shen
University of Pennsylvania
School of Medicine
Department of Microbiology
Philadelphia, PA 19104-6076
USA

Library of Congress Control Number: 2006939779

ISBN-10: 0-387-49373-5   e-ISBN-10: 0-387-49376-X
ISBN-13: 978-0-387-49373-2   e-ISBN-13: 978-0-387-49376-3

Printed on acid-free paper.

Printed in the United States of America.

9 8 7 6 5 4 3 2 1

springer.com

Preface

We are at the beginning of the third decade of studies at the molecular level on the pathogenesis of *Listeria monocytogenes* and the response of the host to its infections. It is a good time to survey the wealth of information that these studies have revealed and to think about perspectives for a more complete understanding of this important pathogen. During the past 20 years, *L. monocytogenes* has emerged from relative obscurity to being one of the most intensely studied bacterial pathogens. In the opening chapter, Daniel A. Portnoy provides a personal and historical account of the development of our understanding at the molecular level of invasion of the host cell, growth in the cytosol, and cell-to-cell spread by means of actin polymerization-powered motility. We are also fortunate to have contributions from Pascale Cossart and Werner Goebel and their colleagues. These pioneering investigators continue to make major contributions to the molecular description of almost every aspect of *L. monocytogenes* pathogenesis. All chapters in this book have been written by experts who have contributed widely to this field. We are extremely grateful for the efforts of all contributors who have provided contemporary accounts of the status of various aspects of *L. monocytogenes* pathogenesis and the host response. Their positive responses to this effort are deeply appreciated.

Any specialist volume in the biological sciences can only provide a snapshot of the field at the time of its publication. In 2001 the complete genomes of *L. monocytogenes* and *L. innocua* were published, and these have provided considerable impetus for ongoing studies that have revealed the importance of proteins beyond those encoded by genes in the PrfA virulence cluster which was described at the beginning of the last decade. We currently anticipate the completion of the genomes of 19 additional strains at the Broad Institute, which will provide a wealth of information about strains that are most often involved in local epidemics and those that are likely to be found in the production chain of foods for human consumption.

Although research over the past 20 years has greatly enlarged our understanding of the exquisite balance achieved by *L. monocytogenes* in promoting its life in the environment and in the tissues of its hosts, we are still a long way from a complete understanding of these mechanisms. We continue to learn more about the transcriptional and translational controls of expression of proteins important for life in these two very different environments. As we learn more, new windows are constantly being opened into the complexity of host cell biology and the interplay of the signals connecting the various cells and organs involved in the host response. Indeed, as revealed in many of the chapters in this book, the study of *L. monocytogenes* has already provided major insights into eukaryotic cell biology. We are hopeful that this volume will attract new investigators into the field and stimulate future studies on *Listeria* that will have similar broadening and cascading effects in the biological sciences.

Howard Goldfine and Hao Shen
Philadelphia
September 2006

Contents

Contributors

Hélène Bierne
Institut Pasteur
Unité des Interactions Bactéries-Cellules
Institut Pasteur, INSERM U604, INRA USC2020
75724 Paris cedex 15, France

Kathryn J. Boor
Cornell University
Department of Food Science
Ithaca, NY 14853

Carmen Buchrieser
Institut Pasteur
Unité de Génomique des Microorganisme Pathogènes — URACNRS 2171
75724 Paris, France

Pascale Cossart
Institut Pasteur
Unité des Interactions Bactéries-Cellules
Paris, F-75015 France

Nancy Freitag
Seattle Biomedical Research Institute
University of Washington
Seattle, WA 98109-5219, USA

Francisco García-del Portillo
Departamento de Biotecnología Microbiana
Centro Nacional de Biotecnología-
Consejo Superior de Investigaciones Científicas (CSIC)
Darwin 3, 28049 Madrid, Spain

Philippe Glaser
Institut Pasteur
Unité de Génomique des Microorganisme Pathogènes - URACNRS 2171
75724 Paris, France

Werner Goebel
Theodor-Boveri-Institut (Biozentrum)
Lehrstuhl für Mikrobiologie
Universität Würzburg
D-97074 Würzburg, Germany

Howard Goldfine
Department of Microbiology
University of Pennsylvania School of Medicine
Philadelphia, PA 19104-6076, USA

Steven Hagens
Institute of Food Science and Nutrition
ETH Zurich, Switzerland

John T. Harty
University of Iowa
Department of Microbiology
Iowa City, IA 52242, USA

Biju Joseph
Theodor-Boveri-Institut (Biozentrum)
Lehrstuhl für Mikrobiologie
Universität Würzburg
D-97074 Würzburg, Germany

Martin J. Loessner
Institute of Food Science and Nutrition
ETH Zurich, Switzerland

Bennett Lorber
Professor of Medicine, Section of Infectious Diseases
Temple University School of Medicine and Hospital
Philadelphia, PA 19140, USA

Hélène Marquis
Dept. of Microbiology and Immunology
Cornell University
Ithaca, NY 14853, USA

Kelly A.N. Messingham
University of Iowa
Department of Microbiology
Iowa City, IA 52242, USA

Maurine D. Miner
Seattle Biomedical Research Institute
University of Washington
Seattle, WA 98109-5219, USA

Haley F Oliver
Cornell University
Department of Food Science
Ithaca, NY 14853, USA

Javier Pizarro-Cerdá
Institut Pasteur
INSERM, U604, Paris, F-75015 France; INRA, USC2020
Paris, F-75015 France

Gary C. Port
Seattle Biomedical Research Institute
University of Washington
Seattle, WA 98109-5219, USA

Daniel A. Portnoy
Depart. of Molecular and Cell Biology
University of California
Berkeley, CA 94729-3202, USA

M. Graciela Pucciarelli
Departamento de Biotecnología Microbiana
Centro Nacional de Biotecnología-
Consejo Superior de Investigaciones Científicas (CSIC)
Darwin 3, 28049 Madrid, Spain

Christophe Rusniok
Institut Pasteur
Unité de Génomique des Microorganisme Pathogènes—URACNRS 2171
75724 Paris, France

Hao Shen
Department of Microbiology
University of Pennsylvania School of Medicine
Philadelphia, PA 19104-6076, USA

Jörg Slaghuis
Theodor-Boveri-Institut (Biozentrum)
Lehrstuhl für Mikrobiologie
Universität Würzburg
D-97074 Würzburg, Germany

Matthew D. Welch
Department of Molecular & Cell Biology
University of California
Berkeley, CA 94720-3200, USA

Martin Wiedmann
Cornell University
Department of Food Science
Ithaca, NY 14853, USA

Lauren A. Zenewicz
Section of Immunobiology
Yale University School of Medicine
New Haven, CT 06520, USA

1
A 20-Year Perspective on *Listeria monocytogenes* Pathogenesis

Daniel A. Portnoy

Department of Molecular and Cell Biology, University of California, Berkeley, CA 94729-3202, USA
e-mail: portnoy@berkeley.edu

1.1. Introduction

Listeria monocytogenes has attracted the attention of a diverse group of investigators, including clinicians, food microbiologists, immunologists, and medical microbiologists. The reason for this broad degree of interest is due, in large part, to the fact that this facultative intracellular pathogen is highly amenable to experimental analysis and has a broad range of relevant biologic activities ranging from its growth in the environment, infection of many different animal species, and as an important human pathogen. Consistent with its broad host range, *L. monocytogenes* infects rodents and is perhaps the most well-characterized bacterial pathogen in a murine model of infection; indeed, *L. monocytogenes* is a darling of the immunologists due to its ease of handling and rapidity of growth, dating back 45 years to the classic work of Mackaness (1962). It took another 25 years for the bacterial pathogenesis community to become enamored with this important pathogen spurred by the realization in the 1980s that *L. monocytogenes* represents a serious public health threat especially to pregnant women (Chap. 2).

1.2. The Beginning of the Modern Era: A Personal Recollection

I began working on *L. monocytogenes* in the fall of 1986, shortly after starting my own laboratory at Washington University in St Louis. I should mention that Pascale Cossart also entered the field about the same time (Mengaud et al. 1987) and remains a monumental presence contributing to all aspects of *L. monocytogenes* pathogenesis. In 1986, basic research on *L. monocytogenes* was mostly in the realm of immunology while ignored by the bacterial pathogenesis community. I was influenced by a number of immunologists, including David Hinrichs in Oregon, Emil Unanue at Washington University, and Ed Havell and Robert North at Trudeau Institute, in the belief that *L. monocytogenes* pathogenesis might represent a fertile area of research. At that time, it was understood that *L. monocytogenes* is a facultative intracellular pathogen, but we knew virtually nothing about its cell biology of infection, bacterial determinants of pathogenesis, and lacked basic tools for genetic analysis. However, in 1986, two seminal papers were published: Havell (1986) showed that *L. monocytogenes* enters fibroblasts, spreads to neighboring cells, and induces high levels of type 1 interferon, while Philippe Sansonetti's lab showed that transposon mutagenesis could be used effectively in *L. monocytogenes* to isolate hemolysin-negative mutants (Gaillard et al. 1986). A second paper from Werner Goebel's lab was published the following year, also using a conjugative transposon to isolate hemolysin-negative mutants (Kathariou et al. 1987). I used similar approaches in my new lab and also isolated hemolysin-negative mutants (Portnoy et al. 1988). Our hypothesis was that the cholesterol-dependent hemolysin, listeriolysin O (LLO), was essential for intracellular growth; indeed, this is the case in the vast majority of cultured cells. However, as luck would have it, the first cell line that we examined was Henle 407 cells, and contrary to our hypothesis, LLO-minus mutants grew fine. Fortunately, we examined a number of primary and cultured cells and soon appreciated that infection of Henle 407 cells represented an exception to the rule that LLO is essential for escape from a vacuole. It turns out that LLO-minus mutants escape from a vacuole, grow, and spread cell-to-cell in human epithelial cells such as HeLa. Later, we and others showed that the broad range phospholipase C (PLC), PlcB, was necessary in these cell types (Marquis et al. 1995; Grundling et al. 2003), but in most cell types, LLO is essential while the PLCs also contribute. A couple of years later, we cloned and expressed LLO in *Bacillus subtilis*, and to our amazement, *B. subtilis* escaped from a vacuole and grew intracellularly (Bielecki et al. 1990). Thus, not only is LLO required to mediate escape, it was sufficient!

It did not take long to notice that *L. monocytogenes* grew readily inside of cultured cells and spread directly from cell to cell even in the presence of high levels of gentamicin. Remarkably, a single cell could be infected, and by 8 h there were 10 cells infected. A clue that led to our current understanding regarding the mechanism of cell-to-cell spread was provided to me by Larry Hale, who worked on *Shigellae*. *Shigellae flexneri* also spreads cell to cell and Larry told me that it could be blocked by cytochalasin D (Pal et al. 1989). Sure enough, cytochalasin D completely blocked the capacity of *L. monocytogenes* to spread cell to cell (Tilney and Portnoy 1989), implicating a role for actin polymerization. This observation was well known in the *Shigella* field, and Sansonetti's group showed that *S. flexneri* was coated with filamentous actin while mutants defective in cell-to-cell spread were not (Bernardini et al. 1989). Sasakawa's group in Japan made similar observations as well (Lett et al. 1989).

My lab moved to the University of Pennsylvania in 1988 where I began to collaborate with the larger-than-life actin cell biologist Lew Tilney. Using electron microscopy, Lew documented all of the stages currently associated with the cell biology of infection (Tilney and Portnoy 1989) (Figure 1.1.). A similar

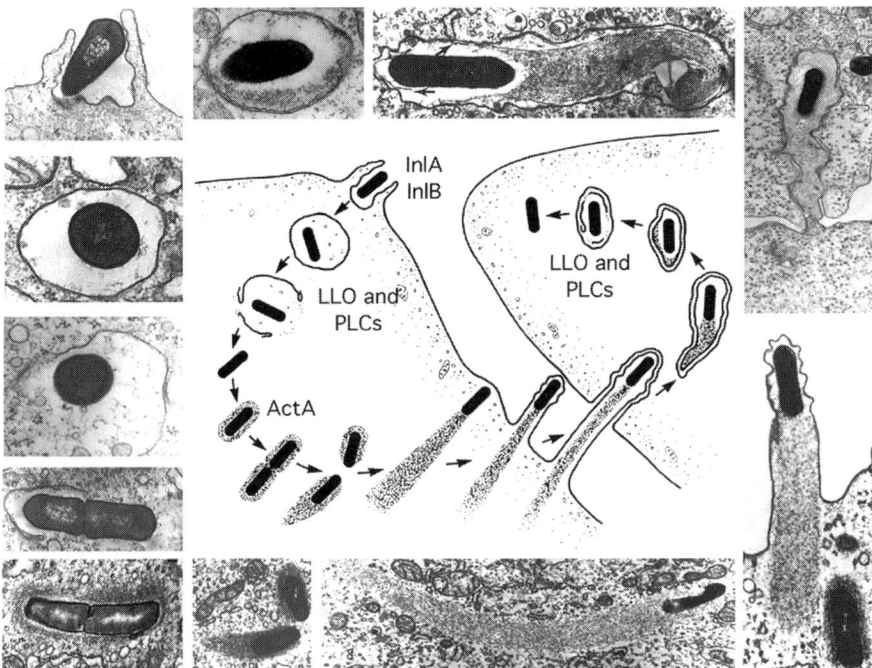

FIGURE 1.1. Stages in the intracellular life cycle of *L. monocytogenes*. *Center*: Cartoon depicting entry, escape from a vacuole, actin nucleation, actin-based motility, and cell-to-cell spread. *Outside*: Representative electron micrographs from which the cartoon was derived. LLO, PLCs, and ActA are all described in the text. Reproduced from *The Journal of Cell Biology*, 2002, 158:409–14; copyright 2002; The Rockefeller University Press.

study was also published from the Sansonetti lab about the time (Mounier et al. 1990), followed by a study by Fred Southwick and Joe and Jean Sanger that used video microscopy to show movement of *L. monocytogenes* in cultured cells (Dabiri et al. 1990). These studies caught the attention of the cell biology community attracting a number of investigators previously interested in the actin cytoskeleton, including Julie Theriot and Matt Welch at Tim Mitchison's lab in UCSF, Jurgen Wehland in Germany, and Marie-France Carlier in France. Indeed, *L. monocytogenes* became a model system with which to study actin-based motility. Lew Tilney et al. (1990) postulated that these studies could lead to the discovery of the actin nucleator, and indeed, the *L. monocytogenes* ActA protein was later shown to activate the actin nucleation properties of the Arp2/3 complex (Welch et al. 1998). Julie Theriot and Tim Mitchison were the first to reconstitute listerial actin-based motility in cell extracts (Theriot et al. 1994), and Marie-France Carlier later reconstituted actin-based motility in vitro using purified components (Loisel et al. 1999).

In 1988, I read an article by David Baltimore urging the scientific community to consider it our responsibility to work on AIDS and/or AIDS vaccines. Since we had just learned that *L. monocytogenes* entered the host cell cytosol, it was obvious that *L. monocytogenes* might be an efficient vector for the induction of cell-mediated immunity, possibly leading to an AIDS vaccine. We thus embarked on a collaborative project with Yvonne Paterson at Penn where we showed that *L. monocytogenes* could be engineered to express and secrete a viral antigen resulting in the induction of antigen-specific CD8+ T cells in vivo (Ikonomidis et al. 1994). Other groups of investigators went on to show the efficacy of *L. monocytogenes* as a live vaccine vector (Goossens et al. 1995; Shen et al. 1995), and later that *L. monocytogenes* vaccines could be used therapeutically to treat tumors as well (Pan et al. 1995). Two biotech companies are currently embarking on clinical trials to test the safety and efficacy of these vaccines in humans, and there has been some progress using *L. monocytogenes*-based vectors for AIDS vaccines (Paterson and Johnson 2004). In this chapter, I will continue with a personal perspective, summarize the current status of the field, and discuss future prospects and unanswered questions.

1.3.   Natural History of Infection

To fully appreciate the pathogenic strategies used by *L. monocytogenes*, or any pathogen for that matter, it is useful to consider how its mechanisms of pathogenesis promote dissemination. *L. monocytogenes* is ubiquitous in nature and associates closely with animals that feed on plant material (Chap. 6). Human infection is generally traced to ingestion of contaminated ready-to-eat food (Chap. 2) but plays little or no role in the natural history of infection. There are two models that may relate listerial pathogenesis to natural history. First, it is reasonable to assume that the gastrointestinal tract is the primary site of bacterial replication, with invasion, intracellular growth, and cell-to-cell spread leading to increased numbers prior to excretion. In this model, infections of the brain or fetus, although clinically relevant, contribute little to dissemination. The second model is that *L. monocytogenes* is primarily a pathogen of pregnant animals causing miscarriage as a means of dissemination. Indeed, the placenta appears to be a highly permissive environment during experimental listeriosis (Hamrick et al. 2003; Bakardjiev et al. 2005). In the second model, dissemination occurs as a result of miscarriage. Recent work from our lab suggests that miscarriage may reflect a host defense to eliminate an uncontrollable nidus of infection (Bakardjiev et al. 2006). It is also possible that gastrointestinal multiplication, excretion, and miscarriage play significant roles in the natural history of infection.

1.4.   Genetics, Genomics, and Gene Regulation

Transposon mutagenesis has been a highly successful approach to identify *L. monocytogenes* determinants of pathogenesis. Indeed, most of the genes located in the PrfA regulon were identified by transposon mutagenesis or by sequencing of genes adjacent to transposon insertion (Portnoy et al. 1992). Use of bacteriophages for genetic analysis (Hodgson 2000) and for the design of phage-based integration vectors (Lauer et al. 2002) has also advanced the field significantly (Chap. 13), as has the development of inducible systems for gene expression (Dancz et al. 2002).

However, with the completion of the *L. monocytogenes* genome sequence, postgenomic approaches including reverse genetics, proteomics, and bacterial microarrays have begun to contribute to the analysis of pathogenesis (Chap. 3). Use of microarrays to examine bacterial gene expression inside of host cells has just begun to be exploited (Chatterjee et al. 2006; Hain et al. 2006; Shen et al. 2006).

1.5.   Models of Infection

One of the major reasons for the advances in the analysis of *L. monocytogenes* pathogenesis has been the development of quantitative tissue culture models of infection using a variety of cultured cells. Unlike many other facultative intracellular pathogens, infection with wild-type *L. monocytogenes* does not result in the death of infected cells until late in infection. Indeed, *L. monocytogenes* mutants that prematurely kill infected host cells, due to a variety of LLO mutations, are avirulent (Glomski et al. 2003). Thus, *L. monocytogenes* has evolved mechanisms to avoid killing its host cell.

Intravenous infection in mice remains the most common animal model to examine infection biology (Chap. 11) and offers very quantitative and reproducible measure of virulence. The murine oral model has been explored but is still unsatisfactory. Pizarro-Cerdá and Cossart (Chap. 8) argue that the oral model is inadequate because mice have a mutation within E-cadherin that prevents the activity of InlA. Thus the guinea pig may be a better model to examine interactions that require E-cadherin (Lecuit et al. 2001). Cossart has shown that E-cadherin is found on the basal and apical plasma membranes of human syncytiotrophoblasts and in villous cytotrophoblasts suggesting that InlA-mediated internalization may lead to placental infection in humans (Lecuit et al. 2004). Indeed, InlA is necessary for the invasion of human trophoblasts (Bakardjiev et al. 2004; Lecuit et al. 2004). However, the InlA mutant has no defect in the pregnant guinea pig model of listeriosis (Bakardjiev et al. 2004). Perhaps there are multiple pathways leading to infection of the placenta.

1.6.   Cell Biology of Infection

1.6.1.   Invasion

Listeria monocytogenes is internalized by the vast majority of adherent cells examined, although the efficiency of uptake can vary by four orders of magnitude.

Although there is a large family of internalin molecules, so far, uptake can be traced to two members of the internalin family (Chap. 8). InlA is necessary for invasion of epithelial cells while InlB is necessary for invasion of hepatocytes. Interestingly, InlB mutants show much less liver toxicity during infection (Brockstedt et al. 2004). The roles of other internalins are not yet appreciated, but it is tempting to speculate that they confer invasion in other animals or in other cell types.

1.6.2. LLO and Escape from Vacuoles

LLO is an essential determinant of *L. monocytogenes* pathogenesis that is largely necessary for escape from the primary vacuole that results upon internalization and the secondary vacuole that results from cell-to-cell spread. The two PLCs (PlcA and PlcB) also contribute to escape from both the primary and the secondary vacuoles (Chap. 9). The precise mechanism of escape has remained elusive, but it seems that LLO has two roles: First, by inserting into a phagosome, LLO may prevent its maturation (Cheng et al. 2005; Henry et al. 2006; Shaughnessy et al. 2006). Secondly, LLO probably acts as a translocation pore for one or both PLCs that may activate host-signaling pathways resulting in vacuolar lysis (Wadsworth and Goldfine 2002). How this leads to disruption of the phagosome is still not understood. The observation that LLO is dispensable in human epithelial cells remains puzzling.

LLO is also essential during cell-to-cell spread (Dancz et al. 2002), and the two PLCs contribute significantly as well (Marquis et al. 1997). Marquis has shown that pro-PlcB is synthesized during intracellular growth and retained within the bacterial cell wall but released upon acidification that presumably occurs during cell-to-cell spread (Marquis and Hager 2000). Metalloprotease (Mpl) also plays a role in the regulated processing and secretion of PlcB (Yeung et al. 2005). LLO mRNA is expressed during intracellular growth, but there is translational regulation that prevents LLO synthesis probably until bacteria enter a vacuole (Schnupf et al. 2006). How the regulated translation of LLO and the activation of the PLCs are coordinated in vivo to mediate cell-to-cell spread represents an important area of future research.

1.6.3. Actin-Based Motility

The capacity of *L. monocytogenes* to exploit a host system of actin-based motility (Chap. 10) to promote its own movement represents one of the best examples of pathogen exploitation of host function. The bacteria need only express ActA, and the host does the rest. This has been a really exciting and fruitful area of research with contributions from many labs. First, ActA was identified as being necessary for the actin-based motility of *L. monocytogenes* (Domann et al. 1992; Kocks et al. 1992; Brundage et al. 1993). Secondly, ActA was shown to be sufficient for actin-based motility (Pistor et al. 1994; Friederich et al. 1995; Smith et al. 1995). Thirdly, the ActA protein was shown to bind to members of the Ena/VASP family

(Pistor et al. 1995). The precise role of Ena/VASP has been difficult to assign as ActA mutants lacking the conserved Ena/VASP recognition sequence still nucleate actin filaments. However, it appears that Ena/VASP proteins contribute by controlling temporal and spatial persistence of bacterial actin-based motility (Auerbuch et al. 2003). Finally, ActA was shown to be a nucleation-promoting factor by binding to and activating the Arp2/3 complex (Welch et al. 1998). Although ActA was the first nucleation-promoting complex to be discovered, it was soon appreciated that mammalian WASP proteins are nucleation-promoting factors and that ActA and WASP share sequence and functional identity (Welch and Mullins 2002).

It is generally assumed that ActA-mediated cell-to-cell spread allows the bacteria to continue to multiply without contacting the extracellular milieu. However, the mechanism of cell-to-cell spread still remains elusive other than one beautiful paper describing the morphological events (Robbins et al. 1999).

1.7.   Relating Pathogenesis to Immunity

There are nearly 3,000 PubMed citations that include *Listeria* and immunity, and the field is nicely summarized in Chaps. 11 and 12. However, one of the most fundamental observations remains that neither heat-killed nor LLO-minus mutants provide protective immunity (von Koenig et al. 1982; Berche et al. 1987; Lauvau et al. 2001). The answer lies at the interface of listerial pathogenesis and cell-mediated immunity (Portnoy et al. 2002). First, most of the known antigens recognized by CD8+ T-cells are derived from secreted proteins. Killed bacteria certainly do not secrete and contain less of these secreted proteins. Similarly, LLO-minus mutants fail to grow and hence probably secrete less of the antigens. Secondly, wild-type bacteria enter the cytosol of cells where they grow and secrete antigens that can readily be processed and presented on MHC Class I molecules. However, even nonsecreted proteins can be processed and presented in vivo (Shen et al. 1998), so secretion into the host cell cytosol cannot be the sole explanation. Finally, the cytokines induced by vacuolar and cytosolic bacteria are quite different (O'Riordan et al. 2002; McCaffrey et al. 2004). Whereas LLO-minus bacteria induce a host cytokine profile that is entirely MyD88-dependent, cytosolic bacteria activate IRF3 leading to production of IFN-β (O'Connell et al. 2004; Stockinger et al. 2004). Thus, the induction of cytokines such as IFN-β may promote acquired immunity, and perhaps the cytokines induced by vacuolar bacteria may not.

1.8.   Future Prospects/Unanswered Questions

Most of the previous 20 years can be viewed from the pregenomic perspective. Now, we are well into the postgenomic era; indeed, another dozen *L. monocytogenes* isolates were just DNA sequenced at the Broad Institute. We also have the complete DNA sequence of the most prevalent animal model. Thus, using microarrays, subsequent to infection, one has the ability to monitor both bacterial and host gene expression. Transposon mutagenesis has led to the identification of bacterial determinants of pathogenesis, but in the future, postgenomic approaches such as TrasH (Sassetti and Rubin 2003) or the construction of comprehensive deletion libraries should lead to a more comprehensive appreciation of the determinants of pathogenesis. By combining comprehensive mutant libraries of bacteria with mouse mutants or RNAi knockdowns in tissue culture, the role of host factors will be further characterized. RNAi knockdown approaches have already been used in cultured Drosophila cells infected with *L. monocytogenes* (Agaisse et al. 2005; Cheng et al. 2005) and similar approaches using mammalian cells are sure to follow.

There are still many questions remaining, regarding the cell biology and bacterial physiology of infection. What bacterial adaptations are required for intracellular growth (Chap. 4)? While many of the PrfA-regulated products have been studied, how do they act in concert to promote vacuolar escape, actin-based motility, and cell-to-cell spread? Does LLO form a translocation pore? What is the cellular process that leads to cell-to-cell spread? How does *L. monocytogenes* avoid or exploit autophagy (Rich et al. 2003), a powerful host process that engulfs bacteria in the cytosol? (Ogawa and Sasakawa 2006).

There are many questions regarding in vivo infection. For example, why does *L. monocytogenes* have so many internalin genes? What are the pathways of gastrointestinal entry during infection? Are there bacterial gene products that affect trafficking to and from the placenta in vivo (Hamrick et al. 2003; Bakardjiev et al. 2006)? What is the significance of bacterial growth in the gall bladder (Hardy et al. 2004)? Regarding immunology, why do you need a live infection to generate cell-mediate immunity, and the related question, how does *L. monocytogenes* avoid and/or manipulate host innate immunity to promote infection? Lastly, will these studies lead to the development of highly attenuated or killed recombinant vaccines for the prevention and/or treatment of infectious disease and malignancies?

References

Agaisse, H., Burrack, L.S., Philips, J.A., Rubin, E.J., Perrimon, N., Higgins, D.E.: Genome-wide RNAi screen for host factors required for intracellular bacterial infection. Science **309**, 1248–51 (2005)

Auerbuch, V., Loureiro, J.J., Gertler, F.B., Theriot, J.A., Portnoy, D.A.: Ena/VASP proteins contribute to *Listeria monocytogenes* pathogenesis by controlling temporal and spatial persistence of bacterial actin-based motility. Mol Microbiol **49**, 1361–75 (2003)

Bakardjiev, A.I., Stacy, B.A., Fisher, S.J., Portnoy, D.A.: Listeriosis in the pregnant guinea pig: a model of vertical transmission. Infect Immun **72**, 489–97 (2004)

Bakardjiev, A.I., Stacy, B.A., Portnoy, D.A.: Growth of *Listeria monocytogenes* in the guinea pig placenta and role of cell-to-cell spread in fetal infection. J Infect Dis **191**, 1889–97 (2005)

Bakardjiev, A.I., Theriot, J.A., Portnoy, D.A.: *Listeria monocytogenes* traffics from maternal organs to the placenta and back. PLoS Pathog **2**, e66 (2006)

Berche, P., Gaillard, J., Sansonetti, P.J.: Intracellular growth of *Listeria monocytogenes* as a prerequisite for in vivo induction of T cell-mediated immunity. J Immunol **138**, 2266–71 (1987)

Bernardini, M.L., Mounier, J., d'Hauterville, H., Coquis-Rondon, M., Sansonetti, P.J.: *ics*A, a plasmid locus of *Shigella flexneri*, governs bacterial intra- and intercellular spread through interaction with F-actin. Proc Natl Acad Sci USA **86**, 3867–71 (1989)

Bielecki, J., Youngman, P., Connelly, P., Portnoy, D.A.: *Bacillus subtilis* expressing a haemolysin gene from *Listeria monocytogenes* can grow in mammalian cells. Nature **345**, 175–6 (1990)

Brockstedt, D.G., Giedlin, M.A., Leong, M.L., Bahjat, K.S., Gao, Y., Luckett, W., Liu, W., Cook, D.N., Portnoy, D.A., Dubensky, T.W., Jr.: Listeria-based cancer vaccines that segregate immunogenicity from toxicity. Proc Natl Acad Sci USA **101**, 13832–7 (2004)

Brundage, R.A., Smith, G.A., Camilli, A., Theriot, J.A., Portnoy, D.A.: Expression and phosphorylation of the *Listeria monocytogenes* ActA protein in mammalian cells. Proc Natl Acad Sci **90**, 11890–4 (1993)

Chatterjee, S.S., Hossain, H., Otten, S., Kuenne, C., Kuchmina, K., Machata, S., Domann, E., Chakraborty, T., Hain, T.: Intracellular gene expression profile of *Listeria monocytogenes*. Infect Immun **74**, 1323–38 (2006)

Cheng, L.W., Viala, J.P., Stuurman, N., Wiedemann, U., Vale, R.D., Portnoy, D.A.: Use of RNA interference in Drosophila S2 cells to identify host pathways controlling compartmentalization of an intracellular pathogen. Proc Natl Acad Sci USA **102**, 13646–51 (2005)

Dabiri, G.A., Sanger, J.M., Portnoy, D.A., Southwick, F.S.: *Listeria monocytogenes* moves rapidly through the host-cell cytoplasm by inducing directional actin assembly. Proc Natl Acad Sci USA **87**, 6068–72 (1990)

Dancz, C.E., Haraga, A., Portnoy, D.A., Higgins, D.E.: Inducible control of virulence gene expression in *Listeria monocytogenes*: temporal requirement of listeriolysin O during intracellular infection. J Bacteriol **184**, 5935–45 (2002)

Domann, E., Wehland, J., Rohde, M., Pistor, S., Hartl, M., Goebel, W., Leimeister-Wachter, M., Wuenscher, M., Chakraborty, T.: A novel bacterial virulence gene in *Listeria monocytogenes* required for host cell microfilament interaction with homology to the proline-rich region of vinculin. EMBO J **11**, 1981–90 (1992)

Friederich, E., Gouin, E., Hellio, R., Kocks, C., Cossart, P., Louvard, D.: Targeting of *Listeria monocytogenes* ActA protein to the plasma membrane as a tool to dissect both actin-based cell morphogenesis and ActA function. Embo J **14**, 2731–44 (1995)

Gaillard, J.L., Berche, P., Sansonetti, P.: Transposon mutagenesis as a tool to study the role of hemolysin in the virulence of *Listeria monocytogenes*. Infect Immun **52**, 50–55 (1986)

Glomski, I.J., Decatur, A.L., Portnoy, D.A.: *Listeria monocytogenes* mutants that fail to compartmentalize listerolysin O activity are cytotoxic, avirulent, and unable to evade host extracellular defenses. Infect Immun **71**, 6754–65 (2003)

Goossens, P.L., Milon, G., Cossart, P., Saron, M.F.: Attenuated *Listeria monocyto-genes* as a live vector for induction of CD8+ T cells in vivo: a study with the nucleoprotein of the lymphocytic choriomeningitis virus. International Immunology **7**, 797–805 (1995)

Grundling, A., Gonzalez, M.D., Higgins, D.E.: Requirement of the *Listeria monocyto-genes* broad-range phospholipase PC-PLC during infection of human epithelial cells. J Bacteriol **185**, 6295–307 (2003)

Hain, T., Steinweg, C., Chakraborty, T.: Comparative and functional genomics of *Listeria* spp. J Biotechnol **126**, 37–51 (2006)

Hamrick, T.S., Horton, J.R., Spears, P.A., Havell, E.A., Smoak, I.W., Orndorff, P.E.: Influence of pregnancy on the pathogenesis of listeriosis in mice inoculated intragastrically. Infect Immun **71**, 5202–9 (2003)

Hardy, J., Francis, K.P., DeBoer, M., Chu, P., Gibbs, K., Contag, C.H.: Extracellular replication of *Listeria monocytogenes* in the murine gall bladder. Science **303**, 851–3 (2004)

Havell, E.A.: Synthesis and secretion of interferon by murine fibroblasts in response to intracellular *Listeria monocytogenes*. Infect Immun **54**, 787–92 (1986)

Henry, R., Shaughnessy, L., Loessner, M.J., Alberti-Segui, C., Higgins, D.E., Swanson, J.A.: Cytolysin-dependent delay of vacuole maturation in macrophages infected with *Listeria monocytogenes*. Cell Microbiol **8**, 107–19 (2006)

Hodgson, D.A.: Generalized transduction of serotype 1/2 and serotype 4b strains of *Listeria monocytogenes*. Mol Microbiol **35**, 312–23 (2000)

Ikonomidis, G., Paterson, Y., Kos, F.J., Portnoy, D.A.: Delivery of a viral antigen to the class I processing and presentation pathway by *Listeria monocytogenes*. J Exp Med **180**, 2209–18 (1994)

Kathariou, S., Metz, P., Hof, H., Goebel, W.: Tn*916*-induced mutations in the hemolysin determinant affecting virulence of *Listeria monocytogenes*. J Bacteriol **169**, 1291–7 (1987)

Kocks, C., Gouin, E., Tabouret, M., Berche, P., Ohayon, H., Cossart, P.: *L. monocytogenes*-induced actin assembly requires the *actA* gene product, a surface protein. Cell **68**, 521–31 (1992)

Lauer, P., Chow, M.Y.N., Loessner, M.J., Portnoy, D.A., Calendar, R.: Construction, characterization and use of two *Listeria monocytogenes* site-specific integration vectors. J Bacteriol **184**, 4177–4186 (2002)

Lauvau, G., Vijh, S., Kong, P., Horng, T., Kerksiek, K., Serbina, N., Tuma, R.A., Pamer, E.G.: Priming of memory but not effector CD8 T cells by a killed bacterial vaccine. Science **294**, 1735–9 (2001)

Lecuit, M., Nelson, D.M., Smith, S.D., Khun, H., Huerre, M., Vacher-Lavenu, M.C., Gordon, J.I., Cossart, P.: Targeting and crossing of the human maternofetal barrier by *Listeria monocytogenes*: role of internalin interaction with trophoblast E-cadherin. Proc Natl Acad Sci USA **101**, 6152–7 (2004)

Lecuit, M., Vandormael-Pournin, S., Lefort, J., Huerre, M., Gounon, P., Dupuy, C., Babinet, C., Cossart, P.: A transgenic model for listeriosis: role of internalin in crossing the intestinal barrier. Science **292**, 1722–5 (2001)

Lett, M., Sasakawa, C., Okada, N., Sakai, T., Makino, S., Yamada, M., Komatsu, K., Yoshikawa, M.: *vir*G, a plasmid-coded virulence gene of *Shigella flexneri*: identification of the virG protein and determination of the complete coding sequence. J Bacteriol **171**, 353–9 (1989)

Loisel, T.P., Boujemaa, R., Pantaloni, D., Carlier, M.F.: Reconstitution of actin-based motility of *Listeria* and *Shigella* using pure proteins. Nature **401**, 613–6 (1999)

Mackaness, M.: Cellular resistance to infection. J Exp Med **116**, 381 (1962)

Marquis, H., Doshi, V., Portnoy, D.A.: The broad-range phospholipase C and a metalloprotease mediate listeriolysin O-independent escape of *Listeria monocytogenes* from a primary vacuole in human epithelial cells. Infect Immun **63**, 4531–4 (1995)

Marquis, H., Goldfine, H., Portnoy, D.A.: Proteolytic pathways of activation and degradation of a bacterial phospholipase C during intracellular infection by *Listeria monocytogenes*. J Cell Biol **137**, 1381–92 (1997)

Marquis, H., Hager, E.J.: pH-regulated activation and release of a bacteria-associated phospholipase C during intracellular infection by *Listeria monocytogenes*. Mol Microbiol **35**, 289–98 (2000).

McCaffrey, R.L., Fawcett, P., O'Riordan, M., Lee, K.D., Havell, E.A., Brown, P.O., Portnoy, D.A.: A specific gene expression program triggered by Gram-positive bacteria in the cytosol. Proc Natl Acad Sci USA **101**, 11386–91 (2004)

Mengaud, J., Chenevert, J., Geoffroy, C., Gaillard, J.L., Cossart, P.: Identification of the structural gene encoding the SH-activated hemolysin of *Listeria monocytogenes*: listeriolysin O is homologous to streptolysin O and pneumolysin. Infect Immun **55**, 3225–7 (1987)

Mounier, J., Ryter, A., Coquis-Rondon, M., Sansonetti, P.J.: Intracellular and cell-to-cell spread of *Listeria monocytogenes* involves interaction with F-actin in the enterocytelike cell line Caco-2. Infect Immun **58**, 1048–58 (1990)

O'Connell, R.M., Saha, S.K., Vaidya, S.A., Bruhn, K.W., Miranda, G.A., Zarnegar, B., Perry, A.K., Nguyen, B.O., Lane, T.F., Taniguchi, T., Miller, J.F., Cheng, G.: Type I interferon production enhances susceptibility to *Listeria monocytogenes* infection. J Exp Med **200**, 437–45 (2004)

Ogawa, M., Sasakawa, C.: Bacterial evasion of the autophagic defense system. Curr Opin Microbiol **9**, 62–8 (2006)

O'Riordan, M., Yi, C.H., Gonzales, R., Lee, K.D., Portnoy, D.A.: Innate recognition of bacteria by a macrophage cytosolic surveillance pathway. Proc Natl Acad Sci USA **99**, 13861–6 (2002)

Pal, T., Newland, J.W., Tall, B.D., Formal, S.B., Hale, T.L.: Intracellular spread of *Shigella flexneri* associated with the kcpA locus and a 140-kilodalton protein. Infect Immun **57**, 477–86 (1989)

Pan, Z.K., Ikonomidis, G., Lazenby, A., Pardoll, D., Paterson, Y.: A recombinant *Listeria monocytogenes* vaccine expressing a model tumour antigen protects mice against lethal tumour cell challenge and causes regression of established tumours. Nature Med **1**, 471–7 (1995)

Paterson, Y., Johnson, R.S.: Progress towards the use of *Listeria monocytogenes* as a live bacterial vaccine vector for the delivery of HIV antigens. Expert Rev Vaccines **3**, S119–34 (2004)

Pistor, S., Chakraborty, T., Niebuhr, K., Domann, E., Wehland, J.: The ActA protein of *Listeria monocytogenes* acts as a nucleator inducing reorganization of the actin cytoskeleton. EMBO J **13**, 758–63 (1994)

Pistor, S., Chakraborty, T., Walter, U., Wehland, J.: The bacterial actin nucleator protein ActA of *Listeria monocytogenes* contains multiple binding sites for host microfilament proteins. Curr Biol **5**, 517–25 (1995)

Portnoy, D.A., Auerbuch, V., Glomski, I.J.: The cell biology of *Listeria monocytogenes* infection: the intersection of bacterial pathogenesis and cell-mediated immunity. J Cell Biol **158**, 409–14 (2002)

Portnoy, D.A., Chakraborty, T., Goebel, W., Cossart, P.: Molecular determinants of *Listeria monocytogenes* pathogenesis. Infect Immun **60**, 1263–67 (1992)

Portnoy, D.A., Jacks, P.S., Hinrichs, D.J.: Role of hemolysin for the intracellular growth of *Listeria monocytogenes*. J Exp Med **167**, 1459–71 (1988)

Rich, K.A., Burkett, C., Webster, P.: Cytoplasmic bacteria can be targets for autophagy. Cell Microbiol **5**, 455–68 (2003)

Robbins, J.R., Barth, A.I., Marquis, H., de Hostos, E.L., Nelson, W.J., Theriot, J.A.: *Listeria monocytogenes* exploits normal host cell processes to spread from cell to cell. J Cell Biol **146**, 1333–50 (1999)

Sassetti, C.M., Rubin, E.J.: Genetic requirements for mycobacterial survival during infection. Proc Natl Acad Sci USA **100**, 12989–94 (2003)

Schnupf, P., Hofmann, J., Norseen, J., Glomski, I.J., Schwartzstein, H., Decatur, A.L.: Regulated translation of listeriolysin O controls virulence of *Listeria monocytogenes*. Mol Microbiol **61**, 999–1012 (2006)

Shaughnessy, L.M., Hoppe, A.D., Christensen, K.A., Swanson, J.A.: Membrane perforations inhibit lysosome fusion by altering pH and calcium in *Listeria monocytogenes* vacuoles. Cell Microbiol **8**, 781–92 (2006)

Shen, A., Higgins, D.E.: The MogR transcriptional repressor regulates nonhierarchal expression of flagellar motility genes and virulence in *Listeria monocytogenes*. PLoS Pathog **2**, e30 (2006)

Shen, H., Miller, J.F., Fan, X., Kolwyck, D., Ahmed, R., Harty, J.T.: Compartmentalization of bacterial antigens: differential effects on priming of CD8 T cells and protective immunity. Cell **92**, 535–45 (1998)

Shen, H., Slifka, M.K., Matloubian, M., Jensen, E.R., Ahmed, R., Miller, J.F.: Recombinant *Listeria monocytogenes* as a live vaccine vehicle for the induction of protective anti-viral cell-mediated immunity. Proc Natl Acad Sci USA **92**, 3987–91 (1995)

Smith, G.A., Portnoy, D.A., Theriot, J.A.: Asymmetric distribution of the *Listeria monocytogenes* ActA protein is required and sufficient to direct actin-based motility. Mol Microbiol **17**, 945–51 (1995)

Stockinger, S., Reutterer, B., Schaljo, B., Schellack, C., Brunner, S., Materna, T., Yamamoto, M., Akira, S., Taniguchi, T., Murray, P.J., Muller, M., Decker, T.: IFN regulatory factor 3-dependent induction of type I IFNs by intracellular bacteria is mediated by a TLR- and Nod2-independent mechanism. J Immunol **173**, 7416–25 (2004)

Theriot, J.A., Rosenblatt, J., Portnoy, D.A., Goldschmidt-Clermont, P.J., Mitchison, T.J.: Involvement of profilin in the actin-based motility of *L. monocytogenes* in cells and in cell-free extracts. Cell **76**, 505–17 (1994)

Tilney, L.G., Connelly, P.S., Portnoy, D.A.: The nucleation of actin filaments by the bacterial intracellular pathogen, *Listeria monocytogenes*. J Cell Biol 111: 2979–88 (1990)

Tilney, L.G., Portnoy, D.A.: Actin filaments and the growth, movement, and spread of the intracellular bacterial parasite, *Listeria monocytogenes*. J Cell Biol **109**, 1597–1608 (1989)

von Koenig, C.H.W., Finger, H., Hof, H.: Failure of killed *Listeria monocytogenes* vaccine to produce protective immunity. Nature **297**, 233–4 (1982)

Wadsworth, S.J., Goldfine, H.: Mobilization of protein kinase C in macrophages induced by *Listeria monocytogenes* affects its internalization and escape from the phagosome. Infect Immun **70**, 4650–60 (2002)

Welch, M.D., Mullins, R.D.: Cellular control of actin nucleation. Annu Rev Cell Dev Biol **18**, 247–88 (2002)

Welch, M.D., Rosenblatt, J., Skoble, J., Portnoy, D.A., Mitchison, T.J.: Interaction of human Arp2/3 complex and the *Listeria monocytogenes* ActA protein in actin filament nucleation. Science **281**, 105–108 (1998)

Yeung, P.S., Zagorski, N., Marquis, H.: The metalloprotease of *Listeria monocytogenes* controls cell wall translocation of the broad-range phospholipase C. J Bacteriol **187**, 2601–8 (2005)

2
Listeriosis

Bennett Lorber

Thomas M. Durant Professor of Medicine
Section of Infectious Diseases
Temple University School of Medicine and Hospital
Philadelphia, PA 19140, USA
e-mail: bennett.lorber@temple.edu

2.1. Introduction

The bacterium *Listeria monocytogenes* is an infrequent cause of illness in the general population. However, in some groups, including neonates, pregnant women, elderly persons, and those with impaired cell-mediated immunity due to underlying disease or immunosuppressive therapy, it is an important cause of life-threatening bacteremia and meningoencephalitis (Lorber 1997; Bucholz and Mascola 2001; Wing and Gregory 2002). Increasing interest in this organism has resulted from food-borne outbreaks, concerns about food safety, and the recognition that food-borne infection may result in self-limited febrile gastroenteritis as well as invasive disease. Separate from its immediate clinical relevance, the study of listeriosis has provided insights into bacterial pathogenesis and the role of cell-mediated immunity in resistance to infection with intracellular pathogens.

2.2. Microbiology

Listeria monocytogenes is a small, facultatively anaerobic, nonsporulating, catalase-positive, oxidase-negative, gram-positive bacillus that grows readily on blood agar, producing incomplete β-hemolysis (Farber and Peterkin 1991; Bille et al. 2003). The bacterium possesses polar flagella and exhibits a characteristic tumbling motility at room temperature (25°C). Optimal growth occurs at 30–37°C, but, unlike most bacteria, *L. monocytogenes* also grows well at refrigerator temperature (4–10°C); and, by so-called cold enrichment, it can be separated from other contaminating bacteria by long incubation in this temperature range. Selective media have been developed to isolate the organism from specimens containing multiple species (food, stool) and appear to be superior to cold enrichment (Hayes et al. 1991). When grown on blood-free agar and viewed with light transmitted at a 45-degree angle (Henry's illumination), listerial colonies appear blue, whereas other bacterial colonies appear yellow or orange.

In clinical specimens, the organisms may be gram-variable and may look like diphtheroids, cocci, or diplococci. Routine growth media are effective for growing *L. monocytogenes* from normally sterile specimens (cerebrospinal fluid (CSF), blood, joint fluid), but media typically used to isolate diarrhea-causing bacteria from stool cultures inhibit listerial growth. Laboratory misidentification as diphtheroids, streptococci, or enterococci is not uncommon, and the isolation of a "diphtheroid" from blood or CSF should always alert one to the possibility that the organism is really *L. monocytogenes* (Buchner and Schneierson 1968; Nieman and Lorber 1980).

Of the six listerial species (*L. monocytogenes, L. seeligeri, L. welshimeri, L. innocua, L. ivanovii,* and *L. grayi*), only *L. monocytogenes* is pathogenic for humans. There are at least 13 serotypes of *L. monocytogenes*, based on cellular O and flagellar H antigens, but almost all diseases are due to types 4b, 1/2a, and 1/2b (Schuchat et al. 1991; Bucholz and Mascola 2001), limiting the utility of serotyping for epidemiological investigations. A number of newer molecular techniques, including pulsed-field gel electrophoresis, ribotyping, and multilocus enzyme electrophoresis, have been employed to separate isolates into distinct groups and have proved useful for investigating epidemics (Czajka and Batt 1994; Gellin et al. 1994; Graves et al. 1994; Louie et al. 1996; Sauders et al. 2003).

2.3.   Epidemiology

Listeria monocytogenes is an important cause of zoonoses, especially in herd animals. It is widespread in nature, being found commonly in soil, decaying vegetation, and as part of the fecal flora of many mammals (Schuchat et al. 1991; Bille et al. 2003). The organism has been isolated from the stool of approximately 5% of healthy adults (Schlech et al. 1983; Schuchat et al. 1991) with higher rates of recovery reported from household contacts of patients with clinical infection (Schuchat et al. 1993). Many foods are contaminated with *L. monocytogenes* and recovery rates of 1–70% or more are common from raw vegetables, raw milk, fish, poultry, and meats, including fresh or processed chicken and beef available at supermarkets or deli counters (Farber and Peterkin 1991). Ingestion of *L. monocytogenes* must be a very common occurrence.

Two active surveillance studies performed in 1980–1982 and 1986 by the Centers for Disease Control and Prevention (CDC) indicated annual infection rates of 7.4 per million population, accounting for approximately 1,850 cases a year in the United States, with 425 deaths (Ciesielski et al. 1988; Gellin et al. 1994). By 1993, after food industry regulations were instituted to minimize the risk of food-borne listeriosis, the annual incidence had declined to 4.4 cases per million, or 1,092 cases, with 248 deaths (Tappero et al. 1995). A similar decline in incidence of human listeriosis was seen in France, following control measures to decrease food contamination (Goulet et al. 2001). Following more recent risk assessment for *L. monocytogenes* in deli meats, regulatory and industry changes have been designed to prevent future contamination of ready-to-eat meat and poultry (Gottlieb et al. 2006).

The highest infection rates are seen in infants < 1 month and in adults > 60 years of age (Ciesielski et al. 1988; Tappero et al. 1995). Pregnant women account for about 30% of all cases and 60% of cases in the 10- to 40-year age group. Almost 70% of nonperinatal infections occur in those with hematologic malignancy, the acquired immunodeficiency syndrome (AIDS), bone marrow or solid organ transplants, or in those receiving corticosteroid therapy (Mylonakis et al. 1988; Blatt and Zajac 1991; Bucholz and Mascola 2001; Safdar et al. 2002; Siegman-Igra et al. 2002) but seemingly healthy persons may develop invasive disease, particularly those over 60 years of age.

Subsequent to the 1983 report (Schlech et al. 1983) of a widespread outbreak of food-borne human listeriosis due to contaminated coleslaw, a number of other food-borne outbreaks resulting in invasive disease (bacteremia, meningitis) have been documented, with vehicles including milk (Fleming et al. 1985), soft cheeses (Linnan et al. 1988; MacDonald et al. 2005), butter (Lyytikaiinen et al. 2000), as well as smoked trout (Miettinen et al. 1999), ready-to-eat pork products (Goulet et al. 1998), hot dogs, and deli-ready turkey (Frye et al. 2002; Olsen et al. 2005; Gottlieb et al. 2006). A 2002 outbreak due to contaminated turkey deli meat involved 54 patients in 9 states and resulted in the recall of more than 30 million pounds of food products, one of the largest meat recalls in US history (Gottlieb et al. 2006). Sporadic cases have been traced to contaminated cheese (Schwartz et al. 1989), turkey franks (Centers for Disease Control and Prevention 1989), and alfalfa tablets (Farber et al. 1990). The importance of food as a source of sporadic listeriosis is illustrated by two CDC studies in which 11% of all refrigerator food samples were contaminated, 64% of patients had at least one contaminated food, and, in 33% of instances, the patient and food isolates had identical strains (Pinner et al. 1992; Schuchat et al. 1992). Delicatessen-style ready-to-eat meats, especially chicken, had the highest rates of contamination. Cases were more likely than were controls to have eaten soft cheeses or deli-counter meats, and 32% of sporadic cases could be attributed to these foods.

Human listeriosis is typically acquired through ingestion of contaminated food but other modes of transmission occur. These include transmission from mother to child transplacentally or through an infected birth canal and cross-infection in neonatal nurseries (Farber et al. 1992; Colodner et al. 2003). Contaminated mineral oil used for bathing infants was the source of one outbreak (Schuchat et al. 1991). Localized cutaneous infections have occurred in veterinarians and farmers after direct contact with aborted calves and infected poultry.

The CDC has established PulseNet (http://www.cdc.gov/pulsenet/), a network of public health and food regulatory laboratories that use pulsed-field gel electrophoresis to subtype food-borne pathogens in order to promptly detect disease clusters that may have a common source (Swaminathan et al. 2001). This system has proved effective in the early detection of listeriosis outbreaks (Centers for Disease Control and Prevention 2002).

2.4. Pathogenesis

Except for vertical transmission from mother to fetus and rare instances of cross-contamination in the delivery suite or neonatal nursery, human-to-human infection has not been documented (Farber et al. 1992; Colodner et al. 2003).

Infection most often begins after ingestion of food contaminated with the organism. The oral inoculum required to produce clinical infection is unknown; experiments in healthy mammals indicate that $\geq 10^9$ organisms are required (Farber et al. 1991). Alkalinization of the stomach by antacids, H_2 blockers, proton pump inhibitors, or ulcer surgery may promote infection (Ho et al. 1986; Schlech et al. 1993). The incubation period for invasive infection is not well established, but evidence from a few cases related to specific ingestions points to a mean incubation period of 31 days, with a range from 11 to 70 days. In one report, two pregnant women, whose only common exposure was attendance at a party, developed listerial bacteremia with the same uncommon enzyme type; incubation periods for illness were 19 and 23 days (Riedo et al. 1994).

Virulent *L. monocytogenes* organisms can cause disease without promoter organisms, but a 1987 outbreak in Philadelphia, for which no particular source was found, suggested that intercurrent gastrointestinal infection with another pathogen may enhance invasion in individuals colonized with *L. monocytogenes* (Schwartz et al. 1989). Evidence for this hypothesis is found in the common history of antecedent gastrointestinal symptoms in patients and household contacts, the long incubation period from ingestion to clinical illness, and two instances in which clinical listeriosis closely followed shigellosis (Schroter and Weil 1977; Lorber 1991). Both listerial meningitis and bacteremia have occurred shortly after colonoscopy or sigmoidoscopy (Sheehan and Galbraith 1993; Witlox et al. 2000).

In the intestine, *L. monocytogenes* crosses the mucosal barrier, aided by active endocytosis of organisms by endothelial cells (Farber and Peterkin 1991; Cossart and Sansonetti 2004). Once in the blood stream, hematogenous dissemination may occur to any site; *L. monocytogenes* has a particular predilection for the central nervous system (CNS) and the placenta. It is generally believed that listeriae reach the CNS by a bacteremic route, but animal experiments suggest that rhombencephalitis may develop by intraaxonal spread of bacteria from peripheral sites to the CNS (Antal et al. 2001; Drevets et al. 2004).

Several virulence factors have been identified that enable *L. monocytogenes* to function as an intracellular organism. The bacterium possesses the cell surface protein, internalin that interacts with E-cadherin, a receptor on macrophages and intestinal lining cells, to induce its own ingestion (Farber and Peterkin 1991; Mengard et al. 1996; Cossart and Sansonetti 2004). A membrane lipoprotein appears to promote entry into nonmacrophage cells (Reglier-Poupet et al., 2003). The major virulence factor, listeriolysin O, along with phospholipases, enables listeriae to escape from the phagosome and avoid intracellular killing (Vasquez-Boland et al. 2001; Portnoy et al. 2002; Dussurget et al. 2004). Once free in the cytoplasm, the bacterium can divide and, by inducing host cell actin polymerization, propel itself to the cell membrane (Sanger et al. 1992; Southwick and Purich 1996). Subsequently, by means of pseudopod-like projections, it can invade adjacent macrophages. The bacterial surface protein Act A is necessary for the induction of actin filament assembly and cell-to-cell spread and, therefore, is a major virulence factor. Thus, through this novel life cycle, L. monocytogenes moves from cell to cell, evading exposure to antibodies, complement, or neutrophils.

Ability to scavenge iron, which is essential for the life of all microorganisms, appears to be an important virulence factor of L. monocytogenes. Siderophores of the organism enable it to take iron from transferrin (Farber and Peterkin 1991). In vitro, iron enhances organism growth (Sword 1966), and, in animal models of listerial infection, iron overload is associated with enhanced susceptibility to infection and iron supplementation with enhanced lethality, whereas iron depletion results in prolonged survival (Ampel et al. 1992; Lorber 1997). The clinical associations of sporadic listerial infection with hemochromatosis (Nieman and Lorber 1980) and of outbreaks with transfusion-induced iron overload in patients receiving dialysis (Mossey and Sondheimer 1985) attest to the importance of iron acquisition as a virulence factor in humans.

2.5. Immunity

Resistance to infection with the intracellular bacterium L. monocytogenes is predominantly cell mediated, as evidenced by experiments (Mackaness 1962; Mackaness 1971) showing that immunity could be transferred by sensitized lymphocytes but not by antibodies. Further evidence is provided by the overwhelming clinical association between listerial infection and conditions of impaired cellular immunity, including lymphoma, pregnancy, AIDS, and corticosteroid immunosuppression (Buchner and Schneierson 1968; Nieman and Lorber 1980; Stamm et al. 1982; Cherubin et al. 1991; Skoberg et al. 1992; Gellin et al. 1994; Lorber 1997; Bucholz and Mascola 2001; Wing and Gregory 2002). Tumor necrosis factor (TNF) alpha-neutralizing agents (e.g., infliximab) are increasingly used to treat rheumatoid arthritis and Crohn's disease; invasive listeriosis has complicated use of these immune modulating agents (Slifman et al. 2003). The production of nitric oxide by activated macrophages may play a role in natural immunity to listeriosis independent of T-cell function (Hibbs 2002). The role of humoral immunity is unknown, although both immunoglobulin M (absent in neonates) and classical complement activity (low in neonates) have been shown to be necessary for efficient opsonization of L. monocytogenes (Bortolussi et al. 1986).

Although listeriosis is 100–1,000 times more common in patients with AIDS compared with the general population (Jurado et al. 1993; Ewert et al. 1995), it is somewhat surprising that it is not seen more commonly, given the ubiquity of the organism (Berenguer et al. 1981; Mascola et al. 1988; Kales and

Holzman 1990; Decker et al. 1991). A partial explanation may lie in the experimental observation that resistance to listeriosis appears to be mediated by
lymphocytes that do not carry CD4 or CD8 markers (Dunn and North 1991a,b).
Additionally, use of trimethoprim–sulfamethoxazole (TMP–SMX) for prophylaxis against *Pneumocystis jiroveci (formerly carinii)* provides protection against
listeriosis (Jurado et al. 1993). Frequency of listeriosis is not increased in those
with deficiencies in neutrophil numbers or function, splenectomy, complement
deficiency, or immunoglobulin disorders; the latter finding is not surprising because
L. monocytogenes can pass from cell to cell without being exposed to antibody.

2.6.   Clinical Manifestations

The species name derives from the fact that an extract of the *L. monocytogenes* cell
membrane has potent monocytosis-producing activity in rabbits, but monocytosis
is a very uncommon feature of human infection (Murray et al. 1926; Stanley 1949).

2.6.1.   Infection in Pregnancy

Mild impairment of cell-mediated immunity occurs during gestation, and
pregnant women are prone to developing listerial bacteremia with an estimated
17-fold increase in risk (Weinberg 1984; Mylonakis et al. 2002). Listeriae
proliferate in the placenta in areas that appear to be unreachable by usual
defense mechanisms, and cell-to-cell spread facilitates maternal–fetal transmission
(Bakardjiev et al. 2005). For unexplained reasons, CNS infection, the most
commonly recognized form of listeriosis in other groups, is extremely rare during
pregnancy in the absence of other risk factors (Ciesielski et al. 1988; Gellin
et al. 1994; Bucholz and Mascola 2001). Bacteremia manifests clinically as an acute
febrile illness, often accompanied by myalgia, arthralgia, headache, and backache.
Illness may occur at any time during pregnancy but usually occurs in the third
trimester, probably related to the major decline in cell-mediated immunity seen at
26–30 weeks of gestation (Weinberg 1984). Twenty-two percent of perinatal infections result in stillbirth or neonatal death; premature labor is common (Bucholz and
Mascola 2001). Untreated bacteremia is generally self-limited; although if there is
a complicating amnionitis, fever may persist in the mother until the fetus is aborted.
Early diagnosis and antimicrobial therapy can result in the birth of a healthy infant
(Evans et al. 1985; Kalstone 1991).

There is no convincing evidence that listeriosis is a cause of habitual abortion in
humans.

2.6.2.   Neonatal Infection

In a pregnant primate model, oral administration of *L. monocytogenes* resulted
in stillbirth with isolation of the bacterium from placental and fetal tissues
(Smith et al. 2003). When in utero infection occurs, it can precipitate spontaneous

abortion. The fetus may be stillborn or die within hours of a disseminated form of listerial infection known as granulomatosis infantiseptica, which is characterized by widespread microabscesses and granulomas that are particularly prevalent in the liver and spleen. In this entity, abundant bacteria are often visible on Gram stain of meconium (Visintine et al. 1977; Larsson and Linell 1979).

More commonly, neonatal infection manifests like group B streptococcal disease in one of two forms (Lorber 1997): (1) early-onset sepsis syndrome, usually associated with prematurity and probably acquired in utero; or (2) late onset meningitis, occurring at about 2 weeks of age in term infants, who most likely acquired organisms from the maternal vagina at parturition. Cases have occurred after cesarean delivery, however, and nosocomial transmission has been suggested.

In early-onset disease, *L. monocytogenes* can be isolated from the conjunctivae, external ear, nose, throat, meconium, amniotic fluid, placenta, blood, and, sometimes, CSF; Gram stain of meconium may show gram-positive rods and provide early diagnosis. The highest concentrations of bacteria are found in the neonatal lung and gut, which suggests that infection is acquired in utero from infected amniotic fluid, rather than via a hematogenous route (Becroft et al. 1971). Purulent conjunctivitis and a disseminated papular rash have rarely been described in neonates with early-onset disease, but clinical infection is otherwise similar to that due to other bacterial pathogens.

2.6.3.   Bacteremia

Bacteremia without an evident focus is the most common manifestation of listeriosis after the neonatal period (Gellin et al. 1994). Clinical manifestations typically include fever and myalgias; a prodromal illness with nausea and diarrhea may occur. Since immunocompromised patients are more likely than healthy persons to have blood cultures during febrile illnesses, transient bacteremias in healthy persons may go undetected.

2.6.4.   Central Nervous System Infection

Organisms that cause bacterial meningitis most frequently (*Streptococcus pneumoniae, Neisseria meningitidis*, and *Haemophilus influenzae*), rarely cause parenchymal brain infections such as cerebritis and brain abscess. By contrast, *L. monocytogenes* has tropism for the brain itself (particularly the brain stem), as well as for the meninges (Nieman and Lorber 1980; Lorber 1997). Many patients with meningitis experience altered consciousness, seizures, or movement disorders and truly have meningoencephalitis.

2.6.4.1.   Meningitis

In an active surveillance study of bacterial meningitis reported by the CDC in 1990, *L. monocytogenes* was the fifth most common pathogen, after *H. influenzae, S. pneumoniae, N. meningitidis*, and group B streptococcus, but it had the highest mortality at 22% (Wenger et al. 1990). By 1995, 5 years after the introduction of *H. influenzae* conjugate vaccines, *H. influenzae* had become a less common cause of meningitis than *L. monocytogenes*, which accounted for 20% of cases in neonates and 20% in those > 60 years of age (Schuchat et al. 1997). Worldwide, *L. monocytogenes* is one of the three major causes of neonatal meningitis, is second only to pneumococcus as a cause of bacterial meningitis in adults >50 years, and is the most common cause of bacterial meningitis in patients with lymphoma, patients with organ transplants, or those receiving corticosteroid immunosuppression for any reason (Mylonakis et al. 1988; Lorber 1997; Siegman-Igra et al. 2002; Safdar and Armstrong 2003).

Clinically, meningitis due to *L. monocytogenes* is usually similar to that due to more common causes; features particular to listerial meningitis are summarized in Table 2.1.

2.6.4.2.   Brain Stem Encephalitis (Rhombencephalitis)

An unusual form of listerial encephalitis involves the brain stem (Armstrong and Fung 1993) and is similar to the unique zoonotic listerial infection known as circling disease of sheep (Gill 1993). In contrast to other listerial CNS infections, this illness usually occurs in healthy older children and adults; neonatal cases have not been reported. The typical clinical picture is one of a biphasic illness with a prodrome of fever, headache, nausea, and vomiting lasting about 4 days, followed by the abrupt onset of asymmetrical cranial nerve deficits, cerebellar signs, and hemiparesis or hemisensory deficits, or both. Nuchal rigidity is present in about 50%, CSF is only mildly abnormal, and CSF culture is positive in about 40%; almost two-thirds are bacteremic. Respiratory failure develops in about 4% of cases. Magnetic resonance imaging is superior to computed tomography for demonstrating rhombencephalitis (Armstrong and Fung 1993; Faidas et al. 1993). Mortality is high, and serious sequelae are common in survivors.

TABLE 2.1. Distinctive features of listerial meningitis compared with more common bacterial etiologies.

Feature	Frequency (%)
Presentation can be subacute (mimics tuberculous meningitis)	~ 10
Stiff neck is less common	15–20
Movement disorders (ataxia, tremors, myoclonus) are more common	15–20
Seizures are more common	~ 25
Fluctuating mental status is common	~ 75
Positive blood culture is more common	75
Cerebrospinal fluid (CSF)	
Positive Gram stain is less common	40
Normal CSF glucose is more common	> 60
Mononuclear cell predominance is more common	~ 30

2.6.4.3. Brain Abscess

Macroscopic brain abscesses account for about 10% of CNS listerial infections. Bacteremia is almost always present, and concomitant meningitis with isolation of *L. monocytogenes* from the CSF is found in 25%; both of these features are rare in other forms of bacterial brain abscess (Eckburg et al. 2001). About 50% of cases occur in known risk groups for listerial infection. Subcortical abscesses located in the thalamus, pons, and medulla are common; these sites are exceedingly rare when abscesses are due to other bacteria. Mortality is high, and survivors usually have serious sequelae (Cone et al. 2003).

2.6.5. Endocarditis

Listerial endocarditis may account for as much as 7.5% of adult listerial infections (Nieman and Lorber 1980), affects the population at risk for viridans streptococcal endocarditis, produces both native valve and prosthetic valve disease, and has a high rate of septic complications and a mortality of 48% (Carvajal and Frederiksen 1988). Listerial endocarditis, but not bacteremia per se, may be an indicator of underlying gastrointestinal tract pathology, including cancer (Lorber 1997). Cases in children are rare.

2.6.6. Localized Infection

Rare reports of focal infections from which *L. monocytogenes* have been isolated include direct inoculation resulting in conjunctivitis (Schwartz et al. 1989), skin infection (Cain and McCann 1986), and lymphadenitis (Nieman and Lorber 1980). Bacteremia can lead to hepatic infection (Yu et al. 1992; Braun et al. 1993), cholecystitis (Gordon and Singer 1986), peritonitis (Myers et al. 1983; Winslow and Steele 1984), splenic abscess (Nieman and Lorber 1980), pleuropulmonary infection (Gradon et al. 1982; Mazzulli and Salit 191; Domingo et al. 1992), septic arthritis (Curosh and Perednia 1989), osteomyelitis (Chirgwin and Gleich 1989), pericarditis (Holoshitz et al. 1984), myocarditis (Stamm et al. 1990), arteritis (Gauto et al. 1992), and endophthalmitis (Nieman and Lorber 1980). Complications including disseminated intravascular coagulation (Plaut and Gardner 1972), adult respiratory distress syndrome (Boucher et al. 1984), and rhabdomyolysis with acute renal failure (Thomas and Ravaud 1988) have been documented. There is nothing clinically unique about these localized infections; many, but not all, have occurred in those known to be at risk for listeriosis.

2.6.7. Febrile Gastroenteritis

Many patients with invasive listeriosis give a history of antecedent gastrointestinal illness, often accompanied by fever. Although isolated cases of gastrointestinal illness due to *L. monocytogenes* appear to be quite rare (Schlech et al. 2005), at least seven outbreaks of food-borne gastroenteritis due to *L. monocytogenes* have been documented (Ooi and Lorber 2005). In the largest outbreak to date, 1,566 individuals, most of them children between the ages of 6 and 10, became ill after eating caterer-provided cafeteria food at two schools, and 19% were hospitalized (Aureli et al. 2000). Illness typically occurs 24 h after ingestion of a large inoculum of bacteria (range from 6 h to 10 days) and usually lasts 1–3 days (range 1–7 days); attack rates have been quite high (52–100%). Common symptoms include fever, watery diarrhea, nausea, headache, and pains in joints and muscles. Vehicles of infection have included chocolate milk, cold corn and tuna salad, cold smoked trout, and delicatessen meat. *L. monocytogenes* should be considered to be a possible etiology in outbreaks of febrile gastroenteritis when routine cultures fail to yield a pathogen.

2.6.8. Complications

Disseminated intravascular coagulation, adult respiratory distress syndrome, and rhabdomyolysis with acute renal failure have been documented as complications of invasive listeriosis. Rare episodes of reinfection have occurred (Van et al. 1994).

2.7. Diagnosis

Listeriosis should be a major consideration as part of the differential diagnosis in any of the following clinical settings:

1. Septicemia or meningitis in infants < 2 months of age.
2. Meningitis or parenchymal brain infection in (1) patients with hematologic malignancy, AIDS, organ transplantation, corticosteroid immunosuppression, or those receiving anti-TNF agents; (2) patients with subacute presentation; (3) adults > 50 years; and (4) those in whom CSF shows gram-positive bacilli.
3. Simultaneous infection of the meninges and brain parenchyma.
4. Subcortical brain abscess.
5. Fever during pregnancy.
6. Blood, CSF, or other normally sterile specimen reported to have "diphtheroids" on Gram stain or culture.
7. Food-borne outbreak of febrile gastroenteritis when routine cultures fail to identify a pathogen.

Diagnosis requires isolation of *L. monocytogenes* from clinical specimens (e.g., CSF, blood) and identification through standard microbiologic techniques. Antibodies to listeriolysin O have not proved useful in invasive disease (Chatzipanagiotou and Hof 1988), nor have polymerase chain reaction probes (Greisen et al. 1994). Antibodies to listeriolysin O may be useful during investigation of outbreaks of febrile gastroenteritis (Dalton et al. 1997). Magnetic resonance imaging is superior to computed tomography for demonstrating parenchymal brain involvement, especially in the brain stem (Armstrong and Fung 1993; Faidas et al. 1993).

2.8.   Treatment

Comprehensive reviews of treatment are available (Hof et al. 1997; Lorber 2002). No controlled trials have established drug of choice or duration of therapy for listerial infection. Ampicillin is generally considered the preferred agent (Nieman and Lorber 1980; Scheld 1983; Lorber 1997; Lorber 2002; Safdar and Armstrong 2003). Based on synergy in vitro and in animal models (Edmiston and Gordon 1979), most authorities suggest adding gentamicin to ampicillin for treatment of bacteremia in those with severely impaired T-lymphocyte function and in all cases of meningitis and endocarditis (Cherubin et al. 1991; Lorber 1997). In one uncontrolled study (Merle-Melet et al. 1996), the combination of TMP–SMX plus ampicillin was associated with a lower failure rate and fewer neurologic sequelae than ampicillin combined with an aminoglycoside.

For those intolerant of penicillins, TMP–SMX is believed to be the best alternative (Meyer and Liu 1987; Spitzer et al. 1986; Winslow and Pankey 1982). Chloramphenicol, at one time regarded as the agent of choice for patients with penicillin allergy, should not be used to treat listerial infection because of unacceptable failure and relapse rates (Cherubin et al. 1985; Cherubin et al. 1991). No currently available cephalosporin should be used; none has adequate activity (Cherubin et al. 1991; Espaze and Reynaud 1998), and meningitis has developed in patients receiving cephalosporins (Lorber et al. 1975; Cherubin et al. 1991). For this reason, ampicillin is always included in empirical therapy for septicemia or meningitis in infants < 2 months of age.

Vancomycin has been used successfully in a few patients with penicillin allergy (Blatt and Zajac 1991; Bonacorsi et al. 1993), but other patients have developed listerial meningitis while receiving the drug (Baldassarre et al. 1991). Rifampin is active in vitro and is known to penetrate phagocytic cells; clinical experience is minimal, however, and in animal models, the addition of rifampin to ampicillin was not more effective than when ampicillin was used alone (Scheld 1983).

Initial dosing of antibiotics as for meningitis is prudent for all patients, even in the absence of CNS or CSF abnormalities, because of the high affinity of this organism for the CNS. Patients with meningitis should be treated for no fewer than 3 weeks; bacteremic patients without CSF abnormalities can be treated for 2 weeks.

No data exist concerning antimicrobial efficacy in listerial gastroenteritis; the illness is self-limited, and treatment is not warranted.

Clinically significant antimicrobial resistance has not been encountered, but vigilance is warranted since transfer of resistance from enterococci to *L. monocytogenes* has been reported (Charpentier and Courvalin 1999).

Because iron is a virulence factor for *L. monocytogenes*, it seems prudent to withhold iron replacement in patients with iron deficiency until the listerial infection is resolved.

Nine neonates with septicemia, pneumonia, and severe respiratory failure due to *L. monocytogenes* who were supported by venoarterial bypass have been reported, with recovery in six infants. The duration of extracorporeal membrane oxygenation was comparatively prolonged (median, 9 days), probably because of the necrotizing nature of listerial lung infection (Hirscal et al. 1994).

2.9. Prevention

Recommendations for prevention of listeriosis from a food-borne source have been developed by the CDC (Broome 1993) and are presented in Table 2.2.

Except from infected mother to fetus, human-to-human transmission of listeriosis does not occur; therefore, patients do not require isolation. Neonatal listerial infection complicating successive pregnancies is virtually unheard of, and intrapartum antibiotics are not recommended for mothers with a history of perinatal listeriosis. There is no vaccine. Listerial infections are effectively prevented by TMP–SMX, given as prophylaxis against *P. jiroveci* to recipients of organ transplants or to individuals with the human immunodeficiency virus (Dworkin et al. 2001). The utility, or even the feasibility, of eradicating gastrointestinal colonization as a means to prevent invasive listeriosis is unknown. However, asymptomatic persons at high risk for listeriosis, known to have ingested a food implicated in an outbreak, could reasonably be given several days of oral ampicillin or trimethoprim-sulfamethoxasole.

TABLE 2.2. Dietary recommendations for preventing food-borne listeriosis.

For all persons

1. Cook raw food from animal sources (e.g., beef, pork, and poultry) thoroughly.
2. Wash raw vegetables thoroughly before eating.
3. Keep uncooked meats separate from vegetables, cooked foods, and ready-to-eat foods.
4. Avoid consumption of raw (unpasteurized) milk or foods made from raw milk.
5. Wash hands, knives, and cutting boards after handling uncooked foods.

Additional recommendations for persons at high risk[a]

1. Avoid soft cheeses (e.g., Mexican-style, feta, Brie, Camembert) and blue-veined cheese; there is no need to avoid hard cheeses, cream cheese, cottage cheese, or yogurt.
2. Leftover foods or ready-to-eat foods (e.g., hot dogs) should be reheated until steaming hot before eating.
3. Consider avoidance of foods in delicatessen counters.[b]

[a] Those immunocompromised by illness or medications, pregnant women, and the elderly.
[b] Although the risk for listeriosis associated with foods from delicatessen counters is relatively low, pregnant women and immunosuppressed persons may choose to avoid these foods or to thoroughly reheat cold cuts before consumption.

References

Ampel NM, Bejarano GC, Saavedra M Jr (1992) Deferoxamine increases the susceptibility of beta-thalassemic, iron-overloaded mice to infection with *Listeria monocytogenes*. Life Sci 50:1327–1332

Antal E-A, Leberg EM, Bracht P et al (2001) Evidence for intraaxonal spread of *Listeria monocytogenes* from the periphery to the central nervous system. Brain Pathol 11:432—438

Armstrong RW, Fung PC (1993) Brainstem encephalitis (rhombencephalitis) due to *Listeria monocytogenes*: Case report and review. Clin Infect Dis 16:689–702

Aureli P, Fiorucci GC, Caroli D et al (2000) An outbreak of febrile gastroenteritis associated with corn contaminated by *Listeria monocytogenes*. N Engl J Med 342:1235–1241

Baldassarre JS, Ingerman MJ, Nansteel J, Santoro J (1991) Development of *Listeria* meningitis during vancomycin therapy: A case report. J Infect Dis 164:221–222

Bakardjiev AI, Stacy BA, Portnoy DA (2005) Growth of *Listeria monocytogenes* in the guinea pig placenta and role of cell-to-cell spread in fetal infection. J Infect Dis 191:1889–1897

Becroft DMO, Farmer K, Seddon RJ et al (1971) Epidemic listeriosis in the newborn. BMJ 3:747–751

Berenguer J, Solera J, Diaz MD et al (1991) Listeriosis in patients infected with human immunodeficiency virus. Rev Infect Dis 13:115–119

Bille J, Rocourt J, Swaminathan B (2003) *Listeria* and *Erysipelothrix*. In: Murray PR, Baron EJ, Jorgensen JH et al (eds) Manual of clinical microbiology, 8th edn. ASM, Washington, DC, pp. 461–471

Blatt SP, Zajac RA (1991) Treatment of listeria bacteremia with vancomycin. Rev Infect Dis 13:181–182

Bonacorsi S, Doit C, Aujard Y et al (1993) Successful antepartum treatment of listeriosis with vancomycin plus netilmicin. Clin Infect Dis 17:139–140

Bortolussi R, Issekutz A, Faulkner G (1986) Opsonization of *Listeria monocytogenes* type 4b by human adult and newborn sera. Infect Immun 52:493–498

Boucher M, Yonekura ML, Wallace RJ, Phelan JP (1984) Adult respiratory distress syndrome: A rare manifestation of *Listeria monocytogenes* infection in pregnancy. Am J Obstet Gynecol 149:686–688

Braun TI, Travis D, Dee RR, Nieman RE (1993) Liver abscess due to *Listeria monocytogenes*: Case report and review. Clin Infect Dis 17:267–269

Broome CV (1993) Listeriosis: Can we prevent it? ASM News 59:444–446

Buchner LH, Schneierson SS (1968) Clinical and laboratory aspects of *Listeria monocytogenes* infections with a report of ten cases. Am J Med 45:904–921

Bucholz U, Mascola L (2001) Transmission, pathogenesis, and epidemiology of *Listeria monocytogenes*. Infect Dis Clin Pract 10:34–41

Cain DB, McCann VL (1986) An unusual case of cutaneous listeriosis. J Clin Microbiol 23:976–977

Carvajal A, Frederiksen W (1988) Fatal endocarditis due to *Listeria monocytogenes*. Rev Infect Dis 10:616–623

Centers for Disease Control and Prevention (1989) Listeriosis associated with consumption of turkey franks. MMWR 38:267–268

Centers for Disease Control and Prevention (2002) Public health dispatch: Outbreak of listeriosis–Northeastern United States, 2002. MMWR 51:950–951

Charpentier E, Courvalin P (1999) Antibiotic resistance in *Listeria* spp. Antimicrob Agents Chemother 43:2103–2108

Chatzipanagiotou S, Hof H (1988) Sera from patients with high titers of antibody to streptolysin 0 react with listeriolysin. J Clin Microbiol 26:1066–1067

Cherubin CE, Appleman MD, Heseltine PNR et al (1991) Epidemiological spectrum and current treatment of listeriosis. Rev Infect Dis 13:1108–1114

Cherubin CE, Marr JS, Sierra MF, Becker S (1981) Listeria and gram-negative bacillary meningitis in New York City, 1972–1979. Frequent cases of meningitis in adults. Am J Med 71:199–209

Chirgwin K, Gleich S (1989) Listeria monocytogenes osteomyelitis. Arch Intern Med 149:931–932

Ciesielski CA, Hightower AW, Parsons SK, Broome CV (1988) Listeriosis in the United States: 1980–1982. Arch Intern Med 148:1416–1419

Colodner R, Sakran W, Miron D et al (2003) Listeria monocytogenes cross-contamination in a nursery. Am J Infect Control 31:322–324

Cone LA, Leung MM, Byrd RG et al (2003) Multiple cerebral abscesses because of Listeria monocytogenes: Three case reports and a literature review of supratentorial listerial brain abscess(es). Surg Neurol 59:320–328

Cossart P, Sansonetti PJ (2004) Bacterial invasion: The paradigms of enteroinvasive pathogens. Science 304:242–248

Curosh NA, Perednia DA (1989) Listeria monocytogenes septic arthritis. A case report and review of the literature. Arch Intern Med 149:1207–1208

Czajka J, Batt CA (1994) Verification of causal relationships between Listeria monocytogenes isolates implicated in food-borne outbreaks of listeriosis by randomly amplified polymorphic DNA patterns. J Clin Microbiol 32:1280–1287

Dalton CB, Austin CC, Sobel J et al (1997) An outbreak of gastroenteritis and fever due to Listeria monocytogenes in milk. N Engl J Med 336:100–105

Decker CF, Simon GL, DiGioia RA, Tuazon CV (1991) Listeria monocytogenes infections in patients with AIDS: Report of five cases and review. Rev Infect Dis 13:413–417

Domingo P, Serra J, Sambeat MA, Ausina V (1992) Pneumonia due to Listeria monocytogenes. Clin Infect Dis 14:787–789

Drevets DA, Leenen PJ, Greenfield RA (2004) Invasion of the central nervous system by intracellular bacteria. Clin Microbiol Rev 17:323–347

Dunn PL, North RJ (1991a) Resolution of primary murine listeriosis and acquired resistance to lethal secondary infection can be mediated predominantly by Th-1+CD4– CD8– cells. J Infect Dis 164:869–877

Dunn PL, North RJ (1991b) Limitations of the adoptive immunity assay for analyzing anti-Listeria immunity. J Infect Dis 164:878–882

Dussurget O, Pizarro-Cerda J, Cossart P (2004) Molecular determinants of Listeria monocytogenes virulence. Annu Rev Microbiol 58:587–610

Dworkin MS, Williamson J, Jones JL et al (2001) Prophylaxis with trimethoprim–sulfamethoxasole for human immunodeficiency virus-infected patients: Impact on risk for infectious diseases. Clin Infect Dis 33:393–398

Eckburg PB, Montoya JG, Vosti KL (2001) Brain abscess due to Listeria monocytogenes. Five cases and a review of the literature. Medicine 80: 223–235

Edmiston CE Jr, Gordon RC (1979) Evaluation of gentamicin and penicillin as a synergistic combination in experimental murine listeriosis. Antimicrob Agents Chemother 16:862–863

Espaze EP, Reynaud AE (1998) Antibiotic susceptibilities of Listeria: In vitro studies. Infection 16:S160–S164

Evans JR, Allen AC, Stinson DA et al (1985) Perinatal listeriosis: Report of an outbreak. Pediatr Infect Dis J 4:237–241

Ewert DP, Lieb L, Hayes PS et al (1995) *Listeria monocytogenes* infection and serotype distribution among HIV-infected persons in Los Angeles County, 1985–92. J Acquir Immune Defic Syndr Hum Retrovirus 8:461–465

Faidas A, Shepard DL, Lim J et al (1993) Magnetic resonance imaging in listerial brain stem encephalitis. Clin Infect Dis 16:186–187

Farber JM, Carter AO, Varughese PV et al (1990) Listeriosis traced to consumption of alfalfa tablets and soft cheese. N Engl J Med 322:338

Farber JM, Daley E, Coates F et al (1991) Feeding trials of *Listeria monocytogenes* with a nonhuman primate model. J Clin Microbiol 29:2606–2068

Farber JM, Peterkin PI (1991) *Listeria monocytogenes*, a food-borne pathogen. Microbiol Rev 55:476–511

Farber JM, Peterkin PI, Carter AO et al. (1991) Neonatal listeriosis due to cross-infection confirmed by isoenzyme typing and DNA fingerprinting. J Infect Dis 163:927–928

Fleming DW, Cochi SL, MacDonald KL et al (1985) Pasteurized milk as a vehicle of infection in an outbreak of listeriosis. N Engl J Med 312:404–407

Frye DM, Zweig R, Sturgeon J et al (2002) An outbreak of febrile gastroenteritis associated with delicatessen meat contaminated with *Listeria monocytogenes*. Clin Infect Dis 35:943–949

Gauto AR, Cone LA, Woodard DR et al (1992) Arterial infections due to *Listeria monocytogenes:* Report of four cases and review of world literature. Clin Infect Dis 14:23–28

Gellin BG, Broome CV, Bibb WF et al (1991) The epidemiology of listeriosis in the United States—1986. Am J Epidemiol 133:392–401

Gill DA (1993) Circling disease. A meningoencephalitis of sheep in New Zealand. Vet J 89:258–270

Gordon S, Singer C (1986) *Listeria monocytogenes* cholecystitis. J Infect Dis 154:918–919

Gottlieb SL, Newbern EC, Griffin PM et al (2006) Multistate outbreak of listeriosis linked to turkey deli meat and subsequent changes in US regulatory policy. Clin Infect Dis 42:29–36

Goulet V, de Valk H, Pierre O et al (2001) Effect of prevention measures on incidence of human listeriosis, France, 1987–1997. Emerg Infect Dis 7:983–989

Goulet V, Rocourt J, Rebiere I et al (1998) Listeriosis outbreak associated with the consumption of rillettes in France in 1993. J Infect Dis 177:155–160

Gradon JD, Chapnick EK, Lutwick LI (1992) Pleuropulmonary listeriosis: Case report and review. Infect Dis Clin Pract 1:39–42

Graves LM, Swaminathan B, Reeves MW et al (1994) Comparison of ribotyping and multilocus enzyme electrophoresis for subtyping *Listeria monocytogenes* isolates. J Clin Microbiol 32:2936–2943

Greisen K, Loeffelholz M, Purohit A, Leong D (1994) PCR primers and probes for the 16S rRNA gene of most species of pathogenic bacteria, including bacteria found in cerebrospinal fluid. J Clin Microbiol 32:335–351

Hayes PS, Graves LM, Ajello GW et al (1991) Comparison of cold enrichment and U.S. Department of Agriculture methods for isolating *Listeria monocytogenes* from naturally contaminated foods. Appl Environ Microbiol 57:2109–2113

Hibbs JB Jr (2002) Infection and nitric oxide. J Infect Dis 185(Suppl):S9–17

Hirscal RB, Butler M, Coburn CE et al (1994) *Listeria monocytogenes* and severe newborn respiratory failure supported with extracorporeal membrane oxygenation. Arch Pediatr Adolesc Med 148:513–517

Ho JL, Shands KN, Friedland G et al (1986) An outbreak of type 4b *Listeria monocytogenes* infection involving patients from eight Boston hospitals. Arch Intern Med 146:520–524

Hof H, Nichterlein T, Kretschmar M (1997) Management of listeriosis. Clin Microbiol Rev 10:345–357

Holoshitz J, Schneider M, Yaretzky A et al (1984) *Listeria monocytogenes* pericarditis in a chronically hemodialyzed patient. Am J Med Sci 288:34–37

Jurado RL, Farley MM, Pereira E et al (1993) Increased risk of meningitis and bacteremia due to *Listeria monocytogenes* in patients with human immunodeficiency virus infection. Clin Infect Dis 17:224–227

Kales CP, Holzman RS (1990) Listeriosis in patients with HIV infection: Clinical manifestations and response to therapy. J Acquir Immune Defic Syndr Hum Retrovirus 3:139–143

Kalstone C (1991) Successful antepartum treatment of listeriosis. Am J Obstet Gynecol 164:57–58

Larsson S, Linell F (1979) Correlations between clinical and postmortem findings in listeriosis. Scand J infect Dis 11:55–58

Linnan MJ, Mascola L, Lou XD (1988) Epidemic listeriosis associated with Mexican-style cheese. N Engl J Med 319:823–828

Lorber B (1991) Listeriosis following shigellosis. Rev Infect Dis 13:865–866

Lorber B (1997) Listeriosis. Clin Infect Dis 24:1–11

Lorber B (2002) *Listeria monocytogenes*. In: Antimicrobial Therapy and Vaccines, 2nd edn. Yu VL, Weber R, Raoult D (eds) Apple Trees Productions, LLC, New York, pp. 429–436

Lorber B, Santoro J, Swenson RM (1975) Listeria meningitis during cefazolin therapy. Ann Intern Med 82:226–229

Louie M, Jayaratne P, Luchsinger I et al (1996) Comparison of ribotyping, arbitrarily primed PCR, and pulsed-field gel electophoresis for molecular typing of *Listeria monocytogenes*. J Clin Microbiol 34:15–19

Lyytikaiinen O, Autio T, Maijala R et al (2000) An outbreak of *Listeria monocytogenes* serotype 3a infections from butter in Finland. J Infect Dis 181:1838–1841

MacDonald PDM, Whitwam RE, Boggs JD et al (2005) Outbreak of listeriosis among Mexican immigrants as a result of consumption of illicitly produced Mexican-style cheese. Clin Infect Dis 40: 677–682

Mackaness GB (1962) Cellular resistance to infection. J Exp Med 116:381–406

Mackaness GB (1971) Resistance to intracellular infection. J Infect Dis 123: 439–445

Mascola L, Lieb L, Chiu J et al (1988) Listeriosis: An uncommon opportunistic infection in patients with acquired immunodeficiency syndrome. A report of five cases and a review of the literature. Am J Med 84:162–164

Mazzulli T, Salit IE (1991) Pleural fluid infection caused by *Listeria monocytogenes*: Case report and review. Rev Infect Dis 13:564–570

Mengard J, Ohayon H, Gounon P et al (1996) E-cadherin is the receptor for internalin, a surface protein required for entry of *L. monocytogenes* into epithelial cells. Cell 84:923–932

Merle-Melet M, Dossou-Gbete L, Meyer P et al (1996) Is amoxicillin–cotrimoxazole the most appropriate antibiotic regimen for *Listeria* meningoencephalitis? Review of 22 cases and the literature. J Infect 33:79–85

Meyer RD, Liu S (1987) Determination of the effect of antimicrobics in combination against *Listeria monocytogenes*. Diagn Microbiol Infect Dis 6:199–206

Miettinen MK, Siitonen A, Heiskanen P et al (1999) Molecular epidemiology of an outbreak of febrile gastroenteritis caused by *Listeria monocytogenes* in cold-smoked rainbow trout. J Clin Microbiol 37:2358–2360

Mossey RT, Sondheimer J (1985) Listeriosis in patients with long-term hemodialysis and transfusional iron overload. Am J Med 79:379–400

Murray EGD, Webb RA, Swann MBR (1926) A disease of rabbits characterized by large mononuclear leukocytosis caused by a hitherto undescribed bacillus *Bacterium monocytogenes* (n sp). J Pathol 29:407–439

Myers JP, Peterson G, Rashid A (1983) Peritonitis due to *Listeria monocytogenes* complicating continuous ambulatory peritoneal dialysis. J Infect Dis 148:1130

Mylonakis E, Hohmann EL, Calderwood SB (1988) Central nervous system infection with *Listeria monocytogenes*. 33 years' experience at a general hospital and review of 776 episodes from the literature. Medicine 77:313

Mylonakis E, Paliou M, Hohmann EL et al (2002) Listeriosis during pregnancy: A case series and review of 222 cases. Medicine 81:260–269

Nieman RE, Lorber B (1980) Listeriosis in adults: A changing pattern. Report of eight cases and review of the literature, 1968–1978. Rev Infect Dis 2:207–227

Olsen SJ, Patrick M, Hunter SB et al (2005) Multistate outbreak of *Listeria monocytogenes* infection linked to delicatessen turkey meat. Clin Infect Dis 40:962–967

Ooi ST, Lorber B (2005) Gastroenteritis due to *Listeria monocytogenes*. Clin Infect Dis 40:1327–1332

Pinner RW, Schuchat A, Swaminathan B et al (1992) Role of foods in sporadic listeriosis. II. Microbiologic and epidemiologic investigation. JAMA 267:2046–2050

Plaut M, Gardner P (1972) *Listeria monocytogenes* sepsis with disseminated intravascular coagulation. South Med J 65:490–492

Portnoy DA, Auerbach V, Glomski IJ (2002) The cell biology of *Listeria monocytogenes* infection: The intersection of bacterial pathogenesis and cell-mediated immunity. J Cell Biol 158:409–414

Reglier-Poupet H, Pellegrini E, Charbit A et al (2003) Identification of LpeA, a PsaA-like membrane protein that promotes cell entry by Listeria monocytogenes. Infect Immun 71:474–482

Riedo FX, Pinner RW, Tosca MdeL et al (1994) A point-source foodborne listeriosis outbreak: Documented incubation period and possible mild illness. J Infect Dis 170:693–696

Safdar A, Armstrong D (2003) Listeriosis in patients at a comprehensive cancer center, 1955–1997. Clin Infect Dis 37:359–364

Safdar A, Armstrong D (2003) Antimicrobial activities against 84 *Listeria monocytogenes* isolates from patients with systemic listeriosis at a comprehensive cancer center (1955–1997). J Clin Microbiol 41:483–485

Safdar A, Papadopoulous EB, Armstrong D (2002) Listeriosis in recipients of allogeneic blood and marrow transplantation: Thirteen year review of disease characteristics, treatment outcomes and a new association with human cytomegalovirus infection. Bone Marrow Transplant 29:913–916

Sanger JM, Sanger JW, Southwick FS (1992) Host cell actin assembly is necessary and likely to provide the propulsive force for intracellular movement of *Listeria monocytogenes*. Infect Immun 60:3609–3619

Sauders BD, Fortes ED, Morse DL et al (2003) Molecular subtyping to detect human listeriosis clusters. Emerg Infect Dis 9:672–680

Scheld WM (1983) Evaluation of rifampin and other antibiotics against *Listeria monocytogenes* in vitro and in vivo. Ref Infect Dis 5:S593–S599

Schlech WF III, Chase DP, Badley A (1993) A model of food-borne *Listeria monocytogenes* infection in the Sprague–Dawley rat using gastric inoculation: Development and effect of gastric acidity on infective dose. Int J Food Microbiol 18:15–24

Schlech WF III, Lavigne PM, Bortolussi RA et al (1983) Epidemic listeriosis—evidence for transmission by food. N Engl J Med 308:203–206

Schlech WF III, Schlech WF IV, Haldane H et al (2005) Does sporadic *Listeria* gastroenteritis exist? A 2-year population-based survey in Nova Scotia, Canada. Clin Infect Dis 41:778–784

Schroter GPJ, Weil R (1977) *Listeria monocytogenes* infection after renal transplantation. Arch Intern Med 137:1395–1399

Schuchat A, Deaver K, Hayes PS et al (1993) Gastrointestinal carriage of *Listeria monocytogenes* in household contacts of patients with listeriosis. J Infect Dis 167:1261–1262

Schuchat A, Deaver KA, Wenger JD et al (1992) Role of foods in sporadic listeriosis. 1. Case-control study of dietary risk factors. JAMA 267:2041–2045

Schuchat A, Lizano C, Broome CV et al (1991) Outbreak of neonatal listeriosis associated with mineral oil. Pediatr Infect Dis 10:183–189

Schuchat A, Robinson K, Wenger JD et al (1997) Bacterial meningitis in the United States in 1995. N Engl J Med 337:970–976

Schuchat A, Swaminathan B, Broome CV (1991) Epidemiology of human listeriosis. Clin Microbiol Rev 1991;4:169–183

Schwartz B, Hexter D, Broome CV et al (1989) Investigation of an outbreak of listeriosis: New hypotheses for the etiology of epidemic *Listeria monocytogenes* infections. J Infect Dis 159:680–685

Sheehan GJ, Galbraith JCT (1993) Colonoscopy-associated listeriosis: Report of a case. Clin Infect Dis 17:1061–1062

Siegman-Igra Y, Levin R, Weinberger M et al (2002) *Listeria monocytogenes* infection in Israel and review of cases worldwide. Emerg Infect Dis 8:305–310

Skoberg K, Syrjanen J, Jahkola M et al (1992) Clinical presentation and outcome of listeriosis in patients with and without immunosuppressive therapy. Clin Infect Dis 14:815–821

Slifman NR, Gershon SK, Lee JH et al (2003) *Listeria monocytogenes* infection as a complication of treatment with tumor necrosis factor alpha-neutralizing agents. Arth Rheum 48:319–324

Smith MA, Takeuchi K, Brackett RE et al (2003) Nonhuman primate model for *Listeria monocytogenes*-induced stillbirths. Infect Immun 71:1574–1579

Southwick FS, Purich DL (1996) Intracellular pathogenesis of listeriosis. N Engl J Med 334:770–776

Spitzer PG, Hammer SM, Karchmer AW (1986) Treatment of *Listeria monocytogenes* infection with trimethoprim–sulfamethoxazole: Case report and review of the literature. Rev Infect Dis 8:427–430

Stamm AM, Dismukes WE, Simmons BP et al (1982) Listeriosis in renal transplant recipients: Report of an outbreak and review of 102 cases. Rev Infect Dis 4:665–682

Stamm AM, Smith SH, Kirklin JK, McGiffin DC (1990) Listerial myocarditis in cardiac transplantation. Rev Infect Dis 12:820–823

Stanley NF (1949) Studies of *Listeria monocytogenes:* I. Isolation of a monocytosis-producing agent (MPA). Aust J Exp Biol Med Sci 27:123–131

Swaminathan B, Barrett TJ, Hunter SB et al (2001) PulseNet: The molecular subtyping network for foodborne bacterial disease surveillance, United States. Emerg Infect Dis 7:382–389

Sword CP (1966) Mechanisms of pathogenesis in *Listeria monocytogenes* infection. I. Influence of iron. J Bacteriol 92:536–542

Tappero JW, Schuchat A, Deaver KA et al (1995) Reduction in the incidence of human listeriosis in the United States. Effectiveness of prevention efforts? JAMA 273:1118–1122

Thomas F, Ravaud Y (1988) Rhabdomyolysis and acute renal failure associated with listeria meningitis. J Infect Dis 158:492–493

Van J-CN, Nguyen L, Guillemam R et al (1994) Relapse of infection or reinfection by *Listeria monocytogenes* in a patient with heart transplant: Usefulness of pulsed-field gel electrophoresis for diagnosis. Clin Infect Dis. 1994;19:208–209

Vasquez-Boland JA, Kuhn M, Berche P et al (2001) *Listeria* pathogenesis and molecular virulence determinants. Clin Microbiol Rev 14:584

Visintine CM, Oleske JM, Nahmias AJ (1977) *Listeria monocytogenes* infection in infants and children. Am J Dis Child 131:393–397

Weinberg ED (1984) Pregnancy-associated depression of cell-mediated immunity. Rev Infect Dis 6:814–831

Wenger JD, Hightower AW, Facklam RR et al (1990) Bacterial meningitis in the United States, 1986: Report of a multistate surveillance study. J Infect Dis 162:1316–1323

Wing EJ, Gregory SH (2002) *Listeria monocytogenes*: Clinical and experimental update. J Infect Dis 185(Suppl):S18–S24

Winslow DL, Pankey GA (1982) In vitro activities of trimethoprim and sulfamethoxazole against *Listeria monocytogenes*. Antimicrob Agents Chemother 22:51–54

Winslow DL, Steele ML (1984) *Listeria* bacteremia and peritonitis associated with a peritoneovenous shunt: Successful treatment with sulfamethoxazole and trimethoprim. J Infect Dis 149:820

Witlox MA, Klinkenberg-Knol EC, Meuwissen SGM (2000) *Listeria* sepsis as a complication of endoscopy. Gastrointest Endosc 51:235–236

Yu VL, Miller WP, Wing EJ et al (1982) Disseminated listeriosis presenting as acute hepatitis. Case reports and review of hepatic involvement in listeriosis. Am J Med 73:773–777

3
Listeria Genomics

Philippe Glaser,* Christophe Rusniok, and Carmen Buchrieser
Unité de Génomique des Microorganisme Pathogènes—URACNRS 2171
Institut Pasteur, 28 Rue du Dr Roux 75724 Paris
*For correspondence, *e-mail: pglaser@pasteur.fr*

Abstract: The publication of the complete genome sequences of four clinical *Listeria monocytogenes* isolates and one of its apathogenic relative, *Listeria innocua*, in 2001 and 2003 has paved the way for major breakthroughs in understanding the biology of *L. monocytogenes*. The availability of these sequences allowed applying postgenomics analysis to understand virulence and ecology of *L. monocytogenes* and evolution of the genus *Listeria*. Comparative genomics now help to unravel the molecular basis of the pathogenesis, the host range, the evolution, and the phenotypic differences among *L. monocytogenes* isolates and within the genus *Listeria*. Based on sequence analysis, many new virulence factors have been identified, foremost among them, surface proteins like internalins, in agreement with the expansion of this protein family revealed by genome sequencing. The discovery of a broad transport capacity, in particular, of carbon sources and of unexpected metabolic pathways like the biosynthesis of vitamin B_{12} and glutathione (GSH) explains the ubiquitous nature of *Listeria*. Some of the many regulators identified in the genome are now analyzed, demonstrating, e.g., the importance of an orphan two-component regulator DegU in virulence of *L. monocytogenes* or leading to the identification of a new class of regulators like MogR, the master regulator of flagella expression. Combined with genome analysis, transcriptome studies using microarrays allow for the first time the global and parallel analysis of *L. monocytogenes* gene expression inside and outside eukaryotic cells. Multiple ongoing sequencing projects allowing multigenome comparisons combined with different postgenomics and functional analyses open the way to a global understanding of the ability of *Listeria* to adapt to a broad range of environments and hosts and of the multiple aspects of the disease it causes, listeriosis.

3.1. Introduction

The term "Genomics", coined by Thomas Roderick in 1986, refers to an at-that-time new scientific discipline of mapping, sequencing, and analyzing genomes (Kuska 1998). Although genomics was the object of controversies

33

in its early times, it was boosted by the human genome project and has meanwhile revolutionized our way of thinking in biological research. Given the small size of their genomes, this is particularly true for bacteria. Genomics also provides, for the first time, the possibility to analyze at the molecular level an organism as an entity, and with the emergence of systems biology it will be possible to integrate all different knowledge coherently. However, genomics is still intricately linked to genetics, a discipline developed in the 19th century. With respect to the genus *Listeria*, genomics started with the first chromosome structure analysis based on pulsed field gel electrophoresis (PFGE) and has led now to what is referred to as postgenomics studies including systematic gene knockouts, proteome, and transcriptome analysis.

The first published description of *L. monocytogenes* was that by Murray et al. in 1926. These researchers observed six cases of rather sudden death of young rabbits in 1924 in the animal-breeding establishment of the Department of Pathology of Cambridge and many more in the succeeding 15 months. The interesting characters presented by the disease and the mortality prompted investigation. In 1927, during investigations of unusual deaths observed in gerbils near Johannesburg (South Africa), Pirie discovered the same organism (Pirie 1927). Since then many studies have been undertaken to characterize *L. monocytogenes* which was until 1966 the only member of the genus. *Listeria* are Gram-positive rods that are not considered as naturally competent and, although phages infecting *Listeria* have been known for a long time, general transduction systems were described only in 2000 (Hodgson 2000). Due to the absence of genetic tools, *L. monocytogenes*, like most pathogenic bacteria, escaped classical genetics. Nevertheless, the development of transposon mutagenesis and genetic screens allowed the isolation of mutants affected in virulence genes. The first gene to be inactivated was *hly* encoding the cytolysin, by using the 26-kb-long transposon Tn1545 (Gaillard et al. 1986). The loss of virulence associated with the inactivation of this gene demonstrated the critical role of hemolysin in virulence. Following *hly*, major virulence determinants, most of them encoded on the so-called "virulence gene cluster" as well as on the *inlAB* operon, were subsequently identified by using various phenotypic tests and transposon mutant collections (Portnoy et al. 1992). Genetic identification of virulence factors depends heavily on the experimental approach used to test for virulence. For example, signature tagged mutagenesis (STM), by testing pools of mutants, allows screening of a large number of mutant strains in an animal model. Recently, several STM screens have been performed with *L. monocytogenes* leading to the identification of additional virulence factors including regulatory systems like the two-component system Lgr, a homolog of the *Staphylococcus aureus* Agr system (Autret et al. 2003) and the VirR response regulator (Mandin et al. 2005a). However, genomics of *L. monocytogenes* was the prerequisite to gain a complete understanding of the genetic basis of the virulence and the ecology of this food-borne pathogen.

3.2. Genome Comparisons

3.2.1. Genome Mapping

In 1984, Schwartz and Cantor developed a procedure allowing for the first time to separate DNA fragments larger than 50 kb in size by using alternately pulsed, electrical fields and chromosomal DNA embedded in agarose. This technique was initially used to separate the 14 *Saccharomyces cerevisiae* chromosomes (Schwartz and Cantor 1984). It was then in 1992 that the first physical map of the chromosome of a *L. monocytogenes* strain, strain LO28 of serotype 1/2c, was obtained (Michel and Cossart 1992). Physical maps of the chromosomes of two other extensively studied *L. monocytogenes* strains were subsequently published – that of strain Scott A in 1997 (He and Luchansky 1997) and that of strain EGD in 1999 (von Both et al. 1999). These physical maps show differences but all indicated a genome size of about 3 Mb for *L. monocytogenes*. In 1991, macrorestriction analysis by PFGE was used for the first time to compare the genome organization of food and clinical isolates of *L. monocytogenes* by using *Apa*I, *Not*I, and *Sma*I (Brosch et al. 1991) or *Apa*I and *Not*I (Carriere et al. 1991) as restriction enzymes. This method, established in 1991, for *Listeria* typing was subsequently validated by an international WHO multicenter study (Brosch et al. 1996). Further studies led to the standardization of PFGE for *L. monocytogenes* (Gerner-Smidt et al. 2006; Graves and Swaminathan 2001). PFGE is still thought to be the most reliable method for typing and differentiating *L. monocytogenes* strains.

Characterization of a large number of *L. monocytogenes* strains by PFGE demonstrated that the previously established two primary phylogenetic divisions using multilocus enzyme electrophoresis (MLEE) present within the species *L. monocytogenes* (Bibb et al. 1989; Piffaretti et al. 1989) are also reflected in the genome organization of this species (Brosch et al. 1994). These analyses together with many further studies led to the identification of three lineages within this species that correlate with serovars: lineage I comprises strains of serovar 1/2a, 1/2c, 3a; lineage II comprises strains of serotypes 1/2b, 4b, 4e; and lineage III comprising strains of serotypes 4a and 4c (Bibb et al. 1989; Brosch et al. 1994; Graves et al. 1994; Jeffers et al. 2001; Piffaretti et al. 1989; Wiedmann et al. 1997). This classification is highly robust, as it is reproduced also when using other molecular typing methods such as ribotyping (Wiedmann et al. 1997), Multi Locus Sequence Typing (Zhang et al. 2004), and DNA-array hybridization (Doumith et al. 2004b). This finding led to the idea that the three lineages are of ancient origin and the congruence of various methods highlights the genomic stability of the species.

3.2.2. Genome Sequencing and Analysis

The first bacterial genome projects concerned thoroughly analyzed bacteria like the model organisms *Escherichia coli* (Blattner et al. 1997) and *Bacillus subtilis* (Kunst et al. 1997) or the pathogenic *Mycobacterium tuberculosis* (Cole et al. 1998). Sequencing of these genomes was based on a top-down approach using subcloning of large chromosomal overlapping fragments in lambda phage, plasmids, or cosmids. However, in 1995 the first complete bacterial genome sequence published, that of *Haemophilus influenzae*, a genetically poorly characterized human pathogen, was obtained by using a bottom-up approach known as whole genome shotgun sequencing (Fleischmann et al. 1995). This method has revolutionized genome sequencing. The procedure is based on random sequencing of a large number of clones and assembly of the individual sequences by using their overlapping regions. Combined with improvement of the DNA sequencing technology, the shotgun sequencing can be nearly fully automated and is nowadays an extremely efficient and reliable technique used broadly for whole genome sequencing. With the determination of the nucleotide sequence of the *H. influenzae* genome, the area of bacterial genomics started. Among the more than 430 freely available complete bacterial genome sequences a large proportion are genomes of human pathogens (http://www.ncbi.nlm.nih.gov/genomes/lproks.cgi).

In *Listeria* research, the area of genomics started with the determination of the complete genome sequences of two members of the genus by a consortium of 10 European laboratories, the pathogen *L. monocytogenes* (strain EGDe), and the closely related apathogenic species *L. innocua* (strain CLIP11626) (Glaser et al. 2001a). These sequences were determined by using the whole genome sequence approach and were published 6 years after that of the genome of *H. influenzae*. For the annotation and analysis of these sequences, a software package (CAAT-box) was developed allowing both, to share the annotation among different research laboratories in different countries and the beginning of the annotation before completion of the finishing phase (Frangeul et al. 2004). The genome sequence of the closely related, nonpathogenic species *L. innocua* was determined in parallel with that of *L. monocytogenes* to allow genome comparisons of pathogenic and nonpathogenic *Listeria*. Strain *L. monocytogenes* EGD was chosen, because it is one of the most studied strains in many different laboratories around the world for a broad range of research questions concerning various aspects of *L. monocytogenes* virulence and ecology. *L. monocytogenes* strain EGD is a derivative of the strain originally isolated by Murray that was then used by Mackaness in his studies on cell-mediated immunity (Mackaness 1964). As since that time genetic differences have appeared among different isolates of EGD kept in different laboratories, the strain selected for sequencing was a well-defined highly virulent isolate, provided by Trinad Chakraborty (University of Giessen) and was named EGDe.

Additional sequence information is now available, as in 2004: the Institute of Genomic Research (TIGR) has published the genome sequences of three *L. monocytogenes* strains isolated from food incriminated as source of listeriosis cases (Nelson et al. 2004). Strain F2365 is a serotype 4b cheese isolate from an outbreak in 1985 in California/USA; strain H7858 is also of serotype 4b and was isolated from contaminated frankfurters responsible for a multistate outbreak in the USA (1998–1999); and strain F6854 is of serotype 1/2a and was isolated from turkey frankfurters thought to be responsible for a sporadic case of human listeriosis. The sequence of the genome of strain F2365 was completed; the two other genome sequences were released nearly complete as a set of 133 (strain F6854) and 181 (strain H7858) concatenated contigs. In addition to these published genome sequences, further *Listeria* genome sequencing is under way in the US and in Europe. The Broad Institute has recently released the draft genome sequences of 18 *L. monocytogenes* strains from various origins (http://www.broad.mit.edu/seq/msc/) including the only serotype 3a strain responsible of an outbreak in Finland in 1999 (Lyytikainen et al. 2000) and the strain responsible of an outbreak of gastro enteritis in Italy in 1997 (Aureli et al. 2000). In Europe the German Pathogenomics network in collaboration with the Institut Pasteur is sequencing one isolate from each of the four remaining species of the genus *Listeria (L. ivanovii, L. seeligeri, L. welshimeri,* and *L. grayi)* (Hain et al. 2006). The outcome of this international effort will provide a unique opportunity to characterize and understand the mechanisms of evolution within the genus *Listeria*, to set the basis for genome-based population genetics and to characterize the specificity of clinical isolates.

3.2.3. Comparative Analysis of L. monocytogenes with Available Genome Sequences

Analysis of the available sequences revealed that the *Listeria* genomes are very homogeneous in size ranging from 2893 to 3,011 kb with conserved general features summarized in Table 3.1. If integrated prophages are not considered, the *L. innocua* genome, which contains five integrated prophages accounting for 216.8 kb, has the smallest size of 2,794 kb containing 2,679 protein-encoding genes (Table 3.1.). Based on the 16S rRNA sequence, the genus *Listeria* is classified into the group of low GC Gram positive bacteria closely related to *Bacillaceae*. Another means to compare organisms is based on gene content of their genomes and on the distribution of the highest similarity score of each protein or on the number of bidirectional best hits (BDBH). In 2001, when the genes predicted in the *L. monocytogenes* strain EGDe were first annotated, 318 genes were scored as coding proteins without similarity to any other protein in the public sequence libraries or as showing similarity only to other *L. monocytogenes* proteins. Since then, an enormous amount of new sequences has been added to these public databases. Thus, it was expected that these scores might have changed. In order to be able to compare the results we had gotten in 2001 with the situation 5 years later, we re-analyzed the similarity scores of all predicted protein sequences of *L. monocytogenes* EGDe to proteins present in the public databases. We recorded the first six non-*Listeria* BLASTp hits in the NCBI-NR protein sequence database with each of the EGDe proteins. This database contains protein sequences from over 500 complete or unfinished bacterial genome sequences. Surprisingly, out of the originally 318 orphan genes, only 60 protein sequences had now a similar sequence among this tremendous amount of new protein

38   Glaser et al.

TABLE 3.1. General features of published *Listeria* genome sequences.

	L. monocytogenes EGDe (1/2a)	*L. monocytogenes* F6854 (1/2a)[c]	*L. monocytogenes* F2365 (4b)	*L.monocytogenes* H7858(4b)[c]	*L.innocua* CLIP11262 (6a)
Size of the chromosome (bp)	2 944 528	~ 2 953 211	2 905 310	~ 2 893 921	3 011 209
G+C content (%)	38	37.8	38	38	37.4
G+C content protein-coding genes (%)	38.4	38.5	38.5	38.4	38.0
Total number of CDS[a]	2853	2973	2847	3024	2973
Percentage coding	89.2%	90.3%	88.4%	89.5%	89.1%
Number of prophage regions	1	3	2	2	5
Monocins	1	1	1	1	1
Plasmid	–	–	–	1 (94 CDS)	1 (79 CDS)
Number of strain-specific genes[b]	61	97	51	69	78
Number of transposons	1 (Tn916 like)	–	–	–	–
Number of rRNA operons	6	6	6	6	6
Number of tRNA genes	67	67[c]	67	65[c]	66

[a]CDS = coding sequence
[b]except prophage genes
[c]Draft genome sequence (eightfold coverage without gap closure)

sequences. Thus 258 genes still encode proteins that are orphans. This result strongly supports the view that these genes are specific of the genus *Listeria*.

As horizontal gene transfer may play a considerable role in the adaptation of a bacterium to its specific niche(s), we have compared the occurrence of best BLASTp and phylogeny. In Figure 3.1., the distribution of best BLASTp hits among bacterial genera or higher order taxonomic groups is presented. In agreement with the phylogeny, the largest number of best hits was detected among *Bacillaceae*, with 1,523 genes, which represent 58% of those that encode a protein that has a similarity hit with a protein in the public database. These proteins constitute the core genome of this group of bacteria, encoding mainly functions like the transcription and translation machinery and metabolic pathways. The remaining genes, the flexible gene pool, are probably responsible for the adaptation of these bacteria to their different environments and niches where they are found. When analyzing the distribution of the best hits of these

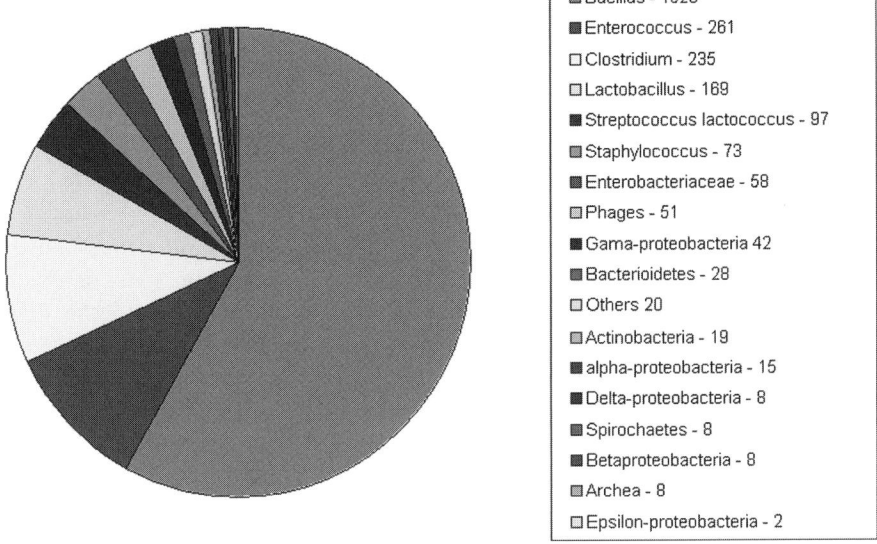

- ■ Bacillus - 1523
- ■ Enterococcus - 261
- □ Clostridium - 235
- □ Lactobacillus - 169
- ■ Streptococcus lactococcus - 97
- ■ Staphylococcus - 73
- ■ Enterobacteriaceae - 58
- □ Phages - 51
- ■ Gama-proteobacteria 42
- ■ Bacterioidetes - 28
- □ Others 20
- ■ Actinobacteria - 19
- ■ alpha-proteobacteria - 15
- ■ Delta-proteobacteria - 8
- ■ Spirochaetes - 8
- ■ Betaproteobacteria - 8
- ■ Archea - 8
- □ Epsilon-proteobacteria - 2

FIGURE 3.1. Distribution of best BLASTp hits of *L. monocytogenes* EGDe proteins among bacterial genera or higher-order taxonomic groups. (A color version of this figure appears between pages 196 and 197.)

remaining genes among the group of low GC Gram positive bacteria, the number of hits to proteins of each of the genera represented in this group depends on both phylogenetic distance and genome size. Accordingly, the highest number of best hits of *Listeria* proteins after those having their best hits with bacilli proteins is found with enterococci proteins, and the third with clostridia proteins. *Clostridia* are third although streptococci are phylogenetically more closely related, because of their larger genome size as compared to streptococci. This could be due to gene loss in particular in the flexible gene pool of streptococci related to the adaptation to the mammalian host environment. Surprisingly only 19 genes have a best BLASTp hit with proteins belonging to the group of Gram-positive *Actinobacteria*. Among the bacterial species more distantly related to *Listeria* than the above mentioned, a high proportion of best hits (58 genes) was observed with proteins similar to proteins encoded by Enterobacteriaceae. These genes or some of them might have been acquired from Enterobacteriaceae by horizontal gene transfer. For example, among these 58 genes, 24 are part of the vitamin B_{12} biosynthesis, propanediol, and ethanolamine utilization gene cluster, which have already been suggested to be horizontally transferred between these two groups of bacteria (Buchrieser et al. 2003). These genes are involved in anaerobic utilization of different carbon sources, which may contribute to the fitness of *Listeria* in the gut, like for some of the *Enterobacteriaceae*. They also could be important for growth in vegetable waste such as silage known to be favorable for *Listeria* growth. Further examples for putative horizontal transfer from *Enterobacteriaceae* are three operons involved in sugar metabolism. The first operon, *lmo2133–2137*, is predicted to be implicated in fructose metabolism as it encodes two type II fructose 1-6 biphospahe aldolases. One is a *Bacillus* type and the second one is highly similar to the *E. coli* one. The second cluster, *lmo2761–2764*, encodes a beta-glucoside utilization operon and the third, *lmo2834–2838*, encodes four genes, probably also involved in sugar import and catabolism. This observation is in agreement with a shared growth environment for these two groups of strains, which is favorable for genetic exchange between strains.

It has been proposed that, in the environment, as for *Legionella*, unicellular eukaryotic organisms, like amoeba, algae, or protozoa, may be hosts for *L. monocytogenes* growth and survival (Gray et al. 2006). In the case of *Legionella pneumophila*, genome analysis revealed the abundance of genes encoding eukaryotic like proteins best explained by gene acquisition from the eukaryotic hosts (Cazalet et al. 2004). In the case of *L. monocytogenes*, we did not identify a single protein with a clear eukaryotic origin (i.e., a best BLASTp with an eukaryotic sequence) clearly indicating that *L. monocytogenes* does not interact so closely and for such a long time with a eukaryotic host as does *Legionella* species. In agreement with this observation, we observed a single gene (*lmo1635*) encoding a protein with a best BLASTp hit with an *L. pneumophila* protein. Overall, 91% of the *Listeria* genes other than phage genes have a best-BLASTp hit with proteins of low GC Gram positive bacteria. This strongly indicates that horizontal gene transfer to *L. monocytogenes* takes place mainly among closely related species.

3.2.4. Intraspecies and Interspecies Comparisons—the Listeria *Genomes*

The initial comparison of the *L. monocytogenes* EGDe and *L. innocua* sequences (Glaser et al. 2001a) revealed several major features of the *Listeria* genomes: a strong conservation of genome organization, a low occurrence of insertion sequences (IS) elements, and the absence of typical pathogenicity islands, but instead the occurrence of multiple insertions and deletions leading to a general organization of a conserved backbone with multiple interspersed smaller islets. An interesting but still unexplained observation is the heterogeneity in the distribution of these islets. The majority of the insertion/deletion events take place in one part of the chromosome ranging from ca 2,600 to 1,200 kb (Figure 3.2.). These first observations from *L. monocytogenes* EGDe and *L. innocua* genome comparison were fully confirmed by the comparison with the three additional genome sequences available. Although the genomes of strains F6854 and H7858 are not finished, a complete synteny of the five genomes was detected, and differences arising mostly by insertions and deletions taking place in one half of the genome (data not shown). According to similarity criteria used by Nelson and colleagues (Nelson et al. 2004) (90% identity over 90% of the gene by BLASTn comparison), 2,499 genes are conserved in the four *L. monocytogenes* genomes, among which 2,394 are also present in *L. innocua*.

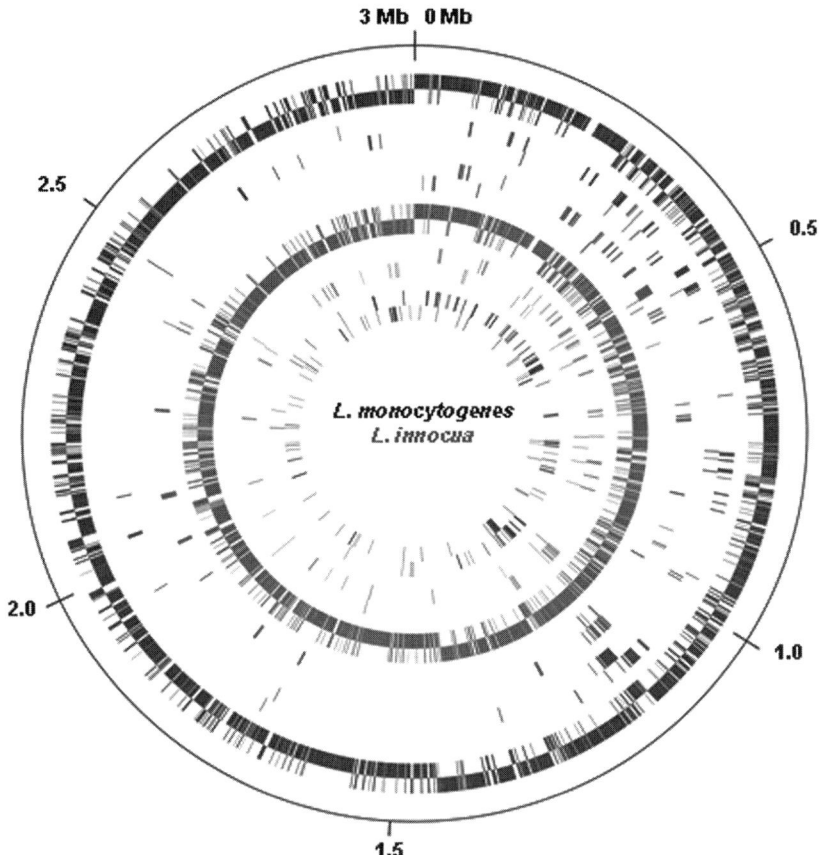

FIGURE 3.2. Circular genome maps of *L. monocytogenes* EGDe and *L. innocua* CLIP11262 showing the position and orientation of genes on the + and − strands, respectively. From the outside: circle 1, *L. monocytogenes* EGDe genes; circle 2, *L. monocytogenes* EGDe genes missing in strain F2365; circle 3, *L. monocytogenes* F2365 missing in EGDe; circle 4, *L. innocua* genes; circle 4, *L. innocua* genes missing in *L. monocytogenes* EGDe; circle 5, *L. monocytogenes* EGDe genes missing in *L. innocua*. The scale in Mb is indicated on the outside with the origin of replication being at position 0. (A color version of this figure appears between pages 196 and 197.)

This conserved genome organization may be related to the low occurrence of IS elements, suggesting that IS transposition or IS-mediated deletions are no key evolutionary mechanisms in *Listeria*. The chromosomes of the serotype 4b strains (F2365 and H7858) lack intact IS but do contain four transposases of the IS3 family that are present in homologous locations in both strains. The serotype 1/2a strains (F6854 and EGDe) each contains three and *L. innocua* contains four of these elements. Furthermore, the serotype 1/2a strains contain an intact IS element named ISLmo1, two are present in strain F6854, and three, one of which is not intact, are present in EGDe. ISLmo1 is missing in the serotype 4b and the *L. innocua* strains. Interestingly, in all five sequenced genomes no gene is inactivated by IS elements.

Despite this high-genome conservation, considerable difference in gene content is present among the different *Listeria* isolates. The combined flexible gene pool (genes which are missing in one or several isolates) of the four *L. monocytogenes* genomes accounts for 1,197 genes highlighting this genome diversity. As extensively discussed by Nelson and colleagues (Nelson et al. 2004), a significant proportion of this flexible gene pool is related to prophages or to remnants of prophages. However, the first comparison of complete *Listeria* genomes, that of *L. monocytogenes* EGDe and of *L. innocua* CLIP 11262, revealed that, if prophages were not taken into account, diversity is also present on the *Listeria* backbone. Two hundred and seventy-two genes clustered in 100 regions are specific to strain EGDe and 144 are specific to the *L. innocua* strain. With the availability of additional genome sequences, multigenome sequence comparisons can be now undertaken to reconstruct the evolutionary history of the *Listeria* genomes as this should allow discriminating between insertions, deletions, and horizontal gene transfer events. Gene clusters that are only missing in one isolate but are present in all the others are best explained by a single deletion event. Multigenome comparisons were also performed to identify strain-specific and serotype-specific genes of the different *L. monocytogenes* isolates. This identified 83 genes specific to the serotype 1/2a strains and 51 genes specific to the serotype 4b strains.

As mentioned above, three different evolutionary-like lineages are present within the species *L. monocytogenes*, eventually even accounting for different disease potential for humans (Wiedmann et al. 1997). Within the species *L. monocytogenes*, comparison of the *L. monocytogenes* EGDe (serotype 1/2a, lineage 1) and of the *L. monocytogenes* F2365 strain (serotype 4b, lineage II) provides a first glimpse of the differences and specificities of these two major lineages. Among the 159 EGDe genes missing in strain F2365, 29 are also present in *L. innocua* corresponding thus probably to deletions in strain F2365. A majority of these 29 genes encodes proteins of unknown function. Conversely, 164 genes annotated in strain F2365 are missing in strain EGDe including 30 genes also present in *L. innocua*. Among these 30 genes, genes encoding functions required for lipoteichoic acid synthesis, glycosyl transferases, probably involved in cell wall modification, a soluble internalin, and six cell-wall-bound proteins were identified. These shared genes may correspond to the similar cell-wall structure and of antigenicity as recognized by serotyping between *L. innocua* and *L. monocytogenes* strains of serotype 4b as previously described (Fiedler 1988; Fiedler and Ruhland 1987).

3.3.   Specific Features of the *L. monocytogenes* Genomes

Functional annotation of the *L. monocytogenes* EGDe genome sequence revealed four major findings: the broad number of membrane-anchored proteins and of secreted proteins, the versatility of carbon source utilization as indicated by the large number of sugar permeation systems involved in sugar or related carbon source utilization, particular metabolic capacities, and a higher proportion than expected of regulatory genes. Genome comparison showed that these classes of genes are variable among the different isolates in agreement with the ability of *Listeria* to adapt to a broad range of environments. These specific features are discussed below.

3.3.1. Virulence Factors and Surface Proteins

The major virulence factors of *L. monocytogenes* identified mainly through different genetic approaches include genes encoding the invasion proteins InlA and InlB, genes encoding proteins that allow escape from the phagocytic vacuole (Listeriolysin O (LLO) and a PI-PLC), intracellular actin-based motility (ActA), and cell-to-cell spread (PlcB). Except *inlAB*, these genes are clustered together with their common transcription regulator, PrfA, on a 9-kb "virulence locus" which is absent from the nonpathogenic species *L. innocua*. Complete genome comparison as well as DNA array hybridization revealed that these genes are present and are generally highly conserved in all *L. monocytogenes* strains.

Other proteins that may have important roles in the interactions of microorganisms with their environments, in particular during host infection are surface proteins. One particular feature of *Listeria*, identified through genome analysis, is the expansion of the surface protein coding gene families. The *Listeria* genomes encode many such proteins, e.g., 4.7% of all predicted genes of *L. monocytogenes* EGDe. The largest surface protein family are lipoproteins and the second largest family are LPXTG proteins including the internalin family, like InlA. Although the major known virulence factors are conserved, there is a pronounced diversity within the surface proteins of the different strains. For example, when comparing *L. monocytogenes* EGDe (serovar 1/2a) to *L. monocytogenes* F2635 (serotype 4b), five lipoproteins, nine LPXTG proteins, and two autolysins are specific to the serotype 4b strain. As another example, among the 41 LPXTG proteins identified in *L. monocytogenes* EGDe, 21 are absent from *L. innocua* CLIP11262. *L. innocua* CLIP11262 codes for 34 LPXTG proteins, 14 of which are absent from *L. monocytogenes* EGDe (Cabanes et al. 2002; Glaser et al. 2001a; Hamon et al. 2006). Analysis of the distribution of these surface protein coding genes in the five sequenced genomes showed that seven internalin/LPXTG proteins, 11 LPXTG, seven GW module-containing proteins, and three secreted internalins are the core set of this class of surface/secreted proteins as they are present in all four sequenced *L. monocytogenes* and the sequenced *L. innocua* strain. Six Internalins, seven LPXTG, one GW module-containing protein (InlB), and two secreted internalins are species specific, as they are present in all *L. monocytogenes* strains sequenced, but are absent from *L. innocua* CLIP11262. In addition, each strain encodes specific ones like one internalin protein in *L. monocytogenes* EGDe (*lmo2026*), two in strain F6854 (*LMOf6854_0338, LMOf6854_0365*), one in strain F2365 (*LMOf2365_0282*), and six in *L. innocua* (*Lin0559, Lin0661, Lin0739, Lin0740, Lin0803, Lin2724*).

The fact that different subgroups of *L. monocytogenes* strains contain different sets of surface proteins may reflect their different potential to cause disease or to multiply in different niches.

3.3.2.   Transport Systems and Sugar Utilization

The *Listeria* genomes encode an abundance of transport proteins (e.g., 11.6% of all predicted genes of *L. monocytogenes* EGDe). These comprise, in particular, proteins dedicated to carbohydrate transport conferring *Listeria* probably in part its ability to colonize a broad range of ecosystems. The overall array of sugar transporters is similar in all *Listeria* genomes, in particular among the four sequenced *L. monocytogenes* strains, but also with *L. innocua*. *Listeria* are predicted to transport and metabolize many simple as well as complex sugars including fructose, rhamnose, rhamnulose, glucose, mannose, chitin, sucrose, cellulose, pullan, trehalose, and tagatose. These sugars are largely associated with the environments where *Listeriae* are found. As in most bacterial genomes the predominant class corresponds to ABC transporters. Interestingly, most of the carbohydrate transport proteins belong to phosphoenolpyruvate-dependent phosphotransferase system (PTS)-mediated carbohydrate transport. The PTS allows the use of different carbon sources and in many bacteria studied so far the PTS is a crucial link between metabolism and regulation of catabolic operons (Barabote and Saier 2005; Kotrba et al. 2001). The *Listeria* genomes contain an unusually large number of PTS loci (e.g., nearly twice as many as *E. coli* and nearly thrice as many as *B. subtilis*). Most of these PTS systems are conserved in the different sequenced genomes; however, subtle differences can be observed, probably allowing niche-specific adaptation. An example is the family of β-glucoside-specific PTSs, of which eight are present in *L. monocytogenes* serotype 1/2a, two of those are missing in the *L. monocytogenes* serotype 4b strains and five are missing from *L. innocua*. As one of these β-glucoside-specific PTS systems named BvrABC was shown to be implicated in virulence of *L. monocytogenes* (Brehm et al. 1999) these differences might play a role in virulence differences among strains.

The different PTS systems should allow *Listeria* to use many different carbon sources during extracellular growth. However, during intracellular growth *L. monocytogenes* also needs a supply of carbon sources. This is in part achieved by a protein similar to UhpT, a hexose phosphate permease found in enteric bacteria, encoded by *hpt*. In addition, *hpt* is absent from the nonpathogenic *L. innocua*. Functional analysis showed that Hpt allows *L. monocytogenes* to utilize phosphorylated sugars such as glucose-1-phosphate within the host cytosol and thus contributes to the bacterial virulence within the mammalian host cell. Deletion of the *hpt* gene resulted in impaired listerial intracytosolic proliferation and attenuated virulence in mice. Thus Hpt is involved in the replicative phase of the intracellular parasitism of *L. monocytogenes* and is an example that adaptation to intracellular parasitism involves exploitation of physiological mechanisms of the eukaryotic host cell (Chico-Calero et al. 2002). Recently it was shown that this transporter can mediate the in vivo uptake of the antibiotic fosfomycin, thus leading to fosfomycin sensitivity of intracellular *L. monocytogenes*, despite the fact that *L. monocytogenes* is resistant to fosfomycin when tested in vitro with conventional methods (Scortti et al. 2006).

3.3.3. Metabolic Capabilities

It is generally accepted that *L. monocytogenes* is an aerobically growing microaerophilic (carbon dioxydophilic) organism. However, *L. monocytogenes* is able to survive and to colonize the mammalian gut where it encounters anaerobic conditions or to multiply in decaying plants, an environment also devoid of oxygen. This is reflected in the genome sequence as many fermentative pathways are predicted (Figure 3.3.). Analysis of the genome sequence identified in *L. monocytogenes* and in *L. innocua* a single continuous locus, coding for the proteins necessary for vitamin B_{12} synthesis. Furthermore, the proteins necessary for degradation of the carbon sources ethanolamine and propanediol in a coenzyme B_{12}-dependent manner are encoded within the same region. Vitamin B_{12} synthesis and degradation of ethanolamine and propanediol may be important for anaerobic growth of *Listeria*.

Not all organisms have the ability to synthesize vitamin B_{12}. A search among genome sequences revealed that vitamin B_{12} biosynthesis genes are found

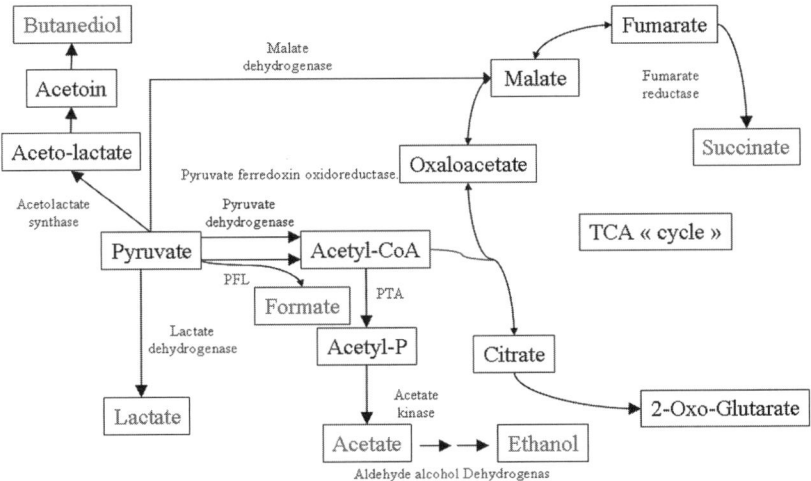

FIGURE 3.3. Fermentative pathways as deduced from the *Listeria monocytogenes* genome analysis. Color code: Black, intermediary metabolites; Red, fermentation end products; Blue, enzymes involved in fermentation. Paralogous genes were detected for several key enzymes: *lmo1917* (*pflA*) and *lmo1406* (*pflB*) encode pyruvate formate lyases; *lmo0210* (*ldh*) and *lmo1057* encode lactate dehydrogenases; *lmo1179* and *lmo1634* encode acetaldehyde dehydrogenase/alcohol dehydrogenase; *lmo1369* and *lmo2103* (*pta*) encode phosphotransacetylases; *lmo1581* (*ackA*) and *lmo1168* (*ackA2*) encode acetate kinases.

in just over one-third of the bacteria sequenced at that time (for a review, see Raux et al. 2000). The proteins deduced from the genome sequence of *L. monocytogenes* and *L. innocua* share the highest homology with those of *S. enterica* serovar Typhimurium (38–65% protein identity). As described above, all three gene clusters seem to have been acquired by *Listeria* en bloc by horizontal gene transfer from Enterobacteriaceae, most probably from *S. enterica* (Buchrieser et al. 2003). Genes coding for CbiD, CbiG, and CbiK, specifically associated with the anaerobic pathway, are present in *Listeria*. This suggests that the two *Listeria* species also contain the oxygen-independent pathway like *Salmonella*. Downstream of the cobalamine biosynthesis genes, *Listeria* contains orthologs of genes necessary in *Salmonella* for the coenzyme B_{12}-dependent degradation of ethanolamine and propanediol. *Salmonella typhimurium* synthesizes vitamin B_{12} anaerobically (Jeter et al. 1984) and can use ethanolamine and 1,2-propanediol as the sole carbon and energy source for growth (Price-Carter et al. 2001). Vitamin B_{12}-dependent anaerobic degradation of ethanolamine and propanediol could enable *L. monocytogenes* to use ethanolamine and 1,2 propanediol as carbon and energy source for growth under anaerobic conditions encountered in the mammalian gut, where both substances are believed to be abundant. It is tempting to speculate that the vitamin B_{12} synthesis genes together with the *pdu* and *eut* operons play a role during listerial infection. This hypothesis is substantiated by a recent report that showed, by studying intracellular gene expression of *L. monocytogenes*, that this bacterium can use ethanolamine as alternative nitrogen sources during replication in epithelial cells (Joseph et al. 2006).

In-depth genome analysis facilitated the identification of a previously unknown and alternative route of gluthatione biosynthesis in *Listeria*, which probably has an important role in protection against oxidative stress (Gopal et al. 2005). GSH is the predominant low-molecular-weight peptide thiol present in living organisms. In bacteria it plays a pivotal role in many metabolic processes including thiol redox homeostasis, protection against reactive oxygen species, protein folding, and provision of electrons via NADPH to reductive enzymes, such as ribonucleotide reductase. A broad survey of the distribution of thiols in microorganisms revealed that several species of gram-positive bacteria, including *Listeria*, streptococci, and enterococci, produce significant amounts of GSH but the source of GSH in these bacteria has remained a puzzle, since their genomes do not contain a canonical *gshB* gene (Fahey et al. 1978; Newton et al. 1996). Copley and Dhillon (Copley and Dhillon 2002) identified in the *L. monocytogenes* genome, a gene containing an N-terminal domain that encodes a molecule significantly related to bacterial Y-glutamylcysteine ligases (GshA) and a C-terminal domain that encodes a molecule that bears little resemblance to typical bacterial glutathione synthetases (GshB) but is clearly related to the ATP-grasp superfamily of proteins. Gopal and colleagues (Gopal et al. 2005) demonstrated that this gene encodes a multidomain protein (termed GshF) that carries out complete synthesis of GSH. Furthermore, gluthatione biosynthesis seems to be involved in virulence of *L. monocytogenes* as a *gsf* deletion mutant is not only defective in gluthatione synthesis but also impaired in growth and survival in mouse macrophages and in Caco-2 enterocyte like cells (Gopal et al. 2005).

3.3.4.   Regulatory Proteins

Given the fact that *L. monocytogenes* is a ubiquitous, opportunistic pathogen that needs a variety of combinatorial pathways to adapt its metabolism to a given niche, an extensive regulatory repertoire is needed. Indeed, a little more than 7% of the *Listeria* genes predicted in the genomes are dedicated to regulatory proteins (Glaser et al. 2001a; Nelson et al. 2004). *Listeriae* have almost twice as many regulators as *Staphylococcus aureus* despite the similar genome size. Only *Pseudomonas aeruginosa* (Stover et al. 2000), another ubiquitous, opportunistic pathogen, encodes with over 8% of its predicted genes a higher proportion of regulatory proteins. Interestingly, diversity among the regulatory genes is not very pronounced, either among the different *L. monocytogenes* genomes or with respect to *L. innocua*, suggesting their implication primarily in features common to the lifestyle of *Listeriae* outside a mammalian host. The most studied regulatory gene of *L. monocytogenes* is *prfA*, encoding the master regulator of virulence. In line with its function in regulating the expression of genes coding proteins necessary for the entry and for intracellular multiplication of *L. monocytogenes*, PrfA is absent from *L. innocua* but conserved in all *L. monocytogenes* strains.

   Although over 7% of the *Listeria* genes were predicted to code transcriptional regulators, this number is probably underestimated, as recently Lmo0674, a new regulator belonging to a not-yet-described class of regulators, was described (Grundling et al. 2004). This protein designated MogR for motility gene repressor was shown to be a transcriptional repressor of flagellar motility and is required for virulence. Deletion of *mogR* reduced the capacity of *L. monocytogenes* for cell-to-cell spread and a 250-fold decrease of virulence in a mouse infection model was observed (Grundling et al. 2004). Most interestingly, MogR is the first known example of a transcriptional repressor functioning as a master regulator controlling nonhierarchical expression of flagellar motility genes (Shen and Higgins 2006). It tightly represses expression of flagellin (FlaA) during extra-cellular growth at 37° C and during intracellular infection. The severe virulence defect of MogR-negative bacteria is due to overexpression of FlaA. Specifically, overproduction of FlaA in MogR-negative bacteria causes pleiotropic defects in bacterial division (chaining phenotype), intracellular spread, and virulence in mice. MogR represses transcription of all known flagellar-motility genes by binding directly to DNA recognition sites. This repression of transcription is antagonized in a temperature-dependent manner by the DegU response regulator, which further regulates FlaA levels through a posttranscriptional mechanism (Shen and Higgins 2006).

3.4.   Diversity Within the Species *L. monocytogenes*

Epidemiological data indicate that not all strains of *L. monocytogenes* are equally capable of causing disease in humans but that differences in virulence among strains seem to exist. Isolates from only four (1/2a; 1/2c; 1/2b; 4b) of the 13 serovars (sv) identified within this species are responsible for over 98% of the human listeriosis cases reported (Jacquet et al. 2002). Furthermore, all major food-borne outbreaks of listeriosis, as well as the majority of sporadic cases, have been caused by serovar 4b strains suggesting that strains of this sv may possess unique virulence properties. Heterogeneity in virulence has also been observed in the mouse infection model (Brosch et al. 1993). Thus, one of the major questions to answer is "What is the genetic basis of virulence differences among different *L. monocytogenes* strains?"

3.4.1. Diversity of L. monocytogenes and Listeria sp.—DNA Array Analysis

The availability of complete genome sequences allows researchers now to use other techniques, like DNA arrays, to investigate the diversity among a large number of *Listeria* strains belonging to these different lineages, to different populations or showing different epidemiological characteristics. This approach allows characterization of the genetic differences among and within these lineages, and to investigate whether these differences can be attributed to virulence differences and/or different niche adaptations.

Different studies aiming to answer these questions have been undertaken. Doumith and colleagues (Doumith et al. 2004b) designed DNA arrays carrying the specific gene pool of three sequenced *Listeria* genomes (*L. monocytogenes* EGDe sv1/2a, *L. monocytogenes* CLIP80459, *L. innocua* CLIP11262) as well as genes coding known virulence and surface proteins, to characterize the variability in gene content of *Listeria* strains belonging to all serovars and being isolated from humans, food, and the environment (93 *L. monocytogenes* and 20 *Listeria sp.*) The results obtained from DNA/DNA hybridization of 113 *Listeria* strains revealed that grouping based on absence and presence of genes clustered all analyzed strains according to their species definition and the *L. monocytogenes* strains according to the three previously defined lineages (I, II, and III). Within each lineage two subdivisions were distinguished, and specific marker genes for the species *L. monocytogenes* and for each subgroup within this species were defined (Table 3.2.).

Nineteen genes were associated specifically with lineage I (Table 3.2., group A). Eight genes allowed the subdivision of lineage I (Table 3.2., group B) and five of the 53-sv 4b specific genes present on the *Listeria* array were identified as markers for lineage II (Table 3.2., group C). Interestingly, 35 of the 53-sv 4b genes spotted on the array were conserved in all 4b strains, suggesting their implication in characteristic features of sv 4b strains (Doumith et al. 2004b). The identification of selective markers for the different subpopulations are an essential contribution for the construction of rapid, accurate identification and subtyping tools, and led to the development of "molecular serotyping" by multiplex PCR (Doumith et al. 2004a). The specificity and reliability of this PCR-based "molecular serotyping" method was validated through an international multicenter study (Doumith et al. 2005).

TABLE 3.2. *L. monocytogenes* lineage-specific marker genes.

Gene name	Lineage I		Lineage II		Lineage III		Functional category
	I .1 (1/2a,3a) (27strains)	I .2 (1/2c,3c) (12 strains)	II .1 (4b,4d,4e) (27 strains)	II .2 (1/2b,3b) (20 strains)	III .1 (4a) (3 strains)	III .2 (4c) (2 strains)	
A *Lmo 0171*	27	12	0	0	0	0	Cell surface proteins
Lmo 0172	27	12	0	0	0	0	Transposon and IS
Lmo 0525	27	12	0	0	0	0	Unknown
Lmo 0734	27	12	0	0	0	0	Regulation
Lmo 0735	27	12	0	0	0	0	Specific pathways
Lmo 0736	27	12	0	0	0	0	Specific pathways
Lmo 0737	27	12	0	0	0	0	Unknown
Lmo 0738	27	12	0	0	0	0	Transport/binding proteins and lipoproteins
Lmo 0739	27	12	0	0	0	0	Specific pathways
Lmo 1060	27	12	0	0	0	0	Regulation
Lmo 1061	27	12	0	0	0	0	Sensors
Lmo 1062	27	12	0	0	0	0	Transport/binding proteins and lipoproteins
Lmo 1063	27	12	0	0	0	0	Transport/binding proteins and lipoproteins
Lmo 1968	27	12	0	0	0	0	Metabolism of amino acids
Lmo 1969	27	12	0	0	0	0	Specific pathways
Lmo 1970	27	12	0	0	0	0	Metabolism of lipids
Lmo 1971	27	12	0	0	0	0	Transport/binding proteins and lipoproteins
Lmo 1973	27	12	0	0	0	0	Transport/binding proteins and lipoproteins
Lmo 1974	27	12	0	0	0	0	Regulation
BvrC	27	12	0	0	0	**2**	Unknown
BvrB	27	12	0	0	0	**2**	Transport/binding proteins and lipoproteins

(Continued)

TABLE 3.2. (*Continued*)

Gene name	Lineage I		Lineage II		Lineage III		Functional category
	I.1 (1/2a,3a) (27 strains)	I.2 (1/2c,3c) (12 strains)	II.1 (4b,4d,4e) (27 strains)	II.2 (1/2b,3b) (20 strains)	III.1 (4a) (3 strains)	III.2 (4c) (2 strains)	
B *Lmo 0151*	3	12	0	0	0	0	Unknown
Lmo 0466	2	12	0	0	0	0	Unknown
Lmo 0467	2	12	0	0	0	0	Unknown
Lmo 0469	2	12	0	0	0	0	Unknown
Lmo 0470	2	12	0	0	0	0	DNA restrictions and modifications
Lmo 0471	2	12	0	0	0	0	Unknown
Lmo 1118	1	12	0	0	0	0	Unknown
Lmo 1119	1	12	0	0	0	0	DNA restrictions and modifications
C ORF2819	0	0	27	20	0	0	Unknown, Similar to hypothetical transcriptional regulator
ORF3840	0	0	27	20	0	0	Unknown, similar to transcriptional regulator
ORF2568	0	0	27	20	0	0	Unknown, Similar to internalin proteins, putative peptidoglycan bound protein (LPXTG)
ORF2017	0	0	27	20	0	0	Unknown, Similar to internalin proteins, putative peptidoglycan bound protein (LPXTG)
ORF0029	0	0	27	19	0	0	Unknown, Similar to internalin proteins, putative peptidoglycan bound protein (LPXTG)
D ORF0799	0	0	27	0	0	0	Unknown
ORF2372	0	0	27	0	0	0	Unknown, Similar to teichoic acid protein precursor C
ORF2110	0	0	27	0	0	0	Unknown, Putative secreted protein

The analysis of the distribution of the known virulence genes (*inlAB, prfA, plcA, hly, mpl, actA, plcB, uhpT,* and *bsh)* among the 113 strains tested revealed that the known virulence factors are present in all *L. monocytogenes* strains. Thus, other not characterized factors must be responsible for these differences (Doumith et al. 2004b). Probably surface proteins play a role. The distribution of 55 genes coding for putative surface proteins belonging to three sequenced *Listeria* genomes (*L. monocytogenes* EGDe sv1/2a, *L. monocytogenes* CLIP80459 and *L. innocua* CLIP11262) was investigated. In agreement with whole genome comparisons, a pronounced heterogeneity among the surface protein coding genes was observed. Twenty five of the 55 surface protein-coding genes of the Internalin/LPXTG/GW motif containing family could be defined as specific for the species *L. monocytogenes* including *inlAB*. These proteins are always present in all *L. monocytogenes* strains but absent from the other *Listeria* sp. (Doumith et al. 2004c). Surface proteins present in all *L. monocytogenes* strains but not in the apathogenic species of this genus (*L. innocua, L. welshimeri, L. seeligeri,* and *L. grayi*) could be proteins involved in virulence. Thus, based on these results, two internalin protein-encoding genes presenting this characteristic were selected for further functional analysis. They were named *inlI* (*lmo0333*) and *inlJ* (*lmo2821*) (Sabet et al. 2005). Indeed, the *inlJ* deletion mutant is significantly attenuated in virulence after intravenous infection of mice or oral inoculation of transgenic hEcad mice (Sabet et al. 2005). *inlJ* encodes a LRR protein that is structurally related to the listerial invasion factor InlA. The consensus sequence of the LRR defined a novel subfamily of cysteine-containing proteins belonging to the internalin family in *L. monocytogenes*. Further studies focusing on strain-specific surface proteins may reveal the genetic basis of strain-specific differences in virulence and in niche adaptation.

The distribution of surface proteins among the *L. monocytogenes* strains also mirrored the three lineages as each lineage, and each subgroup within a lineage is characterized by a specific surface protein combination (Doumith et al. 2004b). Furthermore, according to DNA/DNA array hybridization, epidemic *L. monocytogenes* seem to be characterized by a specific gene content (Doumith et al. 2004b), which may explain in part their higher potential to cause human listeriosis similar to previous reports that attribute specific gene content to epidemic strains (Evans et al. 2004; Yildirim et al. 2004). However, most of these genes are not functionally characterized or their characterization did not allow yet, defining the reason for the higher prevalence of sv 4b strains in human listeriosis. In contrast, the fact that some strains seem less virulent for humans may be related to missing genes. Most interestingly, in the rarely isolated *L. monocytogenes* sv 4a strains (lineage III), which are mostly animal pathogens, 13 of the 25 *L. monocytogenes*-specific surface proteins, including all known internalins except *inlAB*, were missing. The lack of these surface proteins as well as of additional genes of yet unknown function may explain why lineage III strains are mainly found in animals but not in human listeriosis. This is in line with a recent study that investigated *L. monocytogenes* populations present at the farm, in food processing plants, at retail, and in the human population. Analyses of over 400 strains suggested that *L. monocytogenes* populations that are adapted to different niches exist. This study identified among the strains collected from the food processing plants one dominant pork product strain that was not identified among human isolates. DNA array characterization of these strains identified a specific genetic profile and the absence of five genes predicted to encode internalins and other cell surface proteins. These surface protein diversities may explain/contribute to the adaptation to a specific environment (Hong et al. 2006).

Another approach using a shotgun microarray based on clones derived from *L. monocytogenes* strain 10403S (serovar 1/2a) and DNA/DNA hybridization with chromosomal DNA of 44 *L. monocytogenes* strains belonging to serovar 1/2a, 1/2b, and 4b gave similar results identifying genes specific for *L. monocytogenes* serovar 1/2a as compared to *L. monocytogenes* serovar 4b and 1/2b strains (Zhang et al. 2003). Call and colleagues developed a shotgun-clone-based microarray using DNA of 10 different *L. monocytogenes* strains (4 strains of serovar 1/2a, one of serovar 1/2c, one of serovar 3a, and four of serovar 4b). This allowed identifying 29 genes specific to the different lineages. Most of the genes identified in the herein-cited studies as being specific to the different lineages overlap, others were identified as not specific by the other analysis.

Taking these results together, the species *L. monocytogenes* and even the genus *Listeria* show a conserved genome organization with a high number of orthologous genes, but also a considerable number of strain-specific traits most of them organized in many small plasticity zones. Furthermore, the three subgroups present within the species *L. monocytogenes* correspond clearly to distinct evolutionary lineages with specific histories of gene gain and loss by which each lineage has adapted to a primary niche and probably also to a different virulence potential for humans and animals.

3.5. Postgenomics Analysis—Functional Genomics

3.5.1. Systematic Gene Disruption Analysis

A result of the analysis of complete genome sequences was the description of a large number of genes encoding proteins with unknown functions or being similar to poorly characterized proteins. In order to identify functions for these proteins, programs of systematic analysis, including systematic gene knockouts was performed for model organisms like *Saccharomyces cerevisiae* (Oliver 2002) or *B. subtilis* (Kobayashi et al. 2003). No such program has been set up for *L. monocytogenes*. Nevertheless genome-based analysis of several protein families has been undertaken like two component regulatory systems, surface proteins, and the regulators of the Crp-Fnr family. As previously described in EGDe (Glaser et al. 2001b), Uhlich et al. (2006) have identified in the genome sequence of strain F2365 in addition to PrfA 14 genes encoding proteins belonging to the Crp-Fnr family and have systematically inactivated them. However, for 12 strains no phenotype was observed and only for two a weak phenotype was observed as LMOf2365_0171::TN7, which has a slightly reduced growth rate at 30°C and a reduced oxidative stress tolerance, whereas

LMOf2365_0577::TN7 presents only reduced tolerance to oxidative stress. The absence of clear phenotypes for the different mutants is possibly due to partial functional redundancy in large paralogous gene families and the absence of adequate laboratory tests to characterize them. Williams and colleagues have inactivated 15 out of the 16 response regulator genes in *L. monocytogenes* EGDe. The remaining response regulator, *lmo0287*, is homologous to *B. subtilis yycF* gene and is essential for growth of *Listeria*. Among these 15 mutated strains, only one, mutated in the *B. subtilis degU* homolog, had a clear phenotype, including reduced virulence in mice (Williams et al. 2005). In contrast, another study described one of these two component systems, *virR/virS*, identified through STM as being implicated in virulence (Mandin et al. 2005b). Deletion of *virR* severely decreased virulence in mice as well as invasion in cell-culture experiments. Using a transcriptomic approach, 12 genes regulated by VirR, including the *dlt*-operon, were identified. Another VirR-regulated gene is homologous to *mprF*, which encodes a protein that modifies membrane phosphatidylglycerol with L-lysine and that is involved in resistance to human defensins in *Staphylococcus aureus*. VirR thus appears to control virulence by a global regulation of surface components modifications.

However, the often only weak effects of mutations observed for regulatory systems expected to have a strong effect on the physiology of the bacteria suggest that the regulatory network governing the adaptation of the genetic program is highly robust.

3.5.2. Transcriptome Studies

Access to the complete genome sequence is required for the design of whole genome DNA-arrays for exhaustive expression studies. The first study published analyzed the PrfA regulon in three different *L. monocytogenes* strains (EGD, LO28, and PAM55). This analysis indicated that the number of transcription units directly regulated by PrfA was probably much lower than what was predicted from the motif search done on the complete genome sequence. The directly regulated genes are possibly limited to the 10 previously known (*prfA, plcA, hly, mpl, actA, plcB, inlA, inlB, inlC, and hpt*) ones and two new ones (*lmo2219* and *lmo0788*). All of them are preceded by a PrfA binding site. Gene *lmo2219* encodes a protein similar to PrsA of *B. subtilis*, a posttranslocation molecular chaperon (Kontinen et al. 1991; Kontinen and Sarvas 1993) and gene *lmo0788* code probably for a membrane protein of unknown function. Most importantly this study highlights the connections between PrfA induction and the sigma B stress regulon. Indeed, the sigma B regulon is activated in the wt strain compared to the *prfA* strain. Interestingly this induction is abolished in the presence of charcoal, although this addition increases the PrfA induction of PrfA-dependent gene expression in the parental EGDe strain. This in vitro study gave a first image of the set of genes directly required for intracellular multiplication of *Listeria*. However, further insight into the intracellular program of *L. monocytogenes* gene expression profiles requires transcriptome analysis of bacteria isolated during intracellular multiplication. These experiments are difficult due to the fact that the infection needs to be synchronized, that RNA extraction in sufficient amounts is not easily available, and that there is a risk of contamination by host cell RNA. However, two studies characterizing the intracellular gene expression profiles of *L. monocytogenes* strain EGDe were recently published (Chatterjee et al. 2006; Joseph et al. 2006). Both studies used brain–heart infusion (BHI), and exponentially grown *L. monocytogenes* strains as references. Chatterjee et al. (2006) studied both the intraphagosomal (4 h postinfection; Δ *hly* and Δ *plc* mutant strains) and the intracytoplasmic (8 h after infections; wt EGDe) expression profiles, whereas Joseph et al. (2006) analyzed the intracytoplasmic expression profile 6 h postinfection. Both studies reached very similar conclusions. One of the prominent features is that the PrfA and the stress regulons are induced when *L. monocytogenes* is grown intracellularly. Another major finding was the extensive modification of the expression of genes involved in carbon and energetic metabolic pathways together with permeation systems. The study provides evidence that *L. monocytogenes* can use alternative carbon sources like phosphorylated glucose and glycerol, and nitrogen sources like ethanolamine during replication in epithelial cells, and that the pentose phosphate pathway, but not glycolysis, is the predominant pathway of sugar metabolism in the host environment. Additionally, it shows that the synthesis of arginine, isoleucine, leucine, and valine, as well as a species-specific phosphoenolpyruvate-dependent PTS, play a major role in the intracellular growth of *L. monocytogenes*. The gene expression of metabolic genes is in agreement with what is known about the composition of the cytosol of eukaryotic cells. However, it is likely that this composition is modified by the metabolism of *L. monocytogenes* growing in the cell. As an example, lipid degradation by phospholipases releases glycerol available for utilization by *L. monocytogenes*.

These first analyses of the intracellular gene expression program of *L. monocytogenes* provide clues on the metabolic condition encountered by the bacteria during its intracellular growth; however, these studies do not provide hints for a temporal programation of gene expression during cellular growth, as for example observed for *L. pneumophila* (Brüggemann et al. 2006). This is possibly due to the importance of direct cell-to-cell propagation of *L. monocytogenes* during its growth in the host.

3.5.3. *Proteome Analysis*

Functional adaptation relies on protein synthesis, modification, and localization. Proteome analysis based on 2-D gel electrophoresis or on LC–MS–MS (mass spectrometry coupled to high pressure liquid chromatography) allows the systematic description of the protein repertoire present in a cell. Knowledge of the complete genome sequence and therefore of all predicted encoding protein greatly facilitates protein and gene identification by mass spectrometry. Key features identified from whole genome analysis were also addressed by proteome analysis like the cell wall proteome, the secretome, or the carbon catabolism by the analysis of the effect of a mutation in *sigL* (*rpoN*) gene encoding the alternative σ 54 factor (Arous et al. 2004). Proteome analysis is also an experimental validation of the genome annotation. An important tool for further proteome analysis is the establishment of a proteome reference map. Such a map has been established for strain *L. monocytogenes* EGDe, and contains 201 spots corresponding to 126 proteins identified by mass spectrometry (Folio et al. 2004). The detection of multiple spots for a single protein indicates that, as for other organisms, proteins are submitted to posttranslational processing or modification. Based on this study, a database for protein spot identification has been set up (http://www.clermont.inra.fr/proteome.), but will have to be enriched for further analyses. A partial protein map has also been established previously by Ramnath et al. (2003), who compared protein profiles of food isolates to those of a reference map.

Secreted proteins either bound to the cell envelopes or released in the growth medium are important factors in the interaction with the environment and with the host. These two protein classes include the major virulence factors of *Listeria*. The cell wall proteome is discussed in Chapter 5 of this book by Pucciarelli, Bierne, and García-del Portillo and will thus not be discussed here. In line with the *L. monocytogenes–L. innocua* genome comparison, Trost and colleagues have performed a comparative secretome analysis of these two species (Trost et al. 2005). By combining 2-D electrophoresis and LC–MS/MS, 105 secreted proteins were identified for *L. monocytogenes* grown in BHI at late exponential phase. For 54 proteins, their secretion was predicted by *in silico* analysis; however, the remaining 51 proteins had no recognizable signal sequence. Thus the mechanism of secretion remains to be identified. The similar analysis for *L. innocua* identified 96 secreted proteins. When these two sets of secreted proteins are compared, only 46 were shared by the two strains. Among the 59 proteins detected only in *L. monocytogenes* EGDe, 16 are encoded by strain-specific genes including nine virulence factors (Hly, PlcA, PlcB, ActA, Mpl, InlA, InlB, InlC, and InlH). Among the *L. innocua* specific proteins, seven are encoded by genes present only in this strain. Therefore the majority of differences between the two strains rely on differences in gene expression or secretion. The analysis of these two strains from two related species points to the importance of the differences in regulation. Similar comparative analyses will have to be performed among different *L. monocytogenes* isolates in order to characterize the heterogeneity of the secretome within the species and its possible role in the adaptation to different environments and perhaps also to different disease outcomes.

3.6.   Conclusions and Further Perspectives

Genome sequence analyses, comparative genomics, and postgenomics studies provide clues on the ability of *L. monocytogenes* to survive in a broad range of environments, to infect different hosts, and to provoke different kinds of diseases.

Many new genes involved in the different life styles of *Listeria* have been discovered following the publication of the two first *Listeria* genome sequences. However, how all these genes and proteins are interacting together remains still largely unknown but should be put together in the near future like the pieces of a puzzle. This will be made possible through the improvement of transcriptome and proteome analyses together with other global approaches like the study of the interactome that will provide the basis for a systems biology approach for *Listeria* and listeriosis.

Genome analysis indicates how the *Listeria* genomes may have evolved by multiple insertions and deletions of small islets rarely by site-specific recombination mediated by integrases or IS elements. Even the proposed pathogenicity islands LIPI1 and LIPI2 are devoid of integrase genes. Furthermore the genome comparisons highlight the functional diversity between the major lineages present not only within the species *L. monocytogenes* but also within each lineage. These multiple islets constitute the flexible gene pool of the species *L. monocytogenes*. It includes a large proportion of genes encoding functions involved in the adaptation to various environments like permeases, carbon source catabolism enzymes, and surface proteins, and should thus contribute to the fitness of the strains in specific environments. However, we still do not know the complete size of the flexible gene pool and the relative role of vertical transmission and horizontal gene transfer. Further genome sequencing will allow to define this flexible gene pool and to identify the origin of the horizontally acquired DNA-material. Perhaps the diversity within the genus *Listeria* is larger than thought, and minor species, present in unexplored niches or now extinct, may have contributed to the generation of the present six known *Listeria* species and their flexible gene pool.

Many different studies have investigated the virulence gene cluster and its flanking regions in all *Listeria* species by sequencing and subsequent comparison in order to understand its evolution and with it the evolution of virulence (Cai and Wiedmann 2001; Chakraborty et al. 2000; Gouin et al. 1994; Kreft et al. 2002; Schmid et al. 2005). A fully functional virulence gene cluster is only present in *L. monocytogenes* and *L. ivanovii*, and a similar cluster with additional genes is present in *L. seeligeri*. Comparison of the genomic region in *L. monocytogenes, L. innocua*, and *B. subtilis* suggested that this virulence gene cluster was acquired by a common ancestor of *Listeria* and that *L. innocua* subsequently lost most of it (Glaser et al. 2001a). However, this locus is part of a variable region possibly resulting from successive events of gene loss and gene gain.

The ongoing sequencing projects aiming at determining the complete genome sequence of one representative of each species of the genus *Listeria* by the Institut Pasteur and the German PathoGenomiK network (http://www. pasteur.fr/recherche/unites/gmp/; http://www.genomik.uniwuerzburg.de/seq.htm) and the determination of the complete genome sequence of 19 additional *Listeria* strains by the Broad Institute (http://www.broad.mit.edu/seq/msc/) together with new sequencing methods that emerge, which further reduce the cost of and accelerate genome sequencing, will be the driving force for understanding the function of the many factors encoded by the genome. Combining multigenome comparisons and single nucleotide polymorphism (SNP) distribution for a broad range of strains will allow understanding strain-specific differences in niche adaptation and virulence as well as factors involved in organ tropism and host specificity.

Acknowledgments. The authors would like to thank many of the colleagues who have contributed in different ways to this research. We would like to apologize to our colleagues whose work could not be cited. This work received financial support from the Institut Pasteur (GPH 9).

References

Arous S, Buchrieser C, Folio P, Glaser P, Namane A, Hebraud M, Hechard Y (2004) Global analysis of gene expression in an *rpoN* mutant of *Listeria monocytogenes*. Microbiology 150:1581–90

Aureli P, Fiorucci GC, Caroli D, Marchiaro G, Novara O, Leone L, Salmaso S (2000) An outbreak of febrile gastroenteritis associated with corn contaminated by *Listeria monocytogenes*. N Engl J Med 342:1236–41

Autret N, Raynaud C, Dubail I, Berche P, Charbit A (2003) Identification of the agr locus of *Listeria monocytogenes*: role in bacterial virulence. Infect Immun 71: 4463–71

Barabote RD, Saier MH (2005) Comparative genomic analyses of the bacterial phospho-transferase system. Microbiology and Molecular Biology Reviews 69:608–34

Bibb WF, Schwartz B, Gellin BG, Plikaytis BD, Weaver RE (1989) Analysis of *Listeria monocytogenes* by multilocus enzyme electrophoresis and application of the method to epidemiologic investigations. Int J Food Microbiol 8:233–239

Blattner FR, Plunkett GR, Bloch CA, Perna NT, Burland V, Riley M, et al. (1997) The complete genome sequence of *Escherichia coli* K-12. Science 277:1453–74

Brehm K, Ripio MT, Kreft J, Vazquez-Boland JA (1999) The bvr locus of *Listeria monocytogenes* mediates virulence gene repression by beta-glucosides. J Bacteriol 181:5024–32

Brosch R, Brett M, Catimel B, Luchansky JB, Ojeniyi B, Rocourt J (1996) Genomic fingerprinting of 80 strains from the WHO multicenter international typing study of *Listeria monocytogenes* via pulsed-field gel electrophoresis (PFGE). Int J Food Microbiol 32:343–55

Brosch R, Buchrieser C, Rocourt J (1991) Subtyping of *L. monocytogenes* serovar 4b by use of low frequency cleavage restriction endonucleases and pulsed field gel electrophoresis. Res Microbiol 142:667–75

Brosch R, Catimel B, Milon G, Buchrieser C, Vindel E, Rocourt J (1993) Virulence heterogeneity of *Listeria monocytogenes* strains from various sources (Food, Human, Animal) in immunocompetent mice and its association with typing characteristics. J Food Prot 56:296–301

Brosch R, Chen J, Luchansky JB (1994) Pulsed-field fingerprinting of listeriae: identification of genomic divisions for *Listeria monocytogenes* and their correlation with serovar. Appl Environ Microbiol 60:2584–92

Brüggemann H, Hagman A, Jules M, Sismeiro O, Dillies M, Gouyette C, Kunst F, Steinert M, Heuner K, Coppée J, Buchrieser C (2006) Virulence strategies for infecting phagocytes deduced from the *in vivo* transcriptional program of *Legionella pneumophila*. Cell Microbiol 8:1228–40

Buchrieser C, Rusniok C, Kunst F, Cossart P, Glaser P (2003) Comparison of the genome sequences of *Listeria monocytogenes* and *Listeria innocua*: clues for evolution and pathogenicity. FEMS Immunol Med Microbiol 35:207–13

Cabanes D, Dehoux P, Dussurget O, Frangeul L, Cossart P (2002) Surface proteins and the pathogenic potential of *Listeria monocytogenes*. Trends Microbiol 5:238–45

Cai S, Wiedmann M (2001) Characterization of the *prfA* virulence gene cluster insertion site in non-hemolytic *Listeria* spp.: probing the evolution of the *Listeria* virulence gene island. Curr Microbiol 43:271–7

Carriere C, Allardet-Servent A, Bourg G, Audurier A, Ramuz M (1991) DNA polymorphism in strains of *Listeria monocytogenes*. J Clin Microbiol 29:1351–1355

Cazalet C, Rusniok C, Bruggemann H, Zidane N, Magnier A, Ma L, Tichit M, Jarraud S, Bouchier C, Vandenesch F, Kunst F, Etienne J, Glaser P, Buchrieser C (2004) Evidence in the *Legionella pneumophila* genome for exploitation of host cell functions and high genome plasticity. Nat Genet 36:1165–73

Chakraborty T, Hain T, Domann E (2000) Genome organization and the evolution of the virulence gene locus in *Listeria* species. Int J Med Microbiol 2:167–74

Chatterjee SS, Hossain H, Otten S, Kuenne C, Kuchmina K, Machata S, Domann E, Chakraborty T, Hain T (2006) Intracellular gene expression profile of *Listeria monocytogenes*. Infect Immun 74:1323–38

Chico-Calero I, Suarez M, Gonzalez-Zorn B, Scortti M, Slaghuis J, Goebel W, Vazquez-Boland JA (2002) Hpt, a bacterial homolog of the microsomal glucose-6-phosphate translocase, mediates rapid intracellular proliferation in*Listeria*. Proc Natl Acad Sci USA 99:431–36

Cole ST, Brosch R, Parkhill J, Garnier T, Churcher C, Harris D, Gordon SV, et al. (1998) Deciphering the biology of *Mycobacterium tuberculosis* from the complete genome sequence. Nature 393:537–44

Copley SD, Dhillon JK (2002) Lateral gene transfer and parallel evolution in the history of glutathione biosynthesis genes. Genome Biol 3:0025.1–.16

Doumith M, Buchrieser C, Glaser P, Jacquet C, Martin P (2004a) Differentiation of the Major *Listeria monocytogenes* Serovars by Multiplex PCR. J Clin Microbiol 42:3819–3822

Doumith M, Cazalet C, Simoes N, Frangeul L, Jacquet C, Kunst F, Martin P, Cossart P, Glaser P, Buchrieser C (2004b) New aspects regarding evolution and virulence of *Listeria monocytogenes* revealed by comparative genomics and DNA arrays. Infect Immun 72:1072–83

Doumith M, Jacquet C, Gerner-Smidt P, Graves LM, Loncarevic S, Mathisen T, Morvan A, Salcedo C, Torpdahl M, Vazquez JA, Martin P (2005) Multicenter validation of a multiplex PCR assay for differentiating the major *Listeria monocytogenes* serovars 1/2a, 1/2b, 1/2c, and 4b: toward an international standard. J Food Prot 68:2648–50

Evans MR, Swaminathan B, Graves LM, Altermann E, Klaenhammer TR, Fink RC, Kernodle S, Kathariou S (2004) Genetic markers unique to *Listeria monocytogenes* serotype 4b differentiate epidemic clone II (hot dog outbreak strains) from other lineages. Appl Environ Microbiol 70:2383–90

Fahey RC, Brown WC, Adams WB, Worsham MB (1978) Occurrence of glutathione in bacteria. J Bacteriol 133:1126–9

Fiedler F (1988) Biochemistry of the cell surface of *Listeria* strains: a locating general view. Infection 16:92–7

Fiedler F, Ruhland GJ (1987) Structure of *Listeria monocytogenes* cell walls. Bull Inst Past 85:287–300

Fleischmann RD, Adams MD, White O, Clayton RA, Kirkness EF, Kerlavage AR, Bult CJ, Tomb JF, Dougherty BA, Merrick JM, et al. (1995) Whole-genome random sequencing and assembly of *Haemophilus influenzae* Rd. Science 269:496–512

Folio P, Chavant P, Chafsey I, Belkorchia A, Chambon C, Hebraud M (2004) Two-dimensional electrophoresis database of *Listeria monocytogenes* EGDe proteome and proteomic analysis of mid-log and stationary growth phase cells. Proteomics 4:3187–201

Frangeul L, Glaser P, Rusniok C, Buchrieser C, Duchaud E, Dehoux P, Kunst F (2004) CAAT-Box, Contigs-Assembly and annotation tool-box for genome sequencing projects. Bioinformatics 20:790–7

Gaillard JL, Berche P, Sansonetti P (1986) Transposon mutagenesis as a tool to study the role of hemolysin in the virulence of *Listeria monocytogenes*. Infect. Immun. 52:50–55

Gerner-Smidt P, Hise K, Kincaid J, Hunter S, Rolando S, Hyytia-Trees E, Ribot EM, Swaminathan B, Taskforce P (2006) PulseNet USA: a five-year update. Foodborne pathog Dis 3:9–19

Glaser P, Frangeul L, Buchrieser C, Rusniok C, Amend A, Baquero F, Berche P, Bloecker H, Brandt P, Chakraborty T, Charbit A, Chetouani F, Couve E, de Daruvar A, Dehoux P, Domann E, Dominguez-Bernal G, Duchaud E, Durant L, Dussurget O, Entian KD, Fsihi H, Garcia-del Portillo F, Garrido P, Gautier L, Goebel W, Gomez-Lopez N, Hain T, Hauf J, Jackson D, Jones LM, Kaerst U, Kreft J, Kuhn M, Kunst F, Kurapkat G, Madueno E, Maitournam A, Vicente JM, Ng E, Nedjari H, Nordsiek G, Novella S, de Pablos B, Perez-Diaz JC, Purcell R, Remmel B, Rose M, Schlueter T, Simoes N, Tierrez A, Vazquez-Boland JA, Voss H, Wehland J, Cossart P (2001) Comparative genomics of *Listeria* species. Science 294:849–52

Gopal S, Borovok I, Ofer A, Yanku M, Cohen G, Goebel W, Kreft J, Aharonowitz Y (2005) A multidomain fusion protein in *Listeria monocytogenes* catalyzes the two primary activities for glutathione biosynthesis. J Bacteriol 187:3839–47

Gouin E, Mengaud J, Cossart P (1994) The virulence gene cluster of *Listeria monocytogenes* is also present in *Listeria ivanovii*, an animal pathogen, and *Listeria seeligeri*, a nonpathogenic species. Infect Immun 62:3550–3

Graves L, Swaminathan B, Reeves M, et al. (1994) Comparison of ribotyping and multi-locus enzyme electrophoresis for subtyping of *Listeria monocytogenes* isolates. J Clin Microbiol 32:2936–43

Graves LM, Swaminathan B (2001) PulseNet standardized protocol for subtyping *Listeria monocytogenes* by macrorestriction and pulsed-field gel electrophoresis. Int J Food Microbiol 65:55–62

Gray MJ, Freitag NE, Boor KJ (2006) How the bacterial pathogen *Listeria monocytogenes* mediates the switch from environmental Dr. Jekyll to pathogenic Mr. Hyde. Infect Immun 74:2505–12

Grundling A, Burrack LS, Bouwer HG, Higgins DE (2004) *Listeria monocytogenes* regulates flagellar motility gene expression through MogR, a transcriptional repressor required for virulence. Proc Natl Acad Sci USA 101:12318–23. Epub 2004 Aug 9

Hain T, Steinweg C, Chakraborty T (2006) Comparative and functional genomics of *Listeria* spp. J Biotechnol 126:37–51

Hamon M, Bierne H, Cossart P (2006) *Listeria monocytogenes*: a multifaceted model. Nature Rev Microbiol 4:423–34

He W, Luchansky JB (1997) Construction of the temperature-sensitive vectors pLUCH80 and pLUCH88 for delivery of Tn917: NotI/SmaI and use of these vectors to derive a circular map of *Listeria monocytogenes* Scott A, a serotype 4b isolate. Appl Environ Microbiol 63:3480–7

Hodgson DA (2000) Generalized transduction of serotype 1/2 and serotype 4b strains of *Listeria monocytogenes*. Mol Microbiol 2:312–23

Hong E, Doumith M, Duperrier S, Giovannacci I, Morvan A, Glaser P, Buchrieser C, Jacquet C, Martin M (2006) Genetic diversity of *Listeria monocytogenes* populations present in patients and in pork products at the store distribution level in France in 2000–2001. Int J Food Microbiol doi:10.1016/j.ijfoodmicro.2006.09.011

Jacquet C, Gouin E, Jeannel D, Cossart P, Rocourt J (2002) Expression of ActA, Ami, InlB, and listeriolysin O in *Listeria monocytogenes* of human and food origin. Appl Environ Microbiol 68:616–22

Jeffers GT, Bruce JL, McDonough PL, Scarlett J, Boor KJ, Wiedmann M (2001) Comparative genetic characterization of *Listeria monocytogenes* isolates from human and animal listeriosis cases. Microbiology 147:1095–104

Jeter RM, Olivera BM, Roth JR (1984) *Salmonella typhimurium* synthesizes cobalamin (vitamin B12) de novo under anaerobic growth conditions. J Bacteriol 159:206–13

Joseph B, Przybilla K, Stuhler C, Schauer K, Slaghuis J, Fuchs TM, Goebel W (2006) Identification of *Listeria monocytogenes* genes contributing to intracellular replication by expression profiling and mutant screening. J Bacteriol 188:556–68

Kobayashi K, Ehrlich SD, Albertini A, Amati G, Andersen KK, Arnaud M, Asai K, Ashikaga S, Aymerich S, Bessieres P, et al. (2003) Essential *Bacillus subtilis* genes. Proc Natl Acad Sci USA 100:4678–83

Kontinen VP, Saris P, Sarvas M (1991) A gene (*prsA*) of *Bacillus subtilis* involved in a novel, late stage of protein export. Mol Microbiol 5:1273–83

Kontinen VP, Sarvas M (1993) The PrsA lipoprotein is essential for protein secretion in *Bacillus subtilis* and sets a limit for high-level secretion. Mol Microbiol 4:727–737

Kotrba P, Inui M, Yukawa H (2001) Bacterial phosphotransferase system (PTS) in carbohydrate uptake and control of carbon metabolism. J Biosci Bioeng 92:502–17

Kreft J, Vazquez-Boland J-A, Altrock S, Domingueg-Berrnal G, Goebel W (2002) Pathogenicity islands and other virulence elements in *Listeria*. Curr Top Microbiol Immunol 264:109–25

Kunst F, et al. (1997) The complete genome sequence of the gram positive bacterium *Bacillus subtilis*. Nature 390:249

Kuska B (1998) Beer, Bethesda, and biology: how "genomics" came into being. J Natl Cancer Inst 90:3

Lyytikainen O, Autio T, Maijala R, Ruutu P, Honkanen-Buzalski T, Miettinen M, Hatakka M, Mikkola J, Anttila VJ, Johansson T, Rantala L, Aalto T, Korkeala H, Siitonen A (2000) An outbreak of *Listeria monocytogenes* serotype 3a infections from butter in Finland. J Infect Dis 181:1838–41

Mackaness GB (1964) The immunological basis of acquired cellular resistance. J Exp Med 120:105–20

Mandin P, Fsihi H, Dussurget O, Vergassola M, Milohanic E, Toledo-Arana A, Lasa I, Johansson J, Cossart P (2005) VirR, a response regulator critical for *Listeria monocytogenes* virulence. Mol Microbiol 57:1367–80

Michel E, Cossart P (1992) Physical map of the *Listeria monocytogenes* chromosome. J Bacteriol 174:7098–103

Murray EGD, Webb RE, Swann MBR (1926) A disease of rabbits characterized by a large mononuclear leucocytosis, caused by a hitherto undescribed bacillus *Bacterium monocytogenes* (n. sp.). J Pathol Bacteriol 29:407–39

Nelson KE, Fouts DE, Mongodin EF, Ravel J, DeBoy RT, Kolonay JF, Rasko DA, Angiuoli SV, Gill SR, Paulsen IT, Peterson J, White O, Nelson WC, Nierman W, Beanan MJ, Brinkac LM, Daugherty SC, Dodson RJ, Durkin AS, Madupu R, Haft DH, Selengut J, Van Aken S, Khouri H, Fedorova N, Forberger H, Tran B, Kathariou S, Wonderling LD, Uhlich GA, Bayles DO, Luchansky JB, Fraser CM (2004) Whole genome comparisons of serotype 4b and 1/2a strains of the food-borne pathogen *Listeria monocytogenes* reveal new insights into the core genome components of this species. Nucleic Acids Res 32:2386–95

Newton GL, Arnold K, Price MS, Sherrill C, Delcardayre SB, Aharonowitz Y, Cohen G, Davies J, Fahey RC, Davis C (1996) Distribution of thiols in microorganisms: mycothiol is a major thiol in most actinomycetes. J Bacteriol 178:1990–5

Oliver SG (2002) Functional genomics: lessons from yeast. Philos Trans R Soc Lond B Biol Sci 357:17–23

Piffaretti JC, Kressebuch H, Aeschbacher M, Bille J, Bannerman E, Musser JM, Selander RK, Rocourt J (1989) Genetic characterization of clones of the bacterium *Listeria monocytogenes* causing epidemic disease. Proc Natl Acad Sci USA 86:3818–22

Pirie JHH (1927) A new disease of veld rodents. "Tiger River Disease". Publ S Afr Inst Med Res 3

Portnoy DA, Chakraborty T, Goebel W, Cossart P (1992) Molecular determinants of *Listeria monocytogenes* pathogenesis. Infect Immun 60:1263–67

Price-Carter M, Tingey J, Bobik TA, Roth JR (2001) The alternative electron acceptor tetrathionate supports B12-dependent anaerobic growth of *Salmonella enterica* serovar typhimurium on ethanolamine or 1,2-propanediol. J Bacteriol 183:2463–75

Ramnath M, Rechinger KB, Jansch L, Hastings JW, Knochel S, Gravesen A (2003) Development of a *Listeria monocytogenes* EGDe partial proteome reference map and comparison with the protein profiles of food isolates. Appl Environ Microbiol 69:3368–76

Raux E, Schubert HL, Warren MJ (2000) Biosynthesis of cobalamin (vitamin B12): a bacterial conundrum. Cell Mol Life Sci 57:1880–93

Sabet C, Lecuit M, Cabanes D, Cossart P, Bierne H (2005) LPXTG protein InlJ, a newly identified internalin involved in *Listeria monocytogenes* virulence. Infect Immun. 73:6912–22

Schmid MW, Ng EY, Lampidis R, Emmerth M, Walcher M, Kreft J, Goebel W, Wagner M, Schleifer KH (2005) Evolutionary history of the genus *Listeria* and its virulence genes. Syst Appl Microbiol 28:1–18

Schwartz DC, Cantor CR (1984) Separation of yeast chromosome-sized DNAs by pulsed field gradient gel electrophoresis. Cell 37:67–75

Scortti M, Lacharme-Lora L, Wagner M, Chico-Calero I, Losito P, Vazquez-Boland JA (2006) Coexpression of virulence and fosfomycin susceptibility in *Listeria*: molecular basis of an antimicrobial in vitro-in vivo paradox. Nat Med. 12:515–17. Epub 2006 Apr 23

Shen A, Higgins DE (2006) The MogR Transcriptional Repressor Regulates Nonhier-archal Expression of Flagellar Motility Genes and Virulence in *Listeria monocytogenes*. PLoS Pathog 2:e30

Stover CK, Pham XQ, Erwin AL, Mizoguchi SD, Warrener P (2000) Complete genome sequence of *Pseudomonas aeruginosa* PA01, an opportunistic pathogen. Nature 406:959–64

Trost M, Wehmhoner D, Karst U, Dieterich G, Wehland J, Jansch L (2005) Comparative proteome analysis of secretory proteins from pathogenic and nonpathogenic *Listeria* species. Proteomics 5:1544–57

Uhlich GA, Wonderling LD, Luchansky JB (2006) Analyses of the putative Crp/Fnr family of transcriptional regulators of a serotype 4b strain of *Listeria monocytogenes*. Food Microbiol 23:300–6

von Both U, Otten S, Darbouche A, Domann E, Chakraborty T (1999) Physical and genetic map of the *Listeria monocytogenes* EGD serotype 1/2a chromosome. FEMS Microbiol Lett 175:281–9

Wiedmann M, Bruce JL, Keating C, Johnson AE, McDonough PL, Batt CA (1997) Ribotypes and virulence gene polymorphisms suggest three distinct *Listeria monocytogenes* lineages with differences in pathogenic potential. Infect Immun 65:2707–16

Williams T, Bauer S, Beier D, Kuhn M (2005) Construction and characterization of *Listeria monocytogenes* mutants with in-frame deletions in the response regulator genes identified in the genome sequence. Infect Immun 73:3152–9

Yildirim S, Lin W, Hitchins AD, Jaykus LA, Altermann E, Klaenhammer TR, Kathariou S (2004) Epidemic clone I-specific genetic markers in strains of *Listeria monocytogenes* serotype 4b from foods. Appl Environ Microbiol 70:4158–64

Zhang C, Zhang M, Ju J, Nietfeldt J, Wise J, Terry PM, Olson M, Kachman SD, Wiedmann M, Samadpour M, Benson AK (2003) Genome diversification in phyloge-netic lineages I and II of *Listeria monocytogenes*: identification of segments unique to lineage II populations. J Bacteriol 185:5573–84

Zhang W, Jayarao BM, Knabel SJ (2004) Multi-virulence-locus sequence typing of *Listeria monocytogenes*. Appl Environ Microbiol 70:913–20

4
Metabolism and Physiology of *Listeria monocytogenes*

Jörg Slaghuis, Biju Joseph, and Werner Goebel
*Theodor-Boveri-Institut (Biozentrum), Lehrstuhl für Mikrobiologie,
Universität Würzburg, D-97074 Würzburg, Germany*

Abstract: Compared to the rich wealth of knowledge concerning the molecular basis of *Listeria monocytogenes* virulence, little is known on the physiological background necessary for allowing this facultative intracellular human pathogen to survive and replicate in its natural surroundings, particularly in the host cell's cytosol. This cellular compartment appears to be the preferred site of replication, during a systemic infection caused by *L. monocytogenes*. Complementing earlier physiological studies, especially the more recent results obtained by comparative genomics, transcriptome, and proteome analyses, and by ^{13}C-isotopolog perturbation studies, allow us today to draw a first (although still rather incomplete) picture of how the metabolism of these bacteria may function to facilitate efficient growth under extra- and intracellular conditions. In this chapter, we concentrate on the carbon- and nitrogen-metabolism of *L. monocytogenes* as deduced from these studies. Although many carbon- and nitrogen-metabolic pathways of *L. monocytogenes* appear to be similar to those of the extensively studied *Bacillus subtilis*, which like *L. monocytogenes* belongs to the group of low $G+C$ gram-positive (Gp) bacteria, there seem to be some profound differences that are essential for understanding the interplay of the listerial metabolism with that of the host cells and hence may have an important impact on listerial virulence.

4.1. Introduction

The genus *Listeria* to which the human pathogen *L. monocytogenes* belongs comprises six characterized heterotrophic species, including, besides *L. monocytogenes,* L. ivanovii, *L. innocua, L. welshimeri, L. seeligeri*, and *L. grayi*. With the exception of *L. ivanovii*, which is pathogenic to animals, all other *Listeria* species are harmless saprophytes occurring in nature at various sites (Farber and Peterkin, 1991; Gray and Killinger, 1966; Seeliger, 1984;

Seeliger and Jones, 1986). Earlier genetic and biochemical studies had already pointed to a close physiological relationship of the genus *Listeria* to that of other bacteria belonging to the group of Gp bacteria with low G+C genomic DNA, in particular to the members of the genus *Bacillus* (Collins et al., 1991). Recent comparative genome analyses including the genome sequences of other *Listeria* species show indeed a rather close genomic relationship of *L. monocytogenes* and the other *Listeria* species to *B. subtilis* (Glaser et al., 2001; Karlin et al., 2004) (see Chap. 3 in this book). However, unlike *B. subtilis*, *L. monocytogenes* does not grow readily in a defined mineral salt medium with glucose as sole carbon source, consequently growth of *L. monocytogenes* is routinely carried out in a rich brain–heart infusion (BHI) medium. Most molecular studies on virulence genes and virulence gene expression were performed in the past on *L. monocytogenes* cultivated in this complex medium. Several defined minimal media have been designed in the past (Friedman and Roessler, 1961; Jones et al., 1995; Phan-Thanh and Gormon, 1997; Premaratne et al., 1991; Tsai and Hodgson, 2003), which support the (often rather slow) growth of *L. monocytogenes*. These media contain a more or less complex mixture of amino acids and lipoate as essential additional substrates (O'Riordan et al., 2003), and growth efficiency in these media seems to depend to some extent on the *L. monocytogenes* strain used (Tsai and Hodgson, 2003). Our own studies using the sequenced *L. monocytogenes* EGDe strain show that in liquid culture with glucose as carbon source and glutamine as nitrogen source, a strict requirement for the amino acids cysteine, methionine, isoleucine, and leucine (or valine) and the vitamins, biotin, riboflavin, thiamine, and lipoate; growth is stimulated by the addition of arginine. These data already hint to a more complex metabolism of *L. monocytogenes* even under extracellular conditions which cannot be readily explained by the genome sequence of EGDe, which, for example, shows complete sets of genes for all enzymes of the branched chain amino acids (BCAA; Ile, Leu, and Val) of arginine-, cysteine-, and methionine pathways.

Suitable oligopeptides can be utilized by *L. monocytogenes* as source for essential amino acids (e.g., Val) as shown in the *L. monocytogenes* Scott A strain. Two different oligopeptide transporters were identified, one being driven by proton-motive force (PMF) and the other by ATP (Verheul et al., 1998). Furthermore, specific glycine- and proline-containing peptides were shown to stimulate growth at high osmolarity (Amezaga et al., 1995). Protein-bound lipoate is an essential cofactor of several dehydrogenase complexes, e.g., the pyruvate- and the 2-oxoglutarate dehydrogenases. *L. monocytogenes*, like several other bacteria, lacks the genes for the enzymes involved in lipoate biosynthesis and hence listerial growth depends on its external supply. *L. monocytogenes* possesses genes for two lipoate ligases which link lipoate to the dehydrogenases; their differential expression is essential for intracellular growth (O'Riordan et al., 2003) as discussed later. Among the other vitamins, riboflavin, thiamine, and biotin were also found to be essential for growth in minimal media (Tsai and Hodgson, 2003). *L. monocytogenes* is able to grow under aerobic and anaerobic conditions (Gottschalk, 1986; Pine et al., 1989; Romick et al., 1996). Respiration seems to occur only aerobically, and the respiration chain contains menaquinone but not ubiquinone. Since biosynthesis of menaquinone depends on the common branch of the aromatic amino acids pathway, *L. monocytogenes* strains with mutations in this pathway switch to a predominantly anaerobic metabolism even in the presence of oxygen (Stritzker et al., 2004). When grown aerobically in the presence of glucose, *L. monocytogenes* secretes large amounts of acetoin (as overflow product) into the growth medium (Romick et al., 1996; Romick and Fleming, 1998), while lactate (together with acetate and other products) is the major fermentation product under anaerobic conditions. Under these conditions, the gene for pyruvate–formate lyase is highly induced (Joseph et al., 2006; Karlin et al., 2004), suggesting that mixed acid fermentation is the major mode of fermentation in *L. monocytogenes*.

4.2.   Carbon Metabolism

4.2.1.   The Use of Phosphotransferase System Carbohydrates

Glucose like many other sugars and sugar alcohols is taken up by bacteria via the phosphotransferase system (PTS; for a review on PTS-mediated sugar transport in Gp bacteria, see Reizer et al., 1988; Titgemeyer and Hillen, 2002; Vadeboncoeur et al., 2000).

   Glucose and other PTS sugars such as fructose, mannose, and cellobiose are preferred carbon sources for *L. monocytogenes*, when growing in defined liquid minimal media (Tsai and Hodgson, 2003; our own unpublished results). The *L. monocytogenes* genome contains an unusually large number of genes (>40) encoding PTS*s* (Glaser et al., 2001). Among those, four PTS (determined by *lmo0096–0098*, *lmo0781–0784*, and *lmo1997–2002*) are specific for mannose transport, nine PTS for fructose (determined by *fruAB*, *lmo0021–0023*, *lmo0358*, *lmo0399–0400*, *lmo0426–0428*, *lmo0503*, *lmo0631–0633*, *lmo2135–2137*, and *lmo2733*), and seven PTS for cellobiose (*lmo0034*, *lmo0901*, *lmo1095*, *lmo2683–2685*, *lmo2708*, *lmo2762*, *2763*, *2765*, *lmo2780*, *2782*, and *2783*). Surprisingly, however, *ptsG* which encodes the PTS-dependent glucose transport in many low G+C Gp bacteria, including *B. subtilis* (Gonzy-Treboul et al., 1991), is incomplete in *L. monocytogenes*, and only the gene (*lmo1017*) for the EIIA component of the PTS-G system is present in *L. monocytogenes* and is not organized in an operon with *ptsH* and *ptsI* as in *B. subtilis*. Deletion of this gene does not affect the growth rate of *L. monocytogenes* in minimal media with glucose as carbon source suggesting that this residual part of *ptsG* is not involved in glucose uptake (Mertins et al., 2007). Although non-PTS glucose uptake driven by PMF has been previously suggested for *L. monocytogenes* (Christensen and Hutkins, 1994), a *ptsH* mutant which cannot produce a functional HPr protein (essential for all PTS-dependent systems) is unable to grow in glucose-containing minimal medium (Mertins et al., 2007). These data clearly indicate that glucose transport in *L. monocytogenes* is predominantly, if not exclusively, PTS-mediated.

4.2.2. Glucose Catabolism

Listeria monocytogenes growing in complex media (e.g., BHI) catabolizes first glucose (and the other PTS sugars, which are all ultimately converted to glucose-6-phosphate) mainly by the glycolytic pathway. The principal glycolysis genes (*gap, pgk, tpi, pgm,* and *eno*) of *L. monocytogenes* as in most low G+C Gp bacteria belong to the predicted highly expressed genes (Karlin et al., 2004). In glucose-containing minimal medium, these genes are, however, down-regulated and genes of the pentose phosphate pathway (PPP) are induced when compared to BHI (Joseph et al., 2006), indicating the need of the oxidative decarboxylation of glucose by glucose-6-phosphate dehydrogenase (possibly for the production of CO_2—see below) and/or the generation of increased amounts of erythrose-4-phophate (for the biosynthesis of aromatic amino acids which are not present in the minimal medium). Interestingly, a similar down-regulation of most glycolysis genes and up-regulation of PPP genes are also observed when *L. monocytogenes* grows in the cytosol of mammalian host cells (Joseph et al., 2006).

Entry of pyruvate into the citrate cycle affords the oxidative decarboxylation to acetyl-CoA by the lipoate-dependent pyruvate—dehydrogenase; this step seems to be critical for intracellular *L. monocytogenes*, since a mutant defective in the lipoate ligase 1 (LplA1) is strongly impaired in intracellular growth (O'Riordan et al., 2003).

The citrate cycle of *L. monocytogenes* is interrupted due to the lack of 2-oxoglutarate dehydrogenase (Eisenreich et al., 2006; Trivett and Meyer, 1971). Hence oxaloacetate, which is essential for the entry of acetyl-CoA into the "cycle" leading to citrate and an important intermediate for the synthesis of Asp and other amino acids belonging to the Asp family, cannot be regenerated from citrate, and its synthesis becomes a crucial step in *L. monocytogenes* metabolism. As recently shown by ^{13}C-isotopolog perturbation studies with uniformly labeled [13][C]glucose (Eisenreich et al., 2006), oxaloacetate is mainly produced by carboxylation of pyruvate catalyzed by pyruvate carboxylase (determined by *pycA*, while a gene for PEP-carboxylase seems to be missing in *L. monocytogenes*). Oxaloacetate is probably converted into malonate and succinate by the reducing branch of the citrate cycle for the generation of these important intermediates. Thus, for the generation of oxaloacetate by pyruvate carboxylase, CO_2 is an essential substrate, and we suggest that the observed induced oxidative decarboxylation of glucose-6-phosphate (first step in the PPP) may be therefore required for growth of *L. monocytogenes* in glucose-containing minimal medium. The special role of CO_2 for growth of *L. monocytogenes* and *Yersinia pseudotuberculosis* was pointed out earlier (Buzolyova and

Somov, 1999). This CO_2 requirement may also explain the inability of *L. monocytogenes* to grow in minimal media with pentoses like ribose or rhamnose as carbon sources, although fermentation of rhamnose is observed in rich media (Groves and Welshimer, 1977).

4.2.3. Nonphosphotransferase System Carbon Sources

A *ptsH* mutant of *L. monocytogenes* (unable to metabolize PTS sugars) is able to grow in BHI medium, albeit at a reduced growth rate (Mertins et al., 2007), suggesting that other C-components present in BHI besides PTS sugars can serve as efficient carbon sources. BHI is a rich medium containing many poorly defined C-components which may even vary in composition from batch to batch. It contains probably many peptides, and efficient transporters for oligopeptides were identified in *L. monocytogenes* (Verheul et al., 1998). However, based on previous studies (Premaratne et al., 1991; Tsai and Hodgson, 2003), *L. monocytogenes* fails to grow with casamino acids suggesting that amino acid catabolic pathways may not occur in this microorganism, and hence, amino acids probably cannot serve as sole carbon sources.

Listeria monocytogenes is, however, able to grow on phosphorylated hexoses, like glucose-1(6)-phosphate and fructose-6-phosphate (but again not on the phosphorylated C5 sugars, like ribose-5-phosphate or xylose-5-phosphate). A gene (*hpt or uhpT*) for a special transporter for phosphorylated hexoses has been identified in *L. monocytogenes* (Chico-Calero et al., 2002). This listerial Hpt transporter is highly homologous to a similar sugar transporter in *E. coli* (UhpT) (Weston and Kadner, 1988) and is under the control of PrfA, the central virulence regulator of *L. monocytogenes* (see Chap. 7), and all PrfA-dependent genes including *hpt* are highly up-regulated when *L. monocytogenes* replicates in the cytosol of mammalian host cells (Joseph et al., 2006). Mutants lacking *hpt* are less efficient in replication in host cell's cytosol (Chico-Calero et al., 2002), suggesting that phosphorylated hexoses, presumably mainly glucose-6-phosphate but possibly also glucose-1-phosphate (deriving from cellular glycogen—M. Beck, 2005, personal communication) are major carbon sources in mammalian cells. This assumption is supported by the above mentioned highly induced expression of the *hpt* gene in the host cell's cytosol and the strong activation of the *hpt* promoter in this cellular compartment (S. Pilgrim, personal communication). It is also possible that the listerial Hpt can also transport ribose-5-phospate as shown for the highly related Hpt of *E. coli*; however, in contrast to *E. coli* which can grow on this carbon source, *L. monocytogenes* is unable to use ribose-5-phosphate as a sole carbon source. Glycerol can also replace glucose in defined minimal media (Tsai and Hodgson, 2003); our own unpublished observation). Glycerol, probably taken up by *L. monocytogenes* as by most bacteria via facilitated transport (Heller et al., 1980; Lin, 1976), is phosphorylated by glycerol kinases and oxidized by glycerol-3-phosphate dehydrogenase to glyceraldehyde-3-phosphate which can be further metabolized by the enzymes of the glycolytic pathway. Two glycerol kinase genes (*lmo1034* and *lmo1538*) were identified in *L. monocytogenes*. These two genes as well the glycerol-3-phosphate dehydrogenase are significantly induced not only in *L. monocytogenes* growing in minimal media with glycerol as carbon source but also upon growth within the host cell's cytosol, suggesting that glycerol is a possible carbon source for *L. monocytogenes* metabolism inside the host cells. This assumption is supported by the observation that mutants blocked in both glycerol kinase genes and the glycerol dehydrogenase gene are impaired in cytosolic growth (Joseph et al., 2006). One possible source for the supply of glycerol by the host cell may be phospholipids which by degradation via the listerial phospholipase C (PlcB) together with cellular A-type lipases can yield glycerol, fatty acids, and ethanolamine (or choline). Expression of *plcB* is highly induced in the host cell's cytosol (Joseph et al., 2006; Klarsfeld et al., 1994), and phospholipids as substrates may arise by the disruption of the primary phagosome by which *L. monocytogenes* is taken up in the host cells. In this context, it is interesting to note that the genes for ethanolamine–ammonia lyase which convert ethanolamine into ammonia (which may serve as nitrogen source—see later) and acetyl-CoA are also highly induced in cytosolically growing *L. monocytogenes*. Vitamin B_{12} is required for this reaction (Roof and Roth, 1988, 1989) and interestingly, the genes for cobalamin biosynthesis are also induced in cytosolically growing *L. monocytogenes* (Joseph et al., 2006).

Acetyl-CoA alone cannot be used as a sole carbon source by *L. monocytogenes* since the genes of the glyoxlyate shunt are missing in *L. monocytogenes* (Glaser et al., 2001), a fact that also rules out the utilization of fatty acids, which were shown to be important intracellular carbon sources for *Mycobacterium tuberculosis* and *Salmonella typhimurium* (Fang et al., 2005; McKinney et al., 2000).

4.2.4.   Catabolite Repression and Its Impact on PrfA-Dependent Virulence Gene Expression

Previous studies have repeatedly shown that sugars that can be used by *L. monocytogenes* as carbon source, like glucose, fructose, mannose, and cellobiose, have an inhibitory effect on PrfA activity and hence the PrfA-dependent gene expression (Behari and Youngman, 1998b; Milenbachs et al., 1997; Milenbachs et al., 2004). The strongest inhibition is exerted by cellobiose. These sugars are taken up by PTS-mediated transport and result ultimately in the conversion to glucose-6-phosphate and in catabolite repression of many genes and operons in *L. monocytogenes*. The inhibition of PrfA activity by these sugars thus suggests that PrfA may interact either with components involved in carbon catabolite repression (CCR) or with PTS-mediated sugar transport or with both.

The mechanism of CCR control in Gp bacteria of low G+C content (Figure 4.1) depends on the regulator protein CcpA (*catabolite control protein A*), a member of the LacI/GalR family of bacterial regulatory proteins, which affects the expression of genes containing a *catabolite-responsive element* (CRE-box)

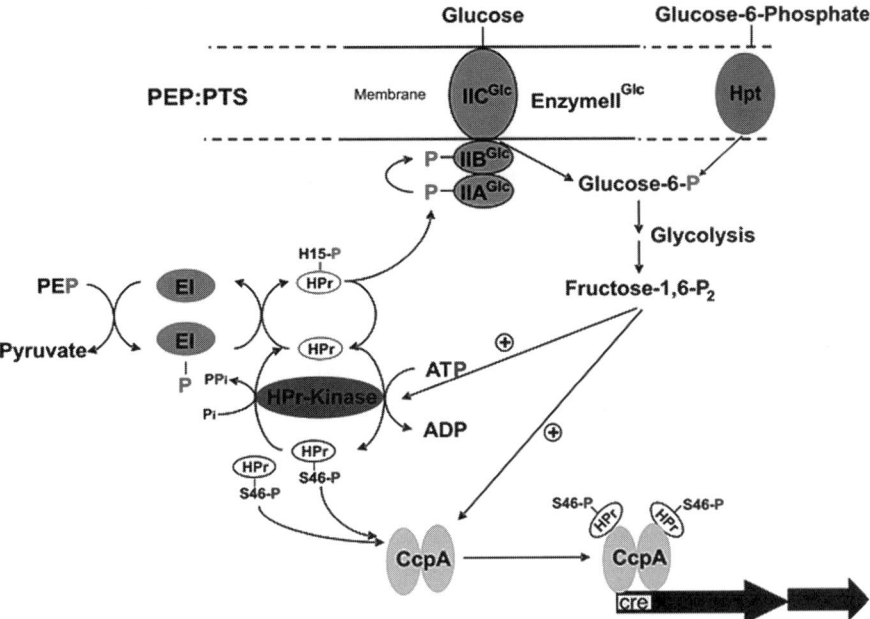

FIGURE 4.1. Schematic representation of CCR in low G+C gram-positive bacteria. *PEP* phosphoenolpyruvate; *PTS* phosphotransferase system; *Hpt* hexose phosphate transporter; *EI*: enzyme I; Enzyme II^{Glc} *A B C* glucose-specific enzyme II A, B, and C; *HPr* PTS phosphocarrier protein HPr; *CcpA* catabolite control protein A; *CRE* catabolisteresponsive *e*lement; *H*15-*P* HPr phosphorylated at His 15; *S*46-*P* HPr phosphorylated at Ser 46; *ATP* adenosine triphosphate; *ADP* adenosine diphosphate; *PP$_i$*: pyrophosphate; *fructose-1,6-P$_2$* fructose-1,6-bisphosphate.

near their regulatory region (about 200 genes in *B. subtilis*) (Hueck and Hillen, 1995; Weickert and Chambliss, 1990). The CcpA activity in itself is dependent on different cofactors, which leads to different modes of gene regulation (Blencke et al., 2003; Gosseringer et al., 1997; Moreno et al., 2001). The major cofactor of CcpA is HPr phosphorylated at position Ser-46. HPr, a component of the general PTS pathway, is phosphorylated at His-24 during PTS sugar transport. This phosphate is transferred to EIIA and further to EIIB of the sugar-specific permeases. During active glycolysis, HPr also becomes phosphorylated at Ser-46 by a specific ATP-dependent HPr-kinase/phosphorylase (HPrK/P) (Reizer et al., 1998). This HPr phosphorylation is stimulated by the intermediates of the glycolytic pathway, especially by fructose-1,6-bisphosphate (FBP), while free phosphate stimulates the phosphorylase activity of HPrK/P (Dossonnet et al., 2000; Galinier et al., 1998; Reizer et al., 1988). HPr-Ser46-P associated with CcpA binds to the CRE sites of CCR-controlled genes and leads to the repression of these genes (type I CcpA-controlled genes) (Deutscher et al., 1995; Jones et al., 1997) as depicted in Figure 4.1.

These genes are therefore up-regulated in *ccpA*- and *HPrK*-deficient mutants (Blencke et al., 2003; Moreno et al., 2001).

A *ccpA* mutant also shows impaired glucose transport and down-regulation of the transcription of several genes that are essential for the C- and the N-metabolism (Blencke et al., 2003). Typical genes of this group (class II CcpA-dependent genes) are involved in the biosynthesis of the BCAA (Ile, Leu, Val) pathway, in glycolysis (*gapA* operon) and glutamate synthesis (*gltAB*). PTS-mediated glucose uptake requires HPr phosphorylation at His15 which is catalyzed by enzyme I in the presence of PEP. EIIA phosphorylated by HPr-His15-P is also involved in other regulatory functions (Titgemeyer and Hillen, 2002). During glucose starvation and by increased inorganic phosphate concentration and low concentrations of glycolytic intermediates, HPr-Ser46-P is dephosphorylated (Fieulaine et al., 2002).

The genomes of *L. monocytogenes* and *L. innocua* contain genes for orthologs of all components involved in CCR of *B. subtilis* (with the exception of *crh* Glaser et al., 2001), suggesting a similar CCR control mechanism in *Listeria* as in *B. subtilis*. This assumption is supported by biochemical and genetic studies on CcpA, HPr, and HPrK/P from *L. monocytogenes* (Behari and Youngman, 1998a; Christensen and Hutkins, 1994; Mertins et al., 2007). CRE sequences highly similar to those of *B. subtilis* were identified in direct proximity to genes and operons which by analogy with the *B. subtilis* counterparts are expected to be under CRR control (Andersson et al., 2005; Joseph et al., 2006). A direct CCR control of PrfA and PrfA-regulated genes and the possible interaction of CcpA with PrfA protein can be ruled out, since a *ccpA* mutant has little effect on *prfA* expression and PrfA activity (Mertins et al., 2007). Recent data (Marr et al., 2006) show, however, that overexpression of PrfA leads to a highly significant growth inhibition of *L. monocytogenes* in glucose-containing media, which seems to be caused by inhibition of PTS-mediated glucose uptake, suggesting that PrfA may interact with components of the PTS-mediated sugar transport rather than with CCR.

4.3. Anabolic Pathways

The genome sequence of *L. monocytogenes* contains all genes for the amino acid-, purine-, pyrimidine-, and several vitamin biosynthetic-pathways (those for biotin, riboflavin, thiamine, and lipoate are absent). The gene for the last enzyme in serine biosynthesis (serine–phosphate phosphatase) has not been annotated in the genome sequence, but all growth studies clearly show that *L. monocytogenes* is not auxotrophic for serine. Surprisingly, previous studies (Phan-Thanh and Gormon, 1997; Premaratne et al., 1991) indicated the requirement of the BCAA (Ile, Val, and Leu) as well as cysteine, methionine, and arginine when *L. monocytogenes* was grown in defined minimal media with glucose as carbon and energy source.

Listeria monocytogenes lacks sulphate and nitrate reductases and hence is dependent on reduced N and S sources, which readily explains the growth requirement for cysteine and methionine. In the presence of cysteine, methionine can be biosynthesized de novo albeit at low rate (J. Slaghuis, personal communication). The ability of *L. monocytogenes* to biosynthesize Arg has been also demonstrated (Tsai and Hodgson, 2003; our own unpublished data), but again, addition of Arg to the minimal medium clearly enhances the growth rate. In the absence of Ile, the growth rate is very low in this minimal medium, and addition of Ile together with one of the other two BCAA (Leu or Val) is required to obtain efficient growth of *L. monocytogenes* in this culture medium. Our recently performed ^{13}[C]-isotopolog perturbation studies (Eisenreich et al., 2006) using uniformly labeled ^{13}[C]glucose show low-level biosynthesis of all three BCAA, even in the presence of externally added BCAA. This synthesis (especially that of Ile) is significantly enhanced in the presence of high PrfA concentration, which as discussed above reduces PTS-mediated glucose uptake and hence may inhibit PMF-dependent BCAA transport. These data clearly show that the biosynthesis of Ile, Leu, and Val is functional, but its efficacy is low in glucose-containing minimal medium.

The BCAA are indicators of the general nutritional status of the bacterial cell because their synthesis depends on several basic catabolic precursors (oxaloacetate, pyruvate, and acetyl-CoA), and hence the rate of BCAA synthesis is an important factor in the overall bacterial physiology. In *B. subtilis*, the central BCAA biosynthesis operon (*ilvB*) is under complex control of the global regulators CcpA, CodY, and TnrA (Shivers and Sonenshein, 2005; Tojo et al., 2005) which also regulate many genes that respond to nutrient availability and growth rate (Molle et al., 2003). All three regulators have binding sites in the *ilvB* regulatory region. CcpA binding to a CRE site within this region activates the transcription starting at the *ilvB* promoter. CodY interacts directly with Ile and GTP, and these two components (indicators of efficient growth and high energy level in the bacterial cell) act as independent corepressors for CodY (Shivers and Sonenshein, 2005).

Listeria monocytogenes contains orthologous genes for CcpA, CodY, and GlnR (highly similar to TnrA), and it is therefore likely that similar control mechanisms may act in *L. monocytogenes* as in *B. subtilis*. The low rate of BCAA biosynthesis in glucose-containing minimal medium may therefore reflect the shortage of necessary catabolic intermediates (especially oxaloacetate) due to the interrupted citrate cycle.

Previous studies indicated that *L. monocytogenes* mutants auxotrophic for some amino acids, like Phe, Gly, and Pro, replicated within host cells like the parental *L. monocytogenes* strain, while a mutant deficient in all three aromatic amino acids was impaired in intracellular replication and virulence (Marquis et al., 1993). Similar results were obtained in our recent investigation with *aro* mutants which are defective in the basic pathway of all aromatic components. These mutants—in addition to exhibiting impaired cytosolic replication, cell-to-cell-spreading, and virulence—showed a predominantly anaerobic metabolism (Stritzker et al., 2004). The reason for this more unexpected result is apparently the lack of synthesis of menaquinone which is the only quinone produced by *L. monocytogenes* and hence its absence strongly impairs aerobic respiration.

Menaquinone biosynthesis involves the condensation of 1,4-dihydroxy-2-naphthoate with polyprenyl-PP which is produced by the isoprenoid biosynthesis pathway. The precursor of all isoprenoids, isopentenyl-PP, is biosynthesized either by the classical mevalonate or by the alternative 2-C-methyl-D-erythritol-4-phosphate (via gyceraldehyde-3-P and pyruvate) pathway. Interestingly, the *L. monocytogenes* genome carries the information for both pathways and both seem to be functional (Begley et al., 2004). It remains to be seen whether there is a preferential activation of one of the pathways when *L. monocytogenes* replicates inside host cells.

Comparative transcript profiling using RNA from extra- and intracellularly grown *L. monocytogenes* shows highly significant up-regulation of the genes for the biosynthesis of all essential amino acids; in particular, the aromatic amino acids and the three BCAA but not the nonessential ones (Joseph et al., 2006), suggesting that the latter ones are provided by the host cell. Strong up-regulation is also observed for the genes involved in purine and pyrimidine biosynthesis, in accordance with previous data (Klarsfeld et al., 1994; Marquis et al., 1993) indicating that nucleotides are not efficiently provided by the host cell to intracellularly growing *L. monocytogenes*.

4.4. Nitrogen Metabolism

For most bacteria including *L. monocytogenes*, glutamine (Gln) is the optimal nitrogen source (Merrick and Edwards, 1995), but in the absence of Gln, *L. monocytogenes* is capable of utilizing alternative nitrogen sources, such as ammonium (Tsai and Hodgson, 2003), arginine, and even ethanolamine (our own unpublished results). These latter N sources might become important, particularly when *L. monocytogenes* replicates in mammalian host cells where the supply of free Gln is limited and its consumption by the intracellular *L. monocytogenes* may strongly impair the "host function" of the invaded cell for the intracellular *L. monocytogenes*. Gln as primary nitrogen source is converted to Glu—the major donor of nitrogen for amino acids and nucleotides—by glutamate synthetase (GOGAT) with 2-oxoglutarate (OG) as additional substrate. The cellular level of Gln, Glu, and OG is stringently controlled in *E. coli* on the transcription—and the glutamine synthetase (GS) activity—level by uridinylation/deuridinylation of the PII protein and by the two-component system NtrB/NtrC (Arcondeguy et al., 2001). Unlike in *E. coli*, the activity of GS of low $G + C$ Gp bacteria (studied extensively in *B. subtilis*) is not modulated by covalent modification, and the global NtrB/C regulatory system is absent here. Synthesis of GS is regulated by the repressor GlnR, by the global regulator TnrA, and possibly by other transcription regulators.

Ammonium as an alternative nitrogen source is transported in *B. subtilis* at low external concentration (at high concentration uptake occurs by diffusion or facilitated transport) by the transporter NrgA which is encoded by the *ngrAB* operon (Detsch and Stulke, 2003), and transcription of the *ngrAB* promoter is activated during nitrogen-limited growth by the global regulator TnrA (Wray et al., 1996; Yoshida et al., 2003). Ammonium is then channeled into glutamine and further to glutamate via glutamine- and glutamate-synthetases as described above. The product of the second gene of the *nrgAB* operon, NrgB belongs to the PII family of regulatory proteins, but there is no indication that NrgB is covalently modified in *B. subtilis*. It is rather believed that NrgB transforms the information of cellular ammonium concentrations by fine-tuning downstream regulatory factors essential for the expression of glutamine- and glutamate-synthetases (Fisher, 1999). Orthologous genes for these regulators of nitrogen metabolism were also identified in the *L. monocytogenes* genome suggesting a similar mode of nitrogen control in *L. monocytogenes* as in *B. subtilis*.

The *nrgAB* operon (*lmo1516–1517*) in *L. monocytogenes* is up-regulated under all conditions which lead to the up-regulation of *glnAR* (operon for GS and the repressor) and *gltAB* (encoding the GOGAT subunits). Up-regulation of these genes is also observed in cytosolically replicating *L. monocytogenes* (Joseph et al., 2006), suggesting that ammonium rather than glutamine might be the major nitrogen source within mammalian host cells.

Ammonium could be provided under the intracellular conditions by excess host cell arginine (normally removed via the urea cycle) or ethanolamine (deriving from phosphatidylethanolamine (PEA)). Indeed the induction of *arpJ*, a gene encoding a specific arginine ABC transporter, has been shown in cytosolically replicating *L. monocytogenes* (Joseph et al., 2006; Klarsfeld et al., 1994). The listerial arginine deiminase (*lmo0043—arcA*) could then degrade this arginine into ammonia and citrulline, two substrates which could serve as nitrogen sources as citrulline can be further converted into another ammonia molecule (together with CO_2 and ATP) and ornithine via the enzymes ornithine carbamoyl transferase and carbamoyl carboxy kinase encoded by the *L. monocytogenes*-specific *arcBCD* operon (*lmo0036-0039*). The possible involvement of the *arcA-D* genes in intracellular ammonium supply does not rule out the participation of these genes in acid resistance of *L. monocytogenes* as well, as recently suggested (Gahan and Hill, 2005). The excess arginine of the host cell would be otherwise removed from the host cell by arginase-catalyzed degradation to ornithine and urea. This urea could not be utilized as nitrogen source by *L. monocytogenes* due to the absence of a listerial urease.

Another possible intracellular nitrogen source provided by the host cell could be ethanolamine generated by degradation of PEA. PEA is an excellent substrate for PlcB, a listerial PlcB encoded by the PrfA-dependent gene *plcB* (Goldfine et al., 1993) that is highly up-regulated inside host cells. Hydrolysis of ethanolamine into ammonia and acetaldehyde occurs by the vitamin B_{12}-dependent ethanolamine–ammonia lyase (Bradbeer, 1965) encoded by the *eutBC* genes. These genes have been shown in *Salmonella enterica* serovar

Typhimurium (Kofoid et al., 1999) to be part of a 55-kb locus which in addition to *eutB* and *eutC* (the genes for the two subunits of the lyase) carry 15 genes that encode a positive regulator (necessary for the induction of the *eut* operon together with B12), enzymes for propandiol degradation, and shell proteins for the carboxysome. A similar gene cluster showing high homology to the *Salmonella eut* operon has been identified in *L. monocytogenes*, and may have been introduced by lateral gene transfer as pointed out previously (Buchrieser et al., 2003). Several genes of the *eut* gene cluster, particularly *eutBC*, but also those for cobalamine synthesis, are highly up-regulated in *L. monocytogenes* replicating inside mammalian host cells (Joseph et al., 2006). Our recent unpublished data indicate that ethanolamine can function as a sole nitrogen source in a minimal medium with glycerol as carbon source, whereas an *eutBC* deletion mutant cannot grow under these conditions.

4.5. Conclusions

Bioinformatic and functional genomic data indicate that *L. monocytogenes* is a heterotrophic and largely prototrophic bacterium belonging to the group of low $G + C$ Gp bacteria. But its metabolism is also optimally adapted for highly efficient growth within the cytosol of many mammalian cells (Goetz et al., 2001). The metabolism of *L. monocytogenes* reveals some unusual features which seem to have profound consequences for extra- and intracellular replication of *L. monocytogenes* and hence for virulence:

1 the inability to use oxidized sulfur and nitrogen sources due to the lack of nitrate and sulfate reductases,
2 the interrupted citrate cycle due to the lack of 2-oxoglutarate dehydrogenase, and
3 the complex PTS-mediated glucose transport.

(1) This feature readily explains the observed auxotrophy of *L. monocytogenes* for Cys (and in the absence of Cys for Met) and indicates that intracellularly replicating *L. monocytogenes* will entirely depend on the host cell for Cys supply. Cystein is a nonessential amino acid for mammalian cells whereas methionine is an essential one. In *L. monocytogenes*, the situation is reverse and the two "partners" could therefore provide each other with the necessary sulphur-containing amino acids.

(2) The incomplete citrate cycle renders the synthesis of oxaloacetate by pyruvate carboxylase to be a critical metabolic step when *L. monocytogenes* grows in environments where glucose (or another carbohydrate) is the sole carbon source. Shortage of this catabolic intermediate (and as a consequence also pyruvate and acetyl-CoA) may be the reason for the unexpected dependency of *L. monocytogenes* on BCAA (Ile and Leu or Val), Met and to a lesser extent Arg for efficient growth in the minimal media containing glucose as a sole carbon

source. Although the biosynthetic pathways for these amino acids are functionally intact, their efficient synthesis depends directly or indirectly on the availability of oxaloacetate. In fact, a *B. subtilis* mutant deficient in 2-oxoglutarate dehydrogenase required Asp (which derives directly from oxaloacetate and represents an important intermediate for the biosynthesis of these amino acids) for growth at wild-type rates in minimal media due to the inability of the mutant to regenerate oxaloacetate from citrate (Fisher and Magasanik, 1984). The syntheses of these amino acids which are needed for protein synthesis and branched-chain fatty acids (especially Ile) (Nichols et al., 2002) in large amounts depend on oxaloacetate (via Asp), pyruvate (needed for synthesis of oxaloacetate), or actyl-CoA (deriving from pyruvate). As mentioned above (see discussion on the ilv-leu operon), nitrogen metabolism and carbon metabolism are coregulated by the global regulators CcpA, TnrA, and CodY (Shivers and Sonenshein, 2005; Tojo et al., 2005). This regulatory network (best studied in *B. subtilis*) guarantees well-balanced intracellular concentrations of the central carbon (especially glucose and its catabolites) and nitrogen (especially Gln, Glu) intermediates essential for the entire cellular metabolism.

The inability of *L. monocytogenes* to regenerate oxaloacetate from citrate may be overcome within the host cell by the supply of malonate. This intermediate, which is directly converted to oxaloacetate in the citrate cycle, could be provided by the host cell via the oxoglutarate/malate shuttle at the expense of the oxoglutarate produced in excess by *L. monocytogenes* due to the absence of oxoglutarate dehydrogenase. It has been shown that a *B. subtilis* mutant deficient in this enzyme secretes considerably larger amounts of OG into the medium than the wild-type strain (Fisher and Magasanik, 1984).

(3) The most important feature is, however, the use of the appropriate carbon source by *L. monocytogenes* within the host cell. The transport of glucose, a preferred carbon source for *L. monocytogenes* metabolism, is achieved in a yet unknown way. Although a large number of (in part *L. monocytogenes* specific) PTS has been identified in the *L. monocytogenes* genome, a functional PTS-G glucose uptake system (characteristic for many bacteria) is missing and glucose may be cotransported by several other PTS permeases (R. Ecke, personal communication).

Uptake of glucose-1-P generated by degradation of host cell's glycogen (a nonessential storage product of the host cell) by the specific transporter (Hpt) avoids the competition for glucose with the host cell, and at the same time, the inactivation of PrfA by PTS-mediated sugar uptake. The efficiency of intracellular replication and virulence may be therefore strongly influenced by the carbon source and its transport.

Phospholipids may act as alternative intracellular carbon source for *L. monocytogenes* (and may even become the primary carbon source in mammalian cells lacking glycogen). Phospholipids are probably generated in sufficient amounts by the disruption of the primary phagosome by which *L. monocytogenes* is internalized. In particular PEA may serve as an important carbon and nitrogen source since it can be converted by cellular lipases of the A-type and especially the listerial PlcB to glycerol, fatty acids, and ethanolamine phosphate, which after dephosphorylation can serve as nitrogen source for *L. monocytogenes* in presence of glycerol as carbon source (Schaffer et al., unpublished results). Unlike *M. tuberculosis* and S. *enterica* which use the fatty acids as major intracellular carbon sources (Fang et al., 2005; McKinney et al., 2000), *L. monocytogenes* could utilize mainly glycerol and ethanolamine as intracellular nutrients.

We hypothesize that these three metabolic features, although unfavourable for growth of *L. monocytogenes* under certain extracellular conditions, are essential for the efficient intracellular replication of *L. monocytogenes* since they lead to an intimate interference between the metabolism of the bacterium and that of the host cell. This metabolic interference will allow an extended survival of the infected cell, which can then serve as a "host cell" for *L. monocytogenes* for a longer period of time.

Acknowledgments. The authors thank Stefanie Müller-Altrock for useful discussions and critical reading of the manuscript.

References

Amezaga MR, Davidson I, McLaggan D, Verheul A, Abee T, and Booth IR (1995) The role of peptide metabolism in the growth of *Listeria monocytogenes* ATCC 23074 at high osmolarity. *Microbiology* **141 (Pt 1)**: 41–49.

Andersson U, Molenaar D, Radstrom P, and de Vos WM (2005) Unity in organisation and regulation of catabolic operons in *Lactobacillus plantarum, Lactococcus lactis* and *Listeria monocytogenes. Syst Appl Microbiol* **28**: 187–195.

Arcondeguy T, Jack R, and Merrick M (2001) P(II) signal transduction proteins, pivotal players in microbial nitrogen control. *Microbiol Mol Biol Rev* **65**: 80–105.

Begley M, Gahan CG, Kollas AK, Hintz M, Hill C, Jomaa H, and Eberl M (2004) The interplay between classical and alternative isoprenoid biosynthesis controls gammadelta T cell bioactivity of *Listeria monocytogenes. FEBS Lett* **561**: 99–104.

Behari J, and Youngman P (1998a) A homolog of CcpA mediates catabolite control in *Listeria monocytogenes* but not carbon source regulation of virulence genes. *J Bacteriol* **180**: 6316–6324.

Behari J, and Youngman P (1998b) Regulation of hly expression in *Listeria monocytogenes* by carbon sources and pH occurs through separate mechanisms mediated by PrfA. *Infect Immun* **66**: 3635–3642.

Blencke HM, Homuth G, Ludwig H, Mader U, Hecker M, and Stulke J (2003) Transcriptional profiling of gene expression in response to glucose in*Bacillus subtilis*: regulation of the central metabolic pathways. *Metab Eng* **5**: 133–149.

Bradbeer C (1965) The clostridial fermentations of choline and ethanolamine. II. Requirement for a cobamide coenzyme by an ethanolamine deaminase. *J Biol Chem* **240**: 4675–4681.

Buchrieser C, Rusniok C, Kunst F, Cossart P, and Glaser P (2003) Comparison of the genome sequences of *Listeria monocytogenes* and *Listeria innocua*: clues for evolution and pathogenicity. *FEMS Immunol Med Microbiol* **35**:207–213.

Buzolyova LS, and Somov GP (1999) Autotrophic assimilation of CO_2 and Cl-compounds by pathogenic bacteria. *Biochemistry (Mosc)* **64**: 1146–1149.

Chico-Calero I, Suarez M, Gonzalez-Zorn B, Scortti M, Slaghuis J, Goebel W, and Vazquez-Boland JA (2002) Hpt, a bacterial homolog of the microsomal glucose-6-phosphate translocase, mediates rapid intracellular proliferation in *Listeria*. *Proc Natl Acad Sci USA* **99**: 431–436.

Christensen DP, and Hutkins RW (1994) Glucose uptake by *Listeria monocytogenes* Scott A and inhibition by pediocin JD. *Appl Environ Microbiol* **60**: 3870–3873.

Collins MD, Wallbanks S, Lane DJ, Shah J, Nietupski R, Smida J, Dorsch M, and Stackebrandt E (1991) Phylogenetic analysis of the genus *Listeria* based on reverse transcriptase sequencing of 16S rRNA. *Int J Syst Bacteriol* **41**: 240–246.

Detsch C, and Stulke J (2003) Ammonium utilization in *Bacillus subtilis*: transport and regulatory functions of NrgA and NrgB. *Microbiology* **149**: 3289–3297.

Deutscher J, Kuster E, Bergstedt U, Charrier V, and Hillen W (1995) Protein kinase-dependent HPr/CcpA interaction links glycolytic activity to carbon catabolite repression in gram-positive bacteria. *Mol Microbiol* **15**: 1049–1053.

Dossonnet V, Monedero V, Zagorec M, Galinier A, Perez-Martinez G, and Deutscher J (2000) Phosphorylation of HPr by the bifunctional HPr Kinase/P-ser-HPr phosphatase from *Lactobacillus casei* controls catabolite repression and inducer exclusion but not inducer expulsion. *J Bacteriol* **182**: 2582–2590.

Eisenreich W, Slaghuis J, Laupitz R, Bussemer J, Stritzker J, Schwarz C, Schwarz R, Dandekar T, Bacher A, and Goebel W (2006) ^{13}C Isotopolog perturbation studies of *Listeria monocytogenes* carbon metabolism and its modulation by the virulence regulator PrfA. *Proc Nath Acad Sci USA* **103**: 2040–2045.

Fang FC, Libby SJ, Castor ME, and Fung AM (2005) Isocitrate lyase (AceA) is required for *Salmonella* persistence but not for acute lethal infection in mice. *Infect Immun* **73**: 2547–2549.

Farber JM, and Peterkin PI (1991) *Listeria monocytogenes*, a food-borne pathogen. *Microbiol Rev* **55**: 476–511.

Fieulaine S, Morera S, Poncet S, Mijakovic I, Galinier A, Janin J, Deutscher J, and Nessler S (2002) X-ray structure of a bifunctional protein kinase in complex with its protein substrate HPr. *Proc Natl Acad Sci USA* **99**: 13437–13441.

Fisher SH, and Magasanik B (1984) Synthesis of oxaloacetate in *Bacillus subtilis* mutants lacking the 2-ketoglutarate dehydrogenase enzymatic complex. *J Bacteriol* **158**: 55–62.

Fisher SH (1999) Regulation of nitrogen metabolism in *Bacillus subtilis*: vive la difference! *Mol Microbiol* **32**: 223–232.

Friedman ME, and Roessler WG (1961) Growth of *Listeria monocytogenes* in defined media. *J Bacteriol* **82**: 528–533.

Gahan CG, and Hill C (2005) Gastrointestinal phase of *Listeria monocytogenes* infection. *J Appl Microbiol* **98**: 1345–1353.

Galinier A, Kravanja M, Engelmann R, Hengstenberg W, Kilhoffer MC, Deutscher J, and Haiech J (1998) New protein kinase and protein phosphatase families mediate signal transduction in bacterial catabolite repression. *Proc Natl Acad Sci USA* **95**: 1823–1828.

Glaser P, Frangeul L, Buchrieser C, Rusniok C, Amend A, Baquero F, Berche P, Bloecker H, Brandt P, Chakraborty T, Charbit A, Chetouani F, Couve E, de Daruvar A, Dehoux P, Domann E, Dominguez-Bernal G, Duchaud E, Durant L, Dussurget O, Entian KD, Fsihi H, Garcia-del Portillo F, Garrido P, Gautier L, Goebel W, Gomez-Lopez N, Hain T, Hauf J, Jackson D, Jones LM, Kaerst U, Kreft J, Kuhn M, Kunst F, Kurapkat G, Madueno E, Maitournam A, Vicente JM, Ng E, Nedjari H, Nordsiek G, Novella S, de Pablos B, Perez-Diaz JC, Purcell R, Remmel B, Rose M, Schlueter T, Simoes N, Tierrez A, Vazquez-Boland JA, Voss H, Wehland J, and Cossart P (2001) Comparative genomics of *Listeria* species. *Science* **294**: 849–852.

Goetz M, Bubert A, Wang G, Chico-Calero I, Vazquez-Boland JA, Beck M, Slaghuis J, Szalay AA, and Goebel W (2001) Microinjection and growth of bacteria in the cytosol of mammalian host cells. *Proc Natl Acad Sci USA* **98**: 12221–12226.

Goldfine H, Johnston NC, and Knob C (1993) Nonspecific phospholipase C of *Listeria monocytogenes*: activity on phospholipids in Triton X-100-mixed micelles and in biological membranes. *J Bacteriol* **175**: 4298–4306.

Gonzy-Treboul G, de Waard JH, Zagorec M, and Postma PW (1991) The glucose permease of the phosphotransferase system of *Bacillus subtilis*: evidence for IIGlc and IIIGlc domains. *Mol Microbiol* **5**: 1241–1249.

Gosseringer R, Kuster E, Galinier A, Deutscher J, and Hillen W (1997) Cooperative and non-cooperative DNA binding modes of catabolite control protein CcpA from *Bacillus megaterium* result from sensing two different signals. *J Mol Biol* **266**: 665–676.

Gottschalk G (1986) *Bacterial Metabolism.* New York: Springer-Verlag.

Gray ML, and Killinger AH (1966) *Listeria monocytogenes* and listeric infections. *Bacteriol Rev* **30**: 309–382.

Groves RD, and Welshimer HJ (1977) Separation of pathogenic from apathogenic *Listeria monocytogenes* by three in vitro reactions. *J Clin Microbiol* **5**: 559–563.

Heller KB, Lin EC, and Wilson TH (1980) Substrate specificity and transport properties of the glycerol facilitator of *Escherichia coli*. *J Bacteriol* **144**: 274–278.

Hueck CJ, and Hillen W (1995) Catabolite repression in*Bacillus subtilis*: a global regulatory mechanism for the gram-positive bacteria? *Mol Microbiol* **15**: 395–401.

Jones BE, Dossonnet V, Kuster E, Hillen W, Deutscher J, and Klevit RE (1997) Binding of the catabolite repressor protein CcpA to its DNA target is regulated by phosphorylation of its corepressor HPr. *J Biol Chem* **272**: 26530–26535.

Jones CE, Shama G, Andrew PW, Roberts IS, and Jones D (1995) Comparative study of the growth of *Listeria monocytogenes* in defined media and demonstration of growth in continuous culture. *J Appl Bacteriol* **78**: 66–70.

Joseph B, Przybilla K, Stühler C, Schauer K, Slaghuis J, Fuchs TM, and Goebel W (2006) Identification of *Listeria monocytogenes* genes contributing to intracellular replication by expression profiling and mutant screening. *J Bacteriol* **188**: 556–568.

Karlin S, Theriot J, and Mrazek J (2004) Comparative analysis of gene expression among low G+C gram-positive genomes. *Proc Natl Acad Sci USA* **101**: 6182–6187.

Klarsfeld AD, Goossens PL, and Cossart P (1994) Five *Listeria monocytogenes* genes preferentially expressed in infected mammalian cells: plcA, purH, purD, pyrE and an arginine ABC transporter gene, arpJ. *Mol Microbiol* **13**: 585–597.

Kofoid E, Rappleye C, Stojiljkovic I, and Roth J (1999) The 17-gene ethanolamine (eut) operon of *Salmonella typhimurium* encodes five homologues of carboxysome shell proteins. *J Bacteriol* **181**: 5317–5329.

Lin EC (1976) Glycerol dissimilation and its regulation in bacteria. *Annu Rev Microbiol* **30**: 535–578.

Marr AK, Joseph B, Mertins S, Ecke R, Müller-Altrock S, and Goebel W (2006) Overexpression of PrfA leads to growth inhibition of *L. monocytogenes* in glucose-containing culture media by interfering with glucose uptake. *J Bacteriol* **188**: 3887–3901.

Marquis H, Bouwer HG, Hinrichs DJ, and Portnoy DA (1993) Intracytoplasmic growth and virulence of *Listeria monocytogenes* auxotrophic mutants. *Infect Immun* **61**: 3756–3760.

McKinney JD, Honer zu Bentrup K, Munoz-Elias EJ, Miczak A, Chen B, Chan WT, Swenson D, Sacchettini JC, Jacobs WR, Jr, and Russell DG (2000) Persistence of *Mycobacterium tuberculosis* in macrophages and mice requires the glyoxylate shunt enzyme isocitrate lyase. *Nature* **406**: 735–738.

Merrick MJ, and Edwards RA (1995) Nitrogen control in bacteria. *Microbiol Rev* **59**: 604–622.

Mertins S, Joseph B, Goetz M, Ecke R, Seidel G, Sprehe M, Hillen W, Goebel W, and Müller-Altrock S (2007) Interference of PrfA with the carbohydrate catabolite repression (CCR) system in *Listeria monocytogenes*. *J Bacteriol* **189**: 473–490.

Milenbachs AA, Brown DP, Moors M, and Youngman P (1997) Carbon-source regulation of virulence gene expression in *Listeria monocytogenes*. *Mol Microbiol* **23**: 1075–1085.

Milenbachs Lukowiak A, Mueller KJ, Freitag NE, and Youngman P (2004) Deregulation of *Listeria monocytogenes* virulence gene expression by two distinct and semi-independent pathways. *Microbiology* **150**: 321–333.

Molle V, Nakaura Y, Shivers RP, Yamaguchi H, Losick R, Fujita Y, and Sonenshein AL (2003) Additional targets of the *Bacillus subtilis* global regulator CodY identified by chromatin immunoprecipitation and genome-wide transcript analysis. *J Bacteriol* **185**: 1911–1922.

Moreno MS, Schneider BL, Maile RR, Weyler W, and Saier MH, Jr (2001) Catabolite repression mediated by the CcpA protein in *Bacillus subtilis*: novel modes of regulation revealed by whole-genome analyses. *Mol Microbiol* **39**: 1366–1381.

Nichols DS, Presser KA, Olley J, Ross T, and McMeekin TA (2002) Variation of branched-chain fatty acids marks the normal physiological range for growth in *Listeria monocytogenes*. *Appl Environ Microbiol* **68**: 2809–2813.

O'Riordan M, Moors MA, and Portnoy DA (2003) Listeria intracellular growth and virulence require host-derived lipoic acid. *Science* **302**: 462–464.

Phan-Thanh L, and Gormon T (1997) A chemically defined minimal medium for the optimal culture of *Listeria*. *Int J Food Microbiol* **35**: 91–95.

Pine L, Malcolm GB, Brooks JB, and Daneshvar MI (1989) Physiological studies on the growth and utilization of sugars by *Listeria* species. *Can J Microbiol* **35**: 245–254.

Premaratne RJ, Lin WJ, and Johnson EA (1991) Development of an improved chemically defined minimal medium for *Listeria monocytogenes*. *Appl Environ Microbiol* **57**: 3046–3048.

Reizer J, Saier MH, Jr, Deutscher J, Grenier F, Thompson J, and Hengstenberg W (1988) The phosphoenolpyruvate: sugar phosphotransferase system in gram-positive bacteria: properties, mechanism, and regulation. *Crit Rev Microbiol* **15**: 297–338.

Reizer J, Hoischen C, Titgemeyer F, Rivolta C, Rabus R, Stulke J, Karamata D, Saier MH, Jr, and Hillen W (1998) A novel protein kinase that controls carbon catabolite repression in bacteria. *Mol Microbiol* **27**: 1157–1169.

Romick TL, Fleming HP, and McFeeters RF (1996) Aerobic and anaerobic metabolism of *Listeria monocytogenes* in defined glucose medium. *Appl Environ Microbiol* **62**: 304–307.

Romick TL, and Fleming HP (1998) Acetoin production as an indicator of growth and metabolic inhibition of *Listeria monocytogenes*. *J Appl Microbiol* **84**: 18–24.

Roof DM, and Roth JR (1988) Ethanolamine utilization in *Salmonella typhimurium*. *J Bacteriol* **170**: 3855–3863.

Roof DM, and Roth JR (1989) Functions required for vitamin B_{12}-dependent ethanolamine utilization in *Salmonella typhimurium*. *J Bacteriol* **171**: 3316–3323.

Seeliger HP (1984) Modern taxonomy of the Listeria group relationship to its pathogenicity. *Clin Invest Med* **7**: 217–221.

Seeliger HP, and Jones D (1986) Genus *Listeria*. In: Sneath PHA, Mair NS, Sharpe ME, and Holt JG (eds) *Bergey's manual of systematic bacteriology*. Vol. 2. Baltimore, MD: Williams & Wilkins, pp.1235–1245.

Shivers RP, and Sonenshein AL (2005) *Bacillus subtilis* ilvB operon: an intersection of global regulons. *Mol Microbiol* **56**: 1549–1559.

Stritzker J, Janda J, Schoen C, Taupp M, Pilgrim S, Gentschev I, Schreier P, Geginat G, and Goebel W (2004) Growth, virulence, and immunogenicity of *Listeria monocytogenes* aro mutants. *Infect Immun* **72**: 5622–5629.

Titgemeyer F, and Hillen W (2002) Global control of sugar metabolism: a gram-positive solution. *Antonie Van Leeuwenhoek* **82**: 59–71.

Tojo S, Satomura T, Morisaki K, Deutscher J, Hirooka K, and Fujita Y (2005) Elaborate transcription regulation of the *Bacillus subtilis* ilv-leu operon involved in the biosynthesis of branched-chain amino acids through global regulators of CcpA, CodY and TnrA. *Mol Microbiol* **56**: 1560–1573.

Trivett TL, and Meyer EA (1971) Citrate cycle and related metabolism of *Listeria monocytogenes*. *J Bacteriol* **107**: 770–779.

Tsai HN, and Hodgson DA (2003) Development of a synthetic minimal medium for *Listeria monocytogenes*. *Appl Environ Microbiol* **69**: 6943–6945.

Vadeboncoeur C, Frenette M, and Lortie LA (2000) Regulation of the pts operon in low G+C Gram-positive bacteria. *J Mol Microbiol Biotechnol* **2**: 483–490.

Verheul A, Rombouts FM, and Abee T (1998) Utilization of oligopeptides by *Listeria monocytogenes* Scott A. *Appl Environ Microbiol* **64**: 1059–1065.

Weickert MJ, and Chambliss GH (1990) Site-directed mutagenesis of a catabolite repression operator sequence in *Bacillus subtilis*. *Proc Natl Acad Sci USA* **87**: 6238–6242.

Weston LA, and Kadner RJ (1988) Role of uhp genes in expression of the *Escherichia coli* sugar-phosphate transport system. *J Bacteriol* **170**: 3375–3383.

Wray LV, Jr, Ferson AE, Rohrer K, and Fisher SH (1996) TnrA, a transcription factor required for global nitrogen regulation in *Bacillus subtilis*. *Proc Natl Acad Sci USA* **93**: 8841–8845.

Yoshida K, Yamaguchi H, Kinehara M, Ohki YH, Nakaura Y, and Fujita Y (2003) Identification of additional TnrA-regulated genes of *Bacillus subtilis* associated with a TnrA box. *Mol Microbiol* **49**: 157–165.

5

The Cell Wall of *Listeria monocytogenes* and its Role in Pathogenicity

M. Graciela Pucciarelli,[1*] Hélène Bierne[2],
and Francisco García-del Portillo[1*]

[1]*Departamento de Biotecnología Microbiana, Centro Nacional
de Biotecnología-Consejo Superior de Investigaciones Científicas (CSIC),
Darwin 3, 28049 Madrid, Spain*
e-mail: mgpuccia@cnb.uam.es, fgportillo@cnb.uam.es
[2]*Unité des Interactions Bactéries-Cellules, Institut Pasteur, INSERM U604,
INRA USC2020, 28 Rue du Docteur Roux, 75724 Paris cedex 15, France
e-mail: hbierne@pasteur.fr*

5.1. Introduction

Listeria monocytogenes contains a cell wall formed by a multilayered cross-linked peptidoglycan decorated with teichoic and lipoteichoic acids. Like in all eubacteria, the cell wall of *L. monocytogenes* plays a critical role in its physiology since it ensures integrity of the cell while maintaining a high internal osmotic pressure. In addition, it also endows the cell with a specific cell shape and provides protection against mechanical stress. As in other noncapsulated gram-positive bacteria, the cell wall of *L. monocytogenes* is the outermost structure of the cell and acts as a scaffold in which different proteins anchor. *L. monocytogenes* is a highly successful pathogen that invades eukaryotic host cells, crosses several natural barriers of the host and survives to extreme environments, and its cell wall must necessarily contain molecules making possible the colonization of these niches. The role in pathogenesis of some of these surface molecules is just starting to be deciphered. Likewise, the genome sequences now known for a few *L. monocytogenes* strains reveal that this pathogen has a large number of genes encoding proteins with domains mediating interactions with cell-wall polymers. Some of these cell-wall-associated proteins are currently subjected to intense investigation. Recent studies have also revisited the structure of the peptidoglycan of *L. monocytogenes* and unravelled new modifications in its structure that may be important for pathogenicity.

In this chapter, we summarize the current knowledge of the biochemistry and enzymology of the *L. monocytogenes* cell wall. Moreover, we discuss on the plethora of proteins that attach to cell-wall components, making emphasis in the distinct modes of protein–cell-wall association and their role in virulence. We also describe recent proteomics studies that have facilitated the identification of novel *L. monocytogenes* surface proteins predicted by the genome data. Finally, we briefly describe what is known on the role of *L. monocytogenes* cell-wall components in the modulation of the host immune response.

5.2.   Biochemistry of the *Listeria monocytogenes* Cell Wall

The first descriptions on the composition and suspected structure of the *L. monocytogenes* peptidoglycan and associated polymers were made more than 30 years ago (Srivastava and Siddique 1973; Ullmann and Cameron 1969). Despite this long period, the fine structure of its peptidoglycan has not been known until very recently (Kloszewska et al. 2006). Likewise, while the biochemistry of teichoic acids (TA) and lipoteichoic acids (LTA) of *L. monocytogenes* was inferred two decades ago (Fiedler 1988; Uchikawa et al. 1986a,b), these polymers are now receiving further attention as they promote the attachment of several virulence proteins and stimulate a large variety of responses in the host.

5.2.1.   The Peptidoglycan of Listeria monocytogenes

Listeria monocytogenes has a peptidoglycan formed by glycan chains containing alternating units of the disaccharide *N*-acetylmuramic acid (MurNAc)-(β-1,4)-*N*-acetyl-D-glucosamine (GlcNAc). Bound to the MurNAc residue is a stem peptide that in *L. monocytogenes* contains L-alanine-γ-D-glutamic acid-*meso*-diaminopimelic acid-D-Ala-D-Ala [L-Ala-γ-D-Glu-*m*-Dap-D-Ala-D-Ala] (Fiedler 1988; Kamisango et al. 1982). The glycan chains are cross-linked by 4→3 linkages between the D-Ala residue of one lateral peptide to the *m*-Dap residue of the other stem peptide. This peptidoglycan structure resembles the reported for many gram-negative bacteria as *Escherichia coli* (Schleifer and Kandler 1972). The biosynthesis of the *L. monocytogenes* peptidoglycan occurs essentially as described in *E. coli* (Holtje 1998), with the formation of the intermediates; lipid I (C_{55}-PP-MurNAc-L-Ala-γ-D-Glu-*m*-Dap-D-Ala-D-Ala) and lipid II [C_{55}-PP-MurNAc-(L-Ala-γ-D-Glu-*m*-Dap-D-Ala-D-Ala)-(β-1,4)-GlcNAc] (Holtje 1998; Navarre and Schneewind 1999). Lipid II is substrate of transglycosylases that bind the disaccharide-pentapeptide molecule to macromolecular peptidoglycan. This reaction leaves the undecaprenyl-pyrophosphate carrier (C_{55}-PP) free to reinitiate the first biosynthetic steps. The transglycosylation reaction is followed by transpeptidation of lateral peptides belonging to adjacent glycan chains, ensuring in this way the incorporation of nascent peptidoglycan.

The first reactions of the peptidoglycan biosynthetic pathway are catalyzed by a conserved number of essential enzymes (Holtje 1998). However, the transglycosylation and transpeptidation reactions are carried out by a variable number of enzymes known as penicillin-binding proteins (PBPs) (Goffin and Ghuysen 2002). All PBPs share the property of binding β-lactam antibiotics, irreversible inhibitors of the transpeptidation (acyl-transferase) reaction involving the rupture of the D-Ala-D-Ala bond. PBPs are multimodular enzymes carrying in their C-terminal module highly conserved SXXK, SXN, and KTG motifs essential for the transpeptidation reaction (Goffin and Ghuysen 2002). Some PBPs also have an N-terminal transglycosylation module. These bi-functional transglycosylase-transpeptidase enzymes are known as class A PBPs. Other PBPs, named as class B, carry an N-module that interacts with components of the morphogenetic apparatus. A third group of PBPs only have the C-module and are named carboxipeptidases or endopeptidases depending on whether they cleave, using water as acceptor molecule, D-Ala-D-Ala or *m*-Dap-D-Ala linkages, respectively. In most bacteria, this third group of PBPs are nonessential and play roles related to peptidoglycan maturation (Goffin and Ghuysen 2002).

Using isotopically labelled β-lactam antibiotics, several studies showed that *L. monocytogenes* has five PBPs, ranging in molecular weight from 95 kDa to 49 kDa (Gutkind et al. 1989; Hakenbeck and Hof 1991; Vicente et al. 1990b). The number of molecules of PBP per cell was estimated in ~ 100 for PBP1, PBP2, PBP3, and PBP4, and ~ 600 for PBP5 (Vicente et al. 1990a). Among these enzymes, PBP4 and PBP3 have the highest and lowest affinity, respectively, for binding of the β-lactam antibiotics (Pierre et al. 1990; Vicente et al. 1990b). The low affinity of PBP3 for β-lactam antibiotics is thought to be the basis of the high intrinsic resistance displayed by *L. monocytogenes* to monobactams and cephalosporins of broad spectrum. *L. monocytogenes* has a distinct PBP profile compared to other *Listeria* species, which is however fairly conserved among *L. monocytogenes* strains (Hakenbeck and Hof 1991). Despite the availability of the complete genome sequence of four *L. monocytogenes* strains (Glaser et al. 2001; Nelson et al. 2004), no study has reported the exact correspondence between each of the PBPs detected by biochemical analysis and their coding genes.

Two PBPs have been recently characterized in the *L. monocytogenes* EGD-e strain: PBP4, a class A PBP encoded by the *lmo2229* gene which displays transglycosylase, transpeptidase, and carboxipeptidase activities (Zawadzka et al. 2006); and, PBP5, a class B PBP with DD-carboxipeptidase activity encoded by *lmo2754* (Korsak et al. 2005). Deletion mutants have been obtained for each of these two PBPs, confirming that none of them is essential for *L. monocytogenes* growth, at least in laboratory conditions. Lack of PBP4 results in a slower growth rate and increased resistance to moenomycin, an antibiotic that inhibits transglycosylase activity (Zawadzka et al. 2006). Phenotypes described for the PBP5 mutant include an increase in the cell-wall thickness and a reduction in the growth rate (Korsak et al. 2005). Whether or not these two PBPs contribute to *L. monocytogenes* pathogenicity remains unknown.

The fine structure of the *L. monocytogenes* peptidoglycan has been recently resolved by reverse-phase high pressure liquid chromatography (HPLC) and mass spectrometry (Kloszewska et al. 2006). Besides the conventional interpeptide linkage (4→3) D-Ala-*m*-Dap, the *L. monocytogenes* peptidoglycan contains (3→3) *m*-Dap-*m*-Dap linkages. This latter interpeptide linkage was previously reported in *Mycobacteria* and *E. coli* (Glauner and Höltje 1990; Wietzerbin et al. 1974). The transpeptidation reaction resulting in the (3→3) linkage is inherently insensitive to β-lactam antibiotics (Goffin and Ghuysen 2002). So, *L. monocytogenes* must have at least one "non-PBP" enzyme responsible for the building of this concrete bridge. The physiological significance of this type of linkage remains to be explored. Other structural features of the *L. monocytogenes* peptidoglycan include the amidation of the free carboxylic group of some *m*-Dap residues and the absence of glucosamine acetylation in certain muropeptides. These types of modifications were previously reported in *Bacillus subtilis* (Atrih et al. 1999). The average cross-linkage of the *L. monocytogenes* peptidoglycan is in the range of ∼ 65% (Kloszewska et al. 2006), slightly higher than that described in the peptidoglycan of *B. subtilis*, ∼ 45%. Unlike most gram-positive bacteria, *L. monocytogenes* does not lyse in the presence of β-lactams, remaining the action of these antibiotics as bacteriostatic. The underlying mechanisms are unknown.

5.2.2. Other Cell-Wall Polyanionic Polymers: Teichoic Acids and Lipoteichoic Acids

As most gram-positive bacteria, *L. monocytogenes* contains two different polyanionic polymers decorating the cell wall: the teichoic acids (TA), covalently bound to the peptidoglycan and the lipoteichoic acids (LTAs), amphipathic molecules that are embedded into the plasma membrane by a diacylglycerolipid (Navarre and Schneewind 1999; Neuhaus and Baddiley 2003). These polymers represent as much as 50–60 % of the total content of isolated dry cell walls of *L. monocytogenes* (Fiedler 1988) and play important functions in metal cation homeostasis, anchoring of surface proteins, and transport of ions, nutrients, and proteins. TA and LTA, which are synthesized by noninterconnecting metabolic pathways (Neuhaus and Baddiley 2003), are main determinants of surface hydrophobicity and immunogenicity. In fact, both TA and LTA confer the basis of the serotype diversity known in *L. monocytogenes*.

Two main types of TA exist in *L. monocytogenes*. The first is formed by a polymer consisting of repeating units (∼20 to ∼45) of 1,5-phopshodiester-linked ribitol residues (Fiedler 1988; Uchikawa et al. 1986a). These ribitol-P units bear variable substitutions. Thus, in serotypes 3a, 3b, and 3c, a GlcNAc residue is linked at position C-4 of the ribitol-P repeating unit whereas in serotypes 1/2a, 1/2b, and 1/2c there is an additional rhamnose residue at position C-2. Remarkably, the serotype 7 has no substitutions in the ribitol-P. The second type of TA includes more complex structures in which the GlcNAc residue incorporates as a part of the poly-ribitol-P chain (Fiedler 1988; Uchikawa et al. 1986b). Thus, the C-1 of GlcNAc binds to hydroxyl groups present at positions C-4 (serotypes 4a and 6) or C-2 (serotypes 4b, 4d, and 4f) of the ribitol-P. The C-4 position of GlcNAc then links to the phosphate of the adjacent ribitol-P unit. In addition, the GlcNAc units are decorated with glucose and/or galactose in some serotypes (case of 4b, 4d, and 4f) (Uchikawa et al. 1986a). A TA structure similar to that of *L. monocytogenes* 4b serotype has been identified in a few *L. innocua* strains. It is possible that *L. innocua* may have acquired from *L. monocytogenes* serotype 4b, the set of genes responsible for these modifications, a hypothesis supported by comparative genome analysis (Doumith et al. 2004). The genome sequences obtained from four *L. monocytogenes* strains, two of serotype 1/2a (EGD-e and F6854) and two of serotype 4b (F2365 and H7858), have revealed the presence of 1/2a serotype-specific genes involved in rhamnose biosynthetic pathway (Nelson et al. 2004). The existence of these serotype-specific genes related to TA biosynthesis was recently confirmed upon genome content analysis of 93 *L. monocytogenes* strains of diverse serotypes (Doumith et al. 2004). Thus, serotypes 1/2, 3, and 7 carry genes involved in TA biosynthesis that are absent in serotype 4. Inversely, a gene annotated with function putatively related to TA synthesis (ORF2372) is present exclusively in serotypes 4b, 4d, and 4e (Doumith et al. 2004). The exact function of ORF2372 has not been elucidated. Another gene named *gtcA* was initially claimed as a 4b serotype-specific gene involved in the decoration (glycosylation) of TA, concretely in the incorporation of galactose and glucose to the GlcNAc residues (Promadej et al. 1999). However, *gtcA* ortholog genes exist in the genome of 1/2a strains (Autret et al. 2001). In fact, insertions in the *gtcA* gene of the serotype 1/2a strain EGD-e impair virulence and its product has been proposed to mediate incorporation of rhamnose to the TA (Autret et al. 2001). Further work is required to unravel whether the role of GtcA in the decoration of TA differs in serotypes 1/2a and 4b. It was later shown that mutants in the *gltA–gltB* gene cassette, which is a truly specific locus of serotype 4b, display a severe reduction or total loss of incorporation of galactose to the GlcNAc residue of TA (Lei et al. 2001). These mutants have unaltered the amount of glucose in the TA, which suggests that GltA and GltB are specifically involved in the linkage of galactose to GlcNAc independently of the glucose substitution. The contribution of GltA and GltB to *L. monocytogenes* pathogenicity has not been yet tested.

The ribitol-P polymer of the *L. monocytogenes* TA is covalently bound to the peptidoglycan by a linkage unit formed by two disaccharides bound by a molecule of phosphoglycerol. This linkage unit has the structure Glc(β1$\rightarrow$3)-Glc(β1$\rightarrow$1/3)Gro-P-(3/4)ManNAc(β1$\rightarrow$4)GlcNAc (Kaya et al. 1985). The disaccharide formed by the two molecules of Glc is bound to the ribitol-P polymer whereas that containing the acetylated amino sugars binds via a 1,6-phosphodiester linkage to the MurNAc residue of the peptidoglycan. The *L. monocytogenes* linkage unit of the TA is more complex than that of the closely related bacteria *B. subtilis* and *Staphylococcus aureus*, which contain only one disaccharide (Navarre and Schneewind 1999; Neuhaus and Baddiley 2003).

Unlike TA, the LTA polymer of *L. monocytogenes* is formed by repeating units of glycerol-P bound by 1,3 linkages. In some strains, the C-2 position of the glycerol-P is decorated with galactose (Fischer et al. 1990). The glycerol-P polymer attaches to the C-6 position of a nonreducing sugar molecule carried by a glycolipid molecule that embeds the LTA into the membrane. In *L. monocytogenes*, the glycolipid molecule has the structure of Gal($\alpha 1 \rightarrow 2$)Glu-1(3),2-diacylglycerol (Fiedler 1988; Fischer et al. 1990). A substitution of a phosphatidic acid in the C-6 position of Glu has been reported in some strains.

A common modification found in TA and LTA of many gram-positive bacteria is D-alanine-esterification at carbons of their respective repeating units (Neuhaus and Baddiley 2003). This modification is accomplished by a unique D-Ala incorporation system encoded in the *dlt* operon. D-Ala esterification has a profound effect on the electromechanical properties of the cell wall since it reduces its global negative charge. This modification modulates distinct cellular functions as the activities of autolysins, the maintenance of cation-homeostasis, and the assimilation of metal cations (Neuhaus and Baddiley 2003). The existence of a *dtl* operon in the genome does not necessarily imply that the TA are D-Ala-esterified. Thus, *Streptococcus pneumoniae* strain R6 harbours an entire *dlt* operon but contains phosphorylcholine-esters instead of D-Ala-esters in both TA and LTA.

The *L. monocytogenes dlt* operon consists of four genes, *dltA*, *dltB*, *dltC*, and *dltD*, encoding all components required for D-Ala esterification (Abachin et al. 2002). D-Ala-esterification is important for *L. monocytogenes* pathogenicity. Thus, a *dltA* mutant displays enhanced sensitivity to antimicrobial cationic peptides and virulence attenuation in the mouse-infection model (Abachin et al. 2002). D-Ala-esterification occurs in laboratory conditions in $\sim$20% of the glycerol-P residues of LTA and is not detected in LTA of the *dltA* mutant. No study has reported the rate of D-Ala-esterification in TA of *L. monocytogenes*. Interestingly, the defect in D-Ala-esterification of LTA displayed by the *dltA* mutant does not alter the relative amount of surface proteins as internalin-A (InlA), InlB, and ActA that are extracted from the cell surface. Based on this result, D-Ala-esterification was postulated to be important for certain surface proteins to reach a functional folding state (Abachin et al. 2002), although it does not discard a potential role of D-Ala-esterification in modulating the anchoring of surface proteins to the cell wall. Noteworthily, VirR, a new response regulator implicated in *L. monocytogenes* virulence, controls among other functions the expression of the *dlt* operon (Mandin et al. 2005). This observation reinforces the idea that D-Ala-esterification is a cell-wall modification playing a prominent role in the *L. monocytogenes* infection process.

5.3. *Listeria monocytogenes* Surface Proteins Anchored to the Cell Wall

One of the most remarkable features of the *L. monocytogenes* genome is the high content of genes encoding surface proteins (Glaser et al. 2001; Nelson et al. 2004). Some of these proteins are not directly associated to the cell wall as they either carry transmembrane domains or N-terminal signals recognized for insertion of a lipid molecule (lipoproteins). None of these two groups will be discussed in this chapter. Instead, we discuss about those proteins known to interact with structures forming part of the cell wall and, therefore, entirely located outside the plasma membrane (Figure 5.1.).

5.3.1. Surface Proteins Anchored Covalently to the Peptidoglycan

5.3.1.1. The LPXTG Protein Family

Pioneering work performed with the *S. aureus* protein A demonstrated that this surface protein is anchored covalently to the peptidoglycan by an enzyme

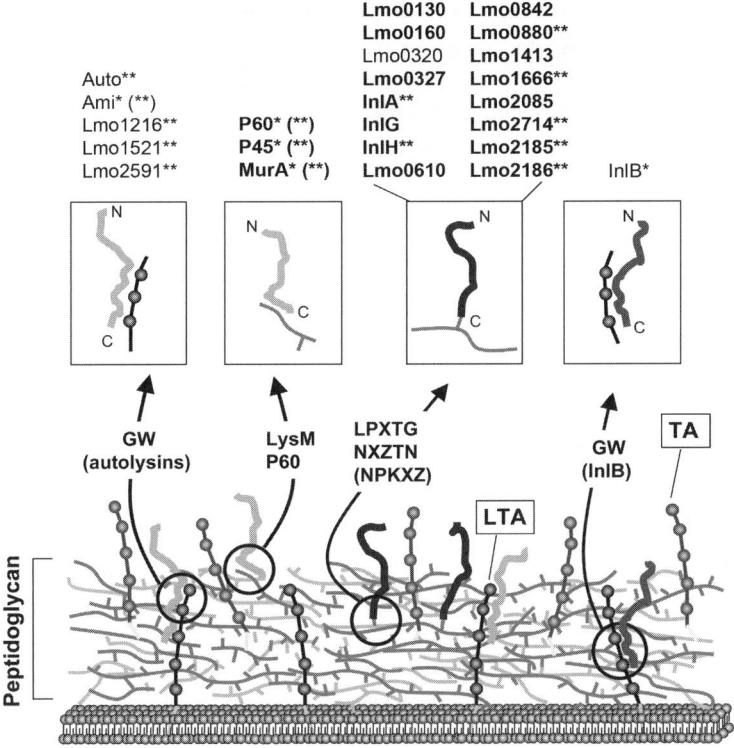

FIGURE 5.1. Global view of the distinct interactions of the *Listeria monocytogenes* surface proteins with the cell wall. Listed in the *upper* part are the proteins identified by proteomics in purified peptidoglycan material (in bold); in extracts obtained from the cell wall upon incubation of bacteria with high concentration of salts (asterisk, *); or, in the extracellular medium as components of the "secretome" (double asterisk, **). Lmo0320 (Vip), although unidentified in these studies, was detected on the cell surface by immunological assays (Cabanes et al. 2005). Other nonlisted surface proteins with mode of association to the cell wall unknown but identified in proteomic analysis include: Lmo1892 (PbpA) in purified peptidoglycan material and Lmo2504, Lmo2522, and Lmo2754 (PBP5) in the extracellular medium. TA: teichoic acid; LTA: lipoteichoic acid.

named sortase (Mazmanian et al. 1999). Protein A contains an N-terminal signal peptide and a C-terminal domain consisting of an LPXTG motif followed by a hydrophobic region of about 20 amino acids that ends in a tail of mostly positively charged residues (Ton-That et al. 2004). This C-terminus configuration, named "sorting signal", is conserved in many surface proteins of gram-positive bacteria (Navarre and Schneewind 1999; Ton-That et al. 2004).

The *in-silico* genome analysis of the *L. monocytogenes* strain EGD-e unravelled 41 genes-encoding surface proteins bearing an LPXTG motif (Glaser et al. 2001). To date, this number of LPXTG proteins is the highest among all the gram-positive bacteria with genome sequence known. A high number of genes-encoding LPXTG proteins, in the average of 45, were also identified in the genome of other three *L. monocytogenes* strains (Nelson et al. 2004). Despite this bulk of information, very few of these proteins have been characterized at the biochemical and/or functional level. The LPXTG protein most extensively studied is InlA, which promotes entry of *L. monocytogenes* into epithelial cells. InlA harbours an LPTTG motif and is anchored covalently by the sortase SrtA to *m*-Dap residues of the peptidoglycan (Bierne et al. 2002; Dhar et al. 2000; Garandeau et al. 2002). In *S. aureus*, the primary acceptor molecule in the anchoring reaction catalyzed by SrtA is the lipid-II precursor (Perry et al. 2002). Whether the *L. monocytogenes* sortase SrtA uses the same mechanism has not yet been formally demonstrated.

The large set of LPXTG proteins of *L. monocytogenes* clusters in two subfamilies that differentiate by the presence in their N-terminal half of a variable number of leucine-rich repeats (LRR) containing 20–22 amino acids each (Cabanes et al. 2002). This domain is a feature shared by all proteins belonging to the "internalin family". The LRR domain is thought to mediate protein–protein interactions, and in the case of InlA is necessary and sufficient to promote bacterial uptake. The *L. monocytogenes* strain EGD-e has 19 LPXTG proteins containing the LRR domain (Cabanes et al. 2002). Of these, only eight are present in the nonpathogenic species *L. innocua*. A similar number of internalins bearing LPXTG motifs (from 14 to 17 proteins) has been reported in the other three *L. monocytogenes* strains with genome sequence known (Nelson et al. 2004). Besides InlA, a few LPXTG proteins of the internalin family have been recently characterized in the EGD-e strain. These include InlH (Lmo0263), InlI (Lmo0333), and InlJ (Lmo2821) (Sabet et al. 2005; Schubert et al. 2001) (see Section 5.4.1).

Proteins containing diverse non-LRR repeat regions preceding the C-terminal sorting region form the second class of LPXTG proteins. The EGD-e strain has 22 LPXTG proteins in this class, of which 14 have orthologs in *L. innocua* (Cabanes et al. 2002; our unpublished data). To date, only one protein of this group, Vip (Lmo0320), has been characterized at a functional level (Cabanes et al. 2005) (see Section 5.4.1).

Genomic comparison studies have revealed that six LPXTG proteins of *L. monocytogenes* serotypes responsible for most cases of listeriosis are absent in the rest of *Listeria* species (Doumith et al. 2004). LPXTG proteins displaying this narrow distribution in the *Listeria* genus are currently investigated for their role in pathogenesis (see Section 5.4.1).

5.3.1.2.   Non-LPXTG Proteins Bearing Cell-Wall Sorting Signals

Staphylococcus aureus has a second sortase, SrtB, which recognizes a sorting motif different than LPXTG. This sortase specifically cleaves an NPQTN motif in a surface protein involved in iron transport, IsdC (Mazmanian et al. 2002). *L. monocytogenes* has also an alternative SrtB sortase (Bierne et al. 2004). Like in *S. aureus*, the *L. monocytogenes srtB* gene maps in an operon containing two genes encoding the Lmo2185 (formerly SvpA) and Lmo2186 (SvpB) proteins, which share homology to *S. aureus* IsdC. Lmo2185 and Lmo2186 bear as putative sorting motifs NAKTN and NKVTN (NPKSS), respectively. Although a direct proof of their covalent anchoring to the peptidoglycan has not been yet shown, both proteins are detected in highly purified peptidoglycan material (Calvo et al. 2005) (see Sect. 5.5.1). Furthermore, Lmo2185 displays a unique migration on gels when extracted from peptidoglycan material (Bierne et al. 2004), suggesting that this species may correspond to the processed form covalently anchored to the peptidoglycan. Detection of Lmo2185 at the cell surface is also abolished in an *srtB* mutant (Bierne et al. 2004). Interestingly, the *L. monocytogenes* operon containing the *lmo2185*, *lmo2186*, and *srtB* genes is induced in iron-deficient conditions, but neither Lmo2185 nor Lmo2186 are required for haemin, haemoglobin, or ferrichrome utilization (Newton et al. 2005). The exact function of these two surface proteins, which are conserved in all *Listeria* species, remains therefore elusive.

5.3.2.   Surface Proteins with Noncovalent Association to the Cell Wall

5.3.2.1.   InlB, a Protein Loosely Associated to the Cell Wall by GW Modules

InlB is a surface protein required for *L. monocytogenes* entry into certain eukaryotic cell types (see Section 4.2). InlB is the only *L. monocytogenes* surface protein carrying a domain organization consisting of a N-terminal LRR domain and a C-terminal domain of three repetitions of 80 amino-acids, called GW modules (Braun et al. 1997). GW modules are also found in autolysins (see below). Domain swapping experiments revealed that the GW-modules mediate binding of InlB to lipoteichoic acids (LTA) (Jonquieres et al. 1999). This association also occur when purified InlB is added externally to intact cells of *L. monocytogenes* serotype 1/2a. InlB is efficiently extracted from the cell surface when bacteria are incubated in the presence of LTA, which indicates that the InlB-LTA association may be rather weak. Noteworthy, InlB does not bind to purified cell walls containing TA (Jonquieres et al. 1999), which supports the idea that the GW-modules specifically interacts with polymers of the LTA type. The strength of the GW domain-LTA association increases with the number of GW-modules. Thus, an InlB variant bearing the 8-GW module region of the autolysin Ami binds more efficiently to cell surface. It has been shown that purified InlB does not attach to either the *L. innocua* surface or that of *S. pneumoniae*, the latter being decorated with LTA devoid of polyglycerol-P. This observation leaves open the possibility that other cell wall components, as TA, may also modulate the association of InlB to the cell wall. This hypothesis could be tested by determining the capacity of purified InlB for binding to *L. monocytogenes* 4b cells, which have a TA structurally different to serotype 1/2a cells.

5.3.2.2. The Autolysin Family

The peptidoglycan is a highly dynamic macromolecule whose structure is continuously modified by hydrolytic enzymes, also known as autolysins (Holtje 1998; Popowska 2004). These enzymes cleave preexisting linkages and act coordinately with biosynthetic activities during the incorporation of nascent peptidoglycan or the separation of daughter cells upon cell division. Hydrolytic enzymes are also responsible for the active release of cell-wall components (up to 30–50% per generation in some bacteria) and, as in the case of gram-negative bacteria, for the active recycling of peptidoglycan turnover products. Hydrolytic enzymes necessarily have to be subjected to tight temporal and spatial control since their indiscriminate activity may lead to cell lysis. The profile of autolytic enzymes in a given bacterium is assessed by zymogram assays using gels loaded with cell walls. Renaturated proteins displaying cell-wall degrading activities are visualized following gel staining. Zymogram assays performed with extracts of *L. monocytogenes* surface proteins have revealed a large number of hydrolytic enzymes (McLaughlan and Foster 1997, 1998; Popowska 2004). In some cases, these assays have proved to be very useful for assigning hydrolytic activity to novel proteins uncovered by genome data (Cabanes et al. 2004; Carroll et al. 2003; McLaughlan and Foster 1998; Milohanic et al. 2001). *L. monocytogenes* contains several types of autolysins, all of them harbouring domains that promote attachment of the protein to the cell wall (Cabanes et al. 2002). These include the "amidase" domain, with similarity to the MurNAc-L-Ala amidase of the Atl autolysin of *S. aureus*; the "LysM" domain; the GW modules; and, the so-called "P60-domain". Some of these enzymes are required for *L. monocytogenes* pathogenicity (see Section 5.4.2). Many of the autolysins containing an amidase domain carry a variable number of GW modules (Milohanic et al. 2001). Domains containing short repetitions are also present in autolysins of *Streptococcus pneumoniae* and *Staphylococci*. In *S. pneumoniae* these modules are responsible for binding of the protein to choline residues that decorate the TA and LTA. Seven proteins with this "amidase-GW" domain organization are known in the *L. monocytogenes* strain EGD-e (Cabanes et al. 2002). Two representative autolysins of this subfamily are Ami (Lmo2558) and Auto (Lmo1076), which contain 8- and 4-GW modules, respectively, in their C-half region (Cabanes et al. 2004; Milohanic et al. 2001). Interestingly, Ami from *L. monocytogenes* strains of serotypes 4b, 4d, and 4e carry only 6-GW modules and, in contrast to its N-half, the C-half region displays low identity (54% at the aminoacid level) to Ami of EGD-e (serotype 1/2a) (Milohanic et al. 2004). Purified Ami protein from serotype 4b binds less efficiently to 1/2a bacterial cells than to 4b cells (Milohanic et al. 2004), suggesting that in addition to LTA, other cell-wall elements may modulate cell wall–protein association mediated by the GW modules. Such an element could be the TA molecule, which is structurally different in serotypes 1/2a and 4b (see Section 5.2.2). This hypothesis fits to the fact that Ami of serotype 1/2a does not bind to the surface of *L. innocua* serotype 6a (Milohanic et al. 2001), which has a TA structure similar to *L. monocytogenes* serotype 4.

Auto is the only autolysin of this subfamily that is absent in the nonpathogenic species *L. innocua*. The *aut* gene is flanked by *lmo1077*, a gene with function related to TA synthesis and also absent in *L. innocua*. Whether Lmo1077 is required for association of Auto to the cell wall is at present unknown.

Another important group of *L. monocytogenes* autolysins is formed by the P60-subfamily. All the members of this subfamily share an NPLC/P60 domain in their C-terminal region. Four proteins of the EGD-e strain have been classified in this subfamily: Spl (P45), Lmo0394, Lmo1104, and P60 itself. Lmo1104 is the only autolysin of the group that is absent in *L. innocua*. In addition to the P60 domain, the P60 protein carries two LysM domains and a bacterial Src-homology 3 (SH3) domain which promote protein association to the cell wall (Cabanes et al. 2002). P60-defective mutants display abnormal morphology, characterized by the presence of filamented cells containing fully formed septa (Gutekunst et al. 1992; Wuenscher et al. 1993). This phenotype links the function of P60 to cell division.

A third type of domain organization is found in a recently characterized autolysin named MurA (Lmo2691). This protein contains an amidase domain in the N-half of the protein followed by 4 LysM domains (Carroll et al. 2003). As in P60-deficient mutants, strains lacking MurA display elongated morphology. The defect in MurA also correlates with increased resistance to detergent-mediated lysis (Carroll et al. 2003), which suggests that MurA could be involved in generalized remodelling of the peptidoglycan. MurA and P60 are secreted, together with other proteins, by a specialized secretion machinery dependent on the SecA2 protein (Lenz et al. 2003). Defects in the secretion of these two autolysins have been linked to the transition to a rough phenotype observed in some *L. monocytogenes* isolates (Machata et al. 2005).

The *L. monocytogenes* EGD-e strain has another two putative autolysins with a unique domain organization. The first, Lmo0849, contains an amidase domain in the middle part of the protein and a transmembrane domain in its C-terminal region. This protein, which would be the only *L. monocytogenes* autolysin embedded in the plasma membrane, has not been characterized yet. The second is Lmo0327, a LPXTG protein containing five LRR domains in the N-half of the protein and 15 repeat regions specific of this protein in its C-half region (Popowska and Markiewicz 2006). The autolytic activity of Lmo0327 was inferred in a screening of autolytic activity using a library of *L. monocytogenes*

EGD-e cloned in *E.coli*. The *L. monocytogenes lmo0327* mutant lacks several bands in zymogram assays, one of them with the expected molecular weight of Lmo0327. This mutant displays an elongated shape, defects in cell separation, and slightly higher resistance to Triton X100-stimulated lysis. Further work is required to confirm whether Lmo0327, which does not contain any domain related to peptidoglycan hydrolysis, is a bona-fide autolysin. In vitro assays with purified protein could provide such evidence. It is worth to mention that to date no study has provided a direct proof of the specific linkage(s) of the pepti-doglycan cleaved by any autolysin of *L. monocytogenes*. Attempts made with purified P60 on peptidoglycan were unsuccessful due to the inherent property of this protein to aggregate (Wuenscher et al. 1993).

5.4.   Role of Cell-Wall-Associated Proteins in *Listeria monocytogenes* Virulence

This section summarizes those studies in which the contribution of *L. monocyto-genes* surface proteins to pathogenicity was tested using diverse infection models. For additional information, the reader is referred to Chap. 8 by Pizarro and Cossart.

5.4.1.   LPXTG Proteins and Sortases

5.4.1.1.   Species-Specific Role of InlA in Crossing of Intestinal and Placental Epithelia

Despite the critical role of InlA as an *L. monocytogenes* invasion protein, it was for long impossible to associate InlA with virulence in mouse models (Gaillard et al. 1996; Gregory et al. 1996; Pron et al. 1998). This unexpected result was explained by the discovery of a species specificity of InlA interaction with its host receptor, the cell adhesion molecule E-cadherin (Ecad). Thus, InlA interacts with human or guinea pig Ecad but does not recognize mouse or rat Ecad. This specificity is due to a single amino acid, a proline at position 16 in the binding site of human Ecad, which is a glutamic acid residue in the mouse or rat Ecad (Lecuit et al. 1999). This change leads to structural modifications that prevent mEcad–InlA interaction (Schubert et al. 2002). The usage of transgenic mice expressing hEcad in the intestine revealed a prominent role of the InlA–hEcad interaction in *L. monocytogenes* invasion of enterocytes (Lecuit et al. 2001). This conclusion was also established upon oral infection of guinea pigs, which naturally possess a permissive Ecad. Altogether, these results demonstrate that in permissive species InlA plays a role in the crossing of the intestinal barrier. Recent data indicate that InlA is also critical for *L. monocytogenes* fetoplacental tropism (Lecuit et al. 2004). Thus, the ability of *L. monocytogenes* to target the placental villi and cross the placental barrier is dependent upon InlA interaction with trophoblast E-cadherin. This observation correlates with epidemiological data showing that

100% of *L. monocytogenes* isolates are obtained from pregnancy-associated listeriosis but only 65% of food isolates express a functional InlA (Jacquet et al. 2004).

5.4.1.2.   Inactivation of Sortases to Assess the Role in Virulence of Cell-Wall Bound Proteins

Inactivation of *L. monocytogenes* SrtA abolishes anchoring of many cell-wall bound proteins to peptidoglycan (Bierne et al. 2002). Interestingly, an *srtA* mutant displays lower organ colonization in mice than an isogenic *inlA* mutant when used by both oral or intravenous routes (Bierne et al. 2002; Garandeau et al. 2002). These data indicate that besides InlA, other LPXTG proteins are required for full virulence. This conclusion was further confirmed in guinea pigs and h-Ecad transgenic mice, in which *L. monocytogenes* efficiently crosses the intestinal barrier in an InlA-dependent-manner. In these models, the *srtA* mutant is also more attenuated for virulence than the *inlA* mutant. These results point to the critical role of SrtA substrates, besides InlA, in bacterial invasion and/or persistence in deeper organs in listeriosis, from the crossing of the intestinal barrier to the hepatic phase of infection (Sabet et al. 2005).

The second *L. monocytogenes* sortase, SrtB, does not play any detectable role in virulence in the mouse model (Bierne et al. 2004). SrtB is also dispensable for virulence following oral infection of guinea pigs (Sabet et al. 2005). Therefore, the two substrates recognized by this sortase, Lmo2185 and Lmo2186, are unlikely to play a major role in food-borne listeriosis. Consistently, an *lmo2185* mutant is not attenuated in virulence in the mouse model (Newton et al. 2005). The inactivation of the *svpA-srtB* operon, containing *srtB*, *lmo2185(svpA)*, *lmo2186*, and genes encoding an iron transporter system, has however a moderate effect on persistence of *L. monocytogenes* in mouse organs (Newton et al. 2005).

5.4.1.3.   New LPXTG proteins as virulence factors

5.4.1.3.1.   InlH

Searching for *inlA*-related genes by southern hybridization using an *inlA* probe revealed the presence of the *inlC2-inlD-inlE* locus in strain EGD (Dramsi et al. 1997). The inactivation of this locus did not result in decreased virulence in the murine model. Surprisingly, three related genes, *inlG-inlH-inlE*, are found at the same locus in strain EGD-e (Raffelsbauer et al. 1998). Multiple deletions of the *inlG-inlH-inlE* genes or a single deletion of *inlH* decreases *L. monocytogenes* virulence in the mouse (Schubert et al. 2001). These results implicate at least *inlH* in pathogenicity although further work is required to discern its exact role in virulence.

5.4.1.3.2.   Vip

The *vip* gene (*lmo0320*), identified as one of the 8 genes encoding LPXTG proteins present in *L. monocytogenes* EGD-e and absent from *L. innocua* is positively regulated by PrfA, the transcriptional activator of the major *L. monocytogenes* virulence factors. Vip is required for bacterial entry into some eukaryotic cells as well as for the infectious process in vivo at both the intestinal level and the later stages of organ colonization (Cabanes et al. 2005). Vip may contribute to pathogenesis by interacting with the host endoplasmic reticulum chaperone Gp96 (Li et al. 2002), inducing either cell invasion and/or signaling events that interfere with the host immune response (Cabanes et al. 2005).

5.4.1.3.3. InlJ

The DNA content and genomic biodiversity of *Listeria* strains of different species and serotypes analyzed by DNA arrays identified *L. monocytogenes*-specific marker genes (Doumith et al. 2004). Among these markers there are five LPXTG proteins of the internalin family, InlA, InlH, InlE, InlI, and InlJ. The contribution to pathogenesis of InlJ (Lmo2821) and InlI (Lmo0333) was recently evaluated. An *inlJ* mutant displays a virulence defect in mice and transgenic hEcad mice after intravenous and oral infection, respectively. In contrast, deletion of the inlI gene has no effect on virulence (Sabet et al. 2005). No phenotype could be attributed to the *inlJ* mutant in tissue culture cells, making its function elusive. Noteworthily, InlJ is not detected in the cell-wall proteome of bacteria grown in brain-heart-infusion (BHI) medium (Pucciarelli et al. 2005) (Section 5.5.1), raising the possibility that expression of *inlJ* might be tightly regulated. InlJ is structurally related to InlA, bearing a new type of cysteine-rich LRR motifs and repetitions related to MucBP domains in the C-terminal region (Sabet et al. 2005). A future challenge will be to assess when and where *inlJ* is expressed and to identify the InlJ eukaryotic binding partner as well as its signaling pathways.

5.4.1.3.4.   L. monocytogenes *serotype 4b-specific LPXTG proteins*

The biodiversity DNA array data revealed two genes encoding LPXTG proteins, *ORF29* and *ORF2568*, present exclusively in serotypes 1/2b, 4a, 4b, and 4c (Doumith et al. 2004). Since *L. monocytogenes* serotype 4b strains are responsible for the majority of epidemic cases of listeriosis, these genes were investigated for their role in infection. Inactivation of these genes does not however alter either infection in vitro or virulence following oral infection of hEcad mice (Sabet et al. 2005). However, the bacterial load of the *ORF2568* deletion mutant in organs, especially spleens, increases compared to the wild type 4b strain. Inactivation of *ORF2568* may somehow affect expression or function of other virulence factors and enhance bacterial fitness in organs.

5.4.2.   Other NonCovalently Bound Proteins

5.4.2.1.   InlB: A Second Species Specificity

The invasion protein InlB is a functional homologue of the human hepatocyte growth factor, h-HGF, acting as an agonist of the hepatocyte growth factor receptor (HGF-R/Met). Met is a widely expressed receptor tyrosine kinase involved in complex cellular processes such as cell proliferation, dissociation, migration, and differentiation (Shen et al. 2000). InlB also interacts with proteoglycans and gClq-R, a ubiquitous glycoprotein (Braun et al. 2000; Jonquieres et al. 2001). InlB–Met interaction is required for *L. monocytogenes* invasion in a variety of cell types in which InlA plays no role, such as hepato-cytes, endothelial cells, and fibroblasts (Dramsi et al. 1995; Gregory et al. 1997; Greiffenberg et al. 1998; Lingnau et al. 1995; Parida et al. 1998) (see also Chap. 8). However, despite in-depth knowledge on InlB in vitro activities, its role in vivo was not extensively explored for long. InlB was first shown to be involved in mouse liver and spleen colonization using a double *inlA-inlB* deletion mutant (Gaillard et al. 1996), or in competition index experiments using wild-type and *inlB* strains (Dramsi et al. 2004). A recent study confirmed that InlB is required for liver and spleen colonization in mice (Khelef et al. 2006). InlB is not however required for the invasion of the intestinal epithelium of transgenic hEcad mice and does not cooperate with InlA for this function. Furthermore, InlB is not involved in *L. monocytogenes* infection of guinea pigs or rabbits due to its inability to stimulate the Met receptors of these two hosts (Khelef et al. 2006). Thus, similar to the InlA–Ecad interaction, InlB mediates an *L. monocytogenes* species-specificity critical for the pathophysiology of listeriosis. These results also emphasize the need for developing new animal models to dissect the exact contribution to pathogenicity of cell wall-associated proteins.

5.4.2.2. AUTOLYSINS

5.4.2.2.1. Ami

Inactivation of Ami alone does not affect *L. monocytogenes* pathogenicity. However, inactivation of Ami in mutants lacking InlA, InlB, or both internalins results in strong reduction of adhesion to hepatocytes and enterocyte-like cell lines (Milohanic et al. 2001). Like in InlB, the GW modules appear to promote Ami adhesion to host cells (Milohanic et al. 2001). Since InlB GW modules bind cellular matrix proteoglycans (Jonquieres et al. 2001), it is possible that Ami GW modules exert a similar function. Thus, Ami may act as a comple-mentary adhesin during infection. This conclusion is supported by the fact that an *ami* mutant is slightly attenuated in the liver of mice infected intravenously (Milohanic et al. 2001).

5.4.2.2.2. Auto

The morphology of an *aut* deletion mutant has been reported similar to those of the wild-type strain, with no defect in septation and cell division (Cabanes et al. 2004). Auto is however required for entry of *L. monocytogenes* into different nonphagocytic eukaryotic cell lines although it is dispensable for efficient adhesion, formation of comet tails, or cell-to-cell spreading. An *aut* mutant displays reduced virulence following intravenous inoculation of mice and oral infection of guinea pigs, which correlates with its low invasiveness (Cabanes et al. 2004). How Auto contributes to pathogenicity is unknown, although its contribution to pathogenicity was tentatively linked to maintenance of the cell surface architecture and/or release of immunologically active cell-wall components (Cabanes et al. 2004).

5.4.2.2.3.   P60

The autolysin P60, encoded by the invasion-associated protein (*iap*) gene, is both secreted and associated with the bacterial cell wall (Kuhn and Goebel 1989; Ruhland et al. 1993; Wuenscher et al. 1993). On the basis of its similarity to the autolysin LytF from *Bacillus subtilis*, P60 is predicted to have a D-*i*Glu-*m*Dap endopeptidase activity (Lenz et al. 2003; Smith et al. 2000). The role of P60 in pathogenicity was first evaluated in rough mutants expressing lower levels of this protein (Kuhn and Goebel 1989). These rough mutants are less virulent and enter less efficiently in certain eukaryotic cells, suggesting a role for P60 in invasion (Gutekunst et al. 1992; Hess et al. 1995; Kuhn and Goebel 1989). A P60-deficient mutant displays similar phenotypes, including virulence attenuation after intravenous infection of mice. This Δ*iap* mutant is also impaired in its intracellular motility process due to mis-localization of the actin-polymerizing factor ActA (Lenz et al. 2003; Pilgrim et al. 2003). P60 plays an important role in the immune response against *L. monocytogenes*. P60-specific antibodies act as opsonins and might play a role in preventing systemic infections in immunocompetent individuals (Kolb-Maurer et al. 2001). P60 is also a major protective antigen that induces both T-CD8 and Th1 protective immune responses (Bouwer and Hinrichs 1996; Geginat et al. 1998; Geginat et al. 1999; Harty and Pamer 1995).

5.5.   Proteomics of the *Listeria monocytogenes* Cell Wall

Listeria monocytogenes surface proteins have been traditionally characterized using extraction methods involving acid treatment, LiCl or detergents like SDS. These methods allow the extraction of surface proteins non-strongly attached to peptidoglycan, although they do not solubilize proteins covalently bound to the peptidoglycan. Unless peptidoglycan-hydrolytic enzymes are used, proteins bound covalently to the peptidoglycan, as those of the LPXTG family, are obtained in minute amounts (Navarre and Schneewind 1999). This is probably one of the reasons of why in *L. monocytogenes*, as in other gram-positive bacteria, relatively few proteins covalently bound to the peptidoglycan have been characterized.

5.5.1.   *The* Listeria monocytogenes *Cell-Wall Proteome*

Based on genome data, a recent study developed a preliminary proteome reference map of the *L. monocytogenes* EGD-e strain (Ramnath et al. 2003). Using total cell extracts, 261 spots were differentiated on two-dimensional (2D) gels. Of these, only 33 distinct proteins were identified, most of them corresponding to abundant proteins (chaperons, translation factors, and enzymes of central metabolism). Noteworthy, no cell-wall-associated protein was identified in this study. A further study, focused on the characterization of the *L. monocytogenes* "secretome" using 2D gels and non-gel proteomics, revealed the presence of numerous cell-wall-associated proteins in the extracellular medium (Trost et al. 2005). These proteins could be released from the cell either as complexes containing fragments of cell-wall polymers or as a result of proteolytic processing. The secretome of the EGD-e strain contains the LPXTG proteins InlA, InlH, Lmo0880, Lmo1666, and Lmo2714; the two NXZTN proteins Lmo2185 and Lmo2186; several autolysins containing GW modules (Auto, Ami, Lmo1216, Lmo1521, and Lmo2591); the autolysin Ami; two members of the P-60 subfamily (P60, P45); two PBPs (Lmo2039 and Lmo2754); and InlB. Two other proteins of unknown function, Lmo2504 and Lmo2522, annotated as "hypothetical-cell-wall binding proteins" were also identified (Trost et al. 2005). Despite the relevant information provided by this study, it did not address the analysis of the protein profile in purified cell-wall material. This was recently made applying a non-gel proteomic approach to peptide mixtures obtained upon incubation of peptidoglycan with trypsin (Calvo et al. 2005). Since peptidoglycan is inherently insensitive to trypsin, this peptide mixture was an ideal source for identifying novel proteins strongly attached to the peptidoglycan. This study performed in *L. monocytogenes* was the first in providing a detailed profile of proteins remaining bound to cell wall upon exhaustive purification of the peptidoglycan employing ionic detergents. Nineteen proteins of the EGD-e strain associated to the peptidoglycan were identified in bacteria growing in BHI medium. All of these proteins correspond to enzymes involved in peptidoglycan metabolism or surface proteins expected to be anchored to this cell-wall component (Calvo et al. 2005). Among these proteins, 13 were LPXTG proteins (including InlA). In comparison with the secretome, a higher number of LPXTG proteins were detected when analyzing purified peptidoglycan material. This result can be explained by the difficulty for detecting in the extracellular medium certain LPXTG proteins that are present in the cell surface only in scarce amounts. It is also possible that some LPXTG proteins locate in surface regions not undergoing massive release of cell-wall fragments, i.e. the polar caps. LPXTG proteins identified in the peptidoglcyan but not in the extracellular medium include Lmo0130, Lmo0160, InlG, Lmo0327, Lmo0610, Lmo0842, Lmo1413, and Lmo2085. Besides LPXTG proteins, the non-gel proteomic analysis performed on peptidoglycan material identified Lmo2185 and Lmo2186 (the two proteins bearing the NXZTN motif); several autolysins (P60, P45, and MurA); and the PBP Lmo1982 (Calvo et al. 2005)(Figure 5.1). No autolysins attaching to the cell wall via GW modules were identified, supporting the idea that the protein–cell wall association mediated by LysM or P60 domains is probably stronger than that mediated by GW modules. This non-gel proteomic study was the first in reporting the identification of a large number of *L. monocytogenes* LPXTG proteins; however, it only covered one-third of LPXTG proteins predicted by the EGD-e genome sequence. Tight regulation of the biosynthesis of concrete LPXTG proteins is conceivable. It is also possible that some LPXTG proteins might be used by *L. monocytogenes* exclusively during the intracellular phase of the infection. In fact, two recent transcriptome analyses have shown up-regulation of a few genes encoding LPXTG proteins in intracellular *L. monocytogenes* growing within epithelial cells (Chatterjee et al. 2006; Joseph et al. 2006). Some of them have not been identified yet in extracellular bacteria. Lastly, in laboratory conditions, some LPXTG proteins may be present on the *L. monocytogenes* cell surface in extremely scarce amounts below the sensitivity threshold of the non-gel proteomics technology.

A proteomic analysis of the *L. monocytogenes* cell wall based on 2D gels has also recently been reported (Schaumburg et al. 2004). In this case, surface proteins were serially extracted from the cell wall with high concentration of salts (Tris and KSCN) and resolved on gels. This work led to the identification of 55 proteins (Schaumburg et al. 2004), but none of them were of the LPXTG family. InlB and the autolysins P60, P45, MurA, and Ami were efficiently extracted from the cell wall with high salt. It becomes clear from these results that the identification of *L. monocytogenes* surface proteins covalently bound to peptidoglycan requires methods involving digestion of either the peptidoglycan itself or the proteins that copurify with it.

5.5.2.   *Identification of* Listeria monocytogenes *Sortase Substrates by Proteomics*

The sortase SrtA of *S. aureus* cleaves the T–G linkage of the LPXTG motif (Ton-That et al. 1999, 2000), whereas SrtB was recently shown to cleave the T–N linkage in the NPQTN sorting motif of IsdC (Marraffini and Schneewind 2005). Sortases contain a conserved TLXTC motif and an H residue involved in catalysis (Ilangovan et al. 2001; Mazmanian et al. 2001). These features have allowed the identification of new putative sortase genes in the genome sequences available in databases. In fact, many gram-positive bacteria contain several sortases to which recognition of distinct sorting motifs has been assigned (Boekhorst et al. 2005; Comfort and Clubb 2004; Dramsi et al. 2005). Like *S. aureus*, *L. monocytogenes* contains two genes encoding sortases, *srtA* (*lmo0929*) and *srtB* (*lmo2181*).

Most of the sortase substrates have been predicted by *in silico* analysis based on the presence of a sorting-signal domain in the C-terminus (Boekhorst et al. 2005; Comfort and Clubb 2004; Dramsi et al. 2005; Ton-That et al. 2004). However, biochemical evidence for recognition of surface proteins by sortases has been reported only in few cases. SrtA and SrtB of *S. aureus* anchor different set of proteins to the cell wall with a specificity that correlates to the presence in the substrate of either a LPXTG or NPQTN motif, respectively (Mazmanian et al. 2001; Ton-That et al. 2004). Sortase substrates have also been identified by 2D SDS-PAGE using cell-wall extracts of sortase-deficient mutants (Osaki et al. 2002). The specificity of the sortases SrtA and SrtB of *L. monocytogenes* strain EGD-e was recently assessed by non-gel proteomics using mutants deficient in these sortases (Pucciarelli et al. 2005). Like in *S.*

aureus, each of the *L. monocytogenes* sortases anchors a distinct subset of proteins to the peptidoglycan. Thus, the LPXTG proteins detected in peptidoglycan of the wild-type strain were all present in the Δ*srtB* mutant but missing in the peptidoglycan of Δ*srtA* or Δ*srtA*Δ*srtB* strains (Pucciarelli et al. 2005). An exception was Lmo0842, which was barely detected in the Δ*srtB* mutant. This result suggests that efficient anchoring of Lmo0842 to the cell wall may require a functional SrtB sortase. Lmo2185 and Lmo2186, both carrying NXZTN sorting motifs, were identified only in strains having a functional StrB sortase. Interestingly, an Lmo2186 peptide covering the first of the two putative sorting motifs predicted in this protein, SDSS*NKVTN*PK, was identified in the peptidoglycan material. This observation supports the hypothesis that SrtB may not cleave the motif NKVTN in certain growth conditions. This hypothesis contemplates that the overlapping motif NPKSS, similar to those described in other gram-positive bacteria (Dramsi et al. 2005), would be preferentially recognized. If demonstrated, Lmo2186 would be the first case of a surface protein covalently anchored to peptidoglycan by a sortase-mediated recognition of two alternative sorting motifs.

Autolysins and PBPs (P60, P45, MurA, and Lmo1892) were identified in the peptidoglycan of all the sortase mutants, an observation that validates the experimental approach and demonstrates that these surface proteins attach to the cell wall by sortase-independent mechanisms.

5.6. The *Listeria monocytogenes* Cell Wall and Inflammation

5.6.1. *Immunogenicity of the* Listeria monocytogenes *Cell Wall*

Bacterial cell-wall components are potent biological effectors of a large variety of stimulatory activities in eukaryotic cells and responsible for severe pathologies like septic shock (Boneca 2005). Early studies performed with distinct cell-wall preparations of *L. monocytogenes* (crude material, peptidoglycan, or LTA) revealed that besides its capacity to activate macrophages, the peptidoglycan has potent adjuvant and antitumor activities (Hether et al. 1983; Paquet et al. 1986; Saiki et al. 1982). Purified peptidoglycan alone is not sufficient to confer protection against *Listeria* infection, requiring priming with crude cell-wall preparations (Hether et al. 1983). Some of the stimulatory effects, such as the adjuvant and mitogenic activities, require both peptidoglycan and TA, whereas others, such as the antitumoral and natural killer activities, are triggered by samples devoid of TA (Hether et al. 1983). These observations indicate that all distinct cell-wall components contribute to stimulate host responses.

Host immunity is also modulated by *L. monocytogenes* cell-wall components. Thus, muropeptide structures found in the peptidoglycan of *L. monocytogenes* are representative of the molecular patterns known to be recognized by Nod1 and Nod2, two members of the intracellular eukaryotic sensor family (Chamaillard et al. 2003a; Inohara et al. 2005; Murray 2005). Nod1 specifically recognizes the disaccharide-tripeptide GlcNAc-MurNAc-L-Ala-γ-D-Glu-m-Dap molecule, whereas Nod2 recognizes the disaccharide-dipeptide GlcNAc-MurNAc-L-Ala-γ-D-Glu (Chamaillard et al. 2003a; Inohara et al. 2005; Murray 2005). Nod1 also recognizes the dipeptide γ-D-Glu-m-Dap (Chamaillard et al. 2003b). Nod1 has been proposed to be involved in sensing of gram-negative bacteria due to the requirement of m-Dap for peptidoglycan recognition and Nod2 would act as a general sensor of bacterial peptidoglycan. Nods have a cytosolic location, which suggests that they have evolved as sensors to respond to infections caused by intracellular bacterial pathogens. Nod2-deficient mice are highly susceptible to bacterial infections initiated at the intestine, including *L. monocytogenes* (Kobayashi et al. 2005). Noteworthy, this increased susceptibility correlates with an impaired secretion of antimicrobial peptides known as cryptdins, an observation that directly links Nod2 function to innate immunity (Kobayashi et al. 2005). Evidence for sensing of *L. monocytogenes* by Nod1 has also been recently found (Opitz et al. 2006). Thus, Nod1 activity is required for both NFκβ-activation and induced secretion of IL-8 in *L. monocytogenes*-infected endothelial cells. Sensing of *L. monocytogenes* by Nod1 also induces the MAPK p38 signaling pathway. Activation of both NFκβ and p38 was shown to mediate IL-8 secretion (Opitz et al. 2006).

5.6.2. Recognition of Listeria monocytogenes Cell-Wall Components by Eukaryotic Molecules

Teichoic acids have been implicated in adhesion of *L. monocytogenes* to host cells. Thus, purified galactose-containing TA, but not those lacking this modification, adhere to HepG-2 epithelial cells (Cowart et al. 1990). Consistently, a *gtcA* mutant deficient in TA glycosylation (see Section 5.2.2) displays a partial defect for adhesion and entry into HepG-2 cells (Autret et al. 2001). These observations favor the hypothesis that TA and/or LTA could act as bacterial "lectins" promoting the interaction of *L. monocytogenes* with membrane-located glycoproteins of nonphagocytic cells. The α-D-galactose receptor present in HepG-2 cells was proposed as the molecule recognizing TA (Cowart et al. 1990), although this interaction has not been formally demonstrated.

In addition to lipopolysaccharide from gram-negative bacteria, the macrophage-scavenger receptor class A (MSR-A) recognizes LTA purified from several gram-positive bacteria, including *L. monocytogenes* (Dunne et al. 1994; Greenberg et al. 1996). MSR-A recognizes preferentially LTA molecules negatively charged since modifications such as D-Ala esterification or positive-charged sugar residues (as those carried by the *S. pneumoniae* LTA) notably reduce LTA–MSRA interaction (Greenberg et al. 1996). MSR-A-deficient mice are highly susceptible to *L. monocytogenes* infection (Ishiguro et al. 2001), which indicates that this receptor is a major effector of the host immune response against *L. monocytogenes*. Toll-like receptors (TLRs), involved in recognition of pathogen-associated molecular patterns (PAMPs) (Takeda and Akira 2005), have also been linked to sensing of *L. monocytogenes*. Thus, mice deficient in TLR2, a receptor that recognizes LTA but not peptidoglycan, are highly susceptible to *L. monocytogenes* infection (Torres et al. 2004). A deficiency in MyD88, an adaptor required for signal transduction from the surface-located TLR receptor, also increases the susceptibility to *L. monocytogenes* (Torres et al. 2004). These data implicate LTA as an important modulator of the host immune response. LTA induce the maturation of dendritic cells and are responsible for the stimulation of IL-18 release observed in *L. monocytogenes*-infected dendritic cells (Kolb-Maurer et al. 2003). *L. monocytogenes* LTA also activate NFκβ in epithelial and macrophage cells (Hauf et al. 1997, 1999).

A recent study showed that *L. monocytogenes dltA* mutants defective in D-Ala esterification (see Section 5.2.2) are less virulent in the mouse model and adhere poorly to phagocytic and non-phagocytic cells (Abachin et al. 2002). It would be of interest to test whether esterification of the TA with D-Ala might alter recognition of LTA by MSR-A or other type of receptor on the surface of epithelial cells.

5.7.   Future Challenges

The *L. monocytogenes* cell wall has revealed as an intricate platform in which several polyanionic polymers, the peptidoglycan and TA precisely dictate the anchoring of a plethora of surface proteins. While some modes of cell wall–protein association have been examined in-depth in a few cases, some unresolved aspects remain. Thus, the association promoted by GW modules, involving the interaction with LTA in the case of InlB, is not yet completely understood for autolysins bearing these modules. The potential role of the TA polymer in modulating anchoring of surface proteins has not been explored, although some observations suggest that it might be the case. Another intriguing issue is the large number of autolysins expressed by *L. monocytogenes* in laboratory conditions. Insights on this phenomenon could be now obtained by determining the specific linkage of the peptidoglycan cleaved by these autolysins. It is expected that these activities do not overlap given the simultaneous presence of these autolysins on the cell surface. These goals can also now be addressed using proteomic techniques. Lastly, an exciting but basically unexplored field deals with the cell-wall physiology of *L. monocytogenes* during the infection process and, more specifically, during the intracellular phase. The changes in expression of genes encoding surface proteins registered when *L. monocytogenes* proliferates inside eukaryotic cells are predictive of significant alterations in the cell wall occurring in intracellular bacteria. These challenges open a new avenue of research that undoubtedly will increase our yet limited knowledge of the *L. monocytogenes* cell wall and its contribution to pathogenicity.

Acknowledgments. We apologize to those colleagues whose work has not been cited due to space limitations. We thank Z. Markiewicz for communicating unpublished data. M.G. Pucciarelli is an investigator of the "Ramón y Cajal" program of the Spanish Ministry of Education and Science. H. Bierne is on the staff of the Institut National de la Recherche Agronomique of France. F. García-del Portillo is an established Investigator of the Consejo Superior de Investigaciones Científicas (CSIC) of Spain.

Note

While this chapter was in the editing process, a study was published by Guinane et al. describing the contribution of distinct pencillin-binding protiens (PBPs) of *Listeria monocytogenes* to antibiotic resistance and virulence (Guinane et al. 2006).

References

Abachin E, Poyart C, Pellegrini E, Milohanic E, Fiedler F, Berche P, and Trieu-Cuot P (2002) Formation of D-alanyl-lipoteichoic acid is required for adhesion and virulence of *Listeria monocytogenes*. Mol Microbiol 43: 1–14.

Atrih A, Bachere G, Allmaier GM, Williamson MP, and Foster SJ (1999) Analysis of peptidoglycan structure from vegetative cells of *Bacillus subtilis* 168 and role of PBP5 in peptidoglycan maturation. J Bacteriol 181: 3956–3966.

Autret N, Dubail I, Trieu-Cuot P, Berche P, and Charbit A (2001) Identification of new genes involved in the virulence of *Listeria monocytogenes* by signature-tagged transposon mutagenesis. Infect Immun 69: 2054–2065.

Bierne H, Mazmanian SK, Trost M, Pucciarelli MG, Liu G, Dehoux P, Jansch L, Garcia-del Portillo F, Schneewind O, and Cossart P (2002) Inactivation of the *srtA* gene in *Listeria monocytogenes* inhibits anchoring of surface proteins and affects virulence. Mol Microbiol 43: 869–881.

Bierne H, Garandeau C, Pucciarelli MG, Sabet C, Newton S, Garcia-del Portillo F, Cossart P, and Charbit A (2004) Sortase B, a new class of sortase in *Listeria monocytogenes*. J Bacteriol 186: 1972–1982.

Boekhorst J, de Been MW, Kleerebezem M, and Siezen RJ (2005) Genome-wide detection and analysis of cell wall-bound proteins with LPxTG-like sorting motifs. J Bacteriol 187: 4928–4934.

Boneca IG (2005) The role of peptidoglycan in pathogenesis. Curr Opin Microbiol 8: 46–53.

Bouwer HG, and Hinrichs DJ (1996) Cytotoxic-T-lymphocyte responses to epitopes of listeriolysin O and p60 following infection with *Listeria monocytogenes*. Infect Immun 64: 2515–2522.

Braun L, Dramsi S, Dehoux P, Bierne H, Lindahl G, and Cossart P (1997) InlB: an invasion protein of *Listeria monocytogenes* with a novel type of surface association. Mol Microbiol 25: 285–294.

Braun L, Ghebrehiwet B, and Cossart P (2000) gC1q-R/p32, a C1q-binding protein, is a receptor for the InlB invasion protein of *Listeria monocytogenes*. Embo J 19: 1458–1466.

Cabanes D, Dehoux P, Dussurget O, Frangeul L, and Cossart P (2002) Surface proteins and the pathogenic potential of *Listeria monocytogenes*. Trends Microbiol 10: 238–245.

Cabanes D, Dussurget O, Dehoux P, and Cossart P (2004) Auto, a surface associated autolysin of *Listeria monocytogenes* required for entry into eukaryotic cells and virulence. Mol Microbiol 51: 1601–1614.

Cabanes D, Sousa S, Cebria A, Lecuit M, Garcia-del Portillo F, and Cossart P (2005) Gp96 is a receptor for a novel *Listeria monocytogenes* virulence factor, Vip, a surface protein. Embo J 24: 2827–2838.

Calvo E, Pucciarelli MG, Bierne H, Cossart P, Albar JP, and Garcia-Del Portillo F (2005) Analysis of the *Listeria* cell wall proteome by two-dimensional nanoliquid chromatography coupled to mass spectrometry. Proteomics 5: 433–443.

Carroll SA, Hain T, Technow U, Darji A, Pashalidis P, Joseph SW, and Chakraborty T (2003) Identification and characterization of a peptidoglycan hydrolase, MurA, of *Listeria monocytogenes*, a muramidase needed for cell separation. J Bacteriol 185: 6801–6808.

Chamaillard M, Girardin SE, Viala J, and Philpott DJ (2003a) Nods, Nalps and Naip: intracellular regulators of bacterial-induced inflammation. Cell Microbiol 5: 581–592.

Chamaillard M, Hashimoto M, Horie Y, Masumoto J, Qiu S, Saab L, Ogura Y, Kawasaki A, Fukase K, Kusumoto S, Valvano MA, Foster SJ, Mak TW, Nunez G, and Inohara N (2003b) An essential role for NOD1 in host recognition of bacterial peptidoglycan containing diaminopimelic acid. Nat Immunol 4: 702–707.

Chatterjee SS, Hossain H, Otten S, Kuenne C, Kuchmina K, Machata S, Domann E, Chakraborty T, and Hain T (2006) Intracellular gene expression profile of *Listeria monocytogenes*. Infect Immun 74: 1323–1338.

Comfort D, and Clubb RT (2004) A comparative genome analysis identifies distinct sorting pathways in gram-positive bacteria. Infect Immun 72: 2710–2722.

Cowart RE, Lashmet J, McIntosh ME, and Adams TJ (1990) Adherence of a virulent strain of *Listeria monocytogenes* to the surface of a hepatocarcinoma cell line via lectin-substrate interaction. Arch Microbiol 153: 282–286.

Dhar G, Faull KF, and Schneewind O (2000) Anchor structure of cell wall surface proteins in *Listeria monocytogenes*. Biochemistry 39: 3725–3733.

Doumith M, Cazalet C, Simoes N, Frangeul L, Jacquet C, Kunst F, Martin P, Cossart P, Glaser P, and Buchrieser C (2004) New aspects regarding evolution and virulence of *Listeria monocytogenes* revealed by comparative genomics and DNA arrays. Infect Immun 72: 1072–1083.

Dramsi S, Biswas I, Maguin E, Braun L, Mastroeni P, and Cossart P (1995) Entry of *Listeria monocytogenes* into hepatocytes requires expression of *inlB*, a surface protein of the internalin multigene family. Mol Microbiol 16: 251–261.

Dramsi S, Dehoux P, Lebrun M, Goossens PL, and Cossart P (1997) Identification of four new members of the internalin multigene family of *Listeria monocytogenes* EGD. Infect Immun 65: 1615–1625.

Dramsi S, Bourdichon F, Cabanes D, Lecuit M, Fsihi H, and Cossart P (2004) FbpA, a novel multifunctional *Listeria monocytogenes* virulence factor. Mol Microbiol 53: 639–649.

Dramsi S, Trieu-Cuot P, and Bierne H (2005) Sorting sortases: a nomenclature proposal for the various sortases of gram-positive bacteria. Res Microbiol 156: 289–297.

Dunne DW, Resnick D, Greenberg J, Krieger M, and Joiner KA (1994) The type I macrophage scavenger receptor binds to gram-positive bacteria and recognizes lipoteichoic acid. Proc Natl Acad Sci U S A 91: 1863–1867.

Fiedler F (1988) Biochemistry of the cell surface of *Listeria* strains: a locating general view. Infection 16 Suppl 2: S92–S97.

Fischer W, Mannsfeld T, and Hagen G (1990) On the basic structure of poly(glycerophosphate) lipoteichoic acids. Biochem Cell Biol 68: 33–43.

Gaillard JL, Jaubert F, and Berche P (1996) The inlAB locus mediates the entry of Listeria monocytogenes into hepatocytes in vivo. J Exp Med 183: 359–369.

Garandeau C, Reglier-Poupet H, Dubail I, Beretti JL, Berche P, and Charbit A (2002) The sortase SrtA of Listeria monocytogenes is involved in processing of internalin and in virulence. Infect Immun 70: 1382–1390.

Geginat G, Lalic M, Kretschmar M, Goebel W, Hof H, Palm D, and Bubert A (1998) Th1 cells specific for a secreted protein of Listeria monocytogenes are protective in vivo. J Immunol 160: 6046–6055.

Geginat G, Nichterlein T, Kretschmar M, Schenk S, Hof H, Lalic-Multhaler M, Goebel W, and Bubert A (1999) Enhancement of the Listeria monocytogenes p60-specific CD4 and CD8 T cell memory by nonpathogenic Listeria innocua. J Immunol 162: 4781–4789.

Glaser P, Frangeul L, Buchrieser C, Rusniok C, Amend A, Baquero F, Berche P, Bloecker H, Brandt P, Chakraborty T, Charbit A, Chetouani F, Couve E, de Daruvar A, Dehoux P, Domann E, Dominguez-Bernal G, Duchaud E, Durant L, Dussurget O, Entian KD, Fsihi H, Garcia-del Portillo F, Garrido P, Gautier L, Goebel W, Gomez-Lopez N, Hain T, Hauf J, Jackson D, Jones LM, Kaerst U, Kreft J, Kuhn M, Kunst F, Kurapkat G, Madueno E, Maitournam A, Vicente JM, Ng E, Nedjari H, Nordsiek G, Novella S, de Pablos B, Perez-Diaz JC, Purcell R, Remmel B, Rose M, Schlueter T, Simoes N, Tierrez A, Vazquez-Boland JA, Voss H, Wehland J, and Cossart P (2001) Comparative genomics of Listeria species. Science 294: 849–852.

Glauner B, and Höltje JV (1990) Structure and metabolism of the murein sacculus. Res Microbiol 141: 75–89.

Goffin C, and Ghuysen JM (2002) Biochemistry and comparative genomics of SxxK superfamily acyltransferases offer a clue to the mycobacterial paradox: presence of penicillin-susceptible target proteins versus lack of efficiency of penicillin as therapeutic agent. Microbiol Mol Biol Rev 66: 702–738, table of contents.

Greenberg JW, Fischer W, and Joiner KA (1996) Influence of lipoteichoic acid structure on recognition by the macrophage scavenger receptor. Infect Immun 64: 3318–3325.

Gregory SH, Sagnimeni AJ, and Wing EJ (1996) Expression of the inlAB operon by Listeria monocytogenes is not required for entry into hepatic cells in vivo. Infect Immun 64: 3983–3986.

Gregory SH, Sagnimeni AJ, and Wing EJ (1997) Internalin B promotes the replication of Listeria monocytogenes in mouse hepatocytes. Infect Immun 65: 5137–5141.

Greiffenberg L, Goebel W, Kim KS, Weiglein I, Bubert A, Engelbrecht F, Stins M, and Kuhn M (1998) Interaction of Listeria monocytogenes with human brain microvascular endothelial cells: InlB-dependent invasion, long-term intracellular growth, and spread from macrophages to endothelial cells. Infect Immun 66: 5260–5267.

Guinane CM, Cotter PD, Ross RP, Hill C (2006) Contribution of penicillin-binding protein homologs to antibiotic resistance, cell morphology, and virulence of Listeria monocytogenes EGDe. Antimicrob Agents Chemother 50: 2824–2828.

Gutekunst KA, Pine L, White E, Kathariou S, and Carlone GM (1992) A filamentous-like mutant of Listeria monocytogenes with reduced expression of a 60-kilodalton extracellular protein invades and grows in 3T6 and Caco-2 cells. Can J Microbiol 38: 843–851.

Gutkind GO, Mollerach ME, and De Torres RA (1989) Penicillin-binding proteins in Listeria monocytogenes. Apmis 97: 1013–1017.

Hakenbeck R, and Hof H (1991) Relatedness of penicillin-binding proteins from various *Listeria* species. FEMS Microbiol Lett 68: 191–195.

Harty JT, and Pamer EG (1995) CD8 T lymphocytes specific for the secreted p60 antigen protect against *Listeria monocytogenes* infection. J Immunol 154: 4642–4650.

Hauf N, Goebel W, Fiedler F, Sokolovic Z, and Kuhn M (1997) *Listeria monocytogenes* infection of P388D1 macrophages results in a biphasic NF-kappaB (RelA/p50) activation induced by lipoteichoic acid and bacterial phospholipases and mediated by IkappaBalpha and IkappaBbeta degradation. Proc Natl Acad Sci U S A 94: 9394–9399.

Hauf N, Goebel W, Fiedler F, and Kuhn M (1999) *Listeria monocytogenes* infection of Caco-2 human epithelial cells induces activation of transcription factor NF-kappa B/Rel-like DNA binding activities. FEMS Microbiol Lett 178: 117–122.

Hess J, Gentschev I, Szalay G, Ladel C, Bubert A, Goebel W, and Kaufmann SH (1995) *Listeria monocytogenes* p60 supports host cell invasion by and in vivo survival of attenuated *Salmonella typhimurium*. Infect Immun 63: 2047–2053.

Hether NW, Campbell PA, Baker LA, and Jackson LL (1983) Chemical composition and biological functions of *Listeria monocytogenes* cell wall preparations. Infect Immun 39: 1114–1121.

Holtje JV (1998) Growth of the stress-bearing and shape-maintaining murein sacculus of Escherichia coli. Microbiol Mol Biol Rev 62: 181–203.

Ilangovan U, Ton-That H, Iwahara J, Schneewind O, and Clubb RT (2001) Structure of sortase, the transpeptidase that anchors proteins to the cell wall of *Staphylococcus aureus*. Proc Natl Acad Sci U S A 98: 6056–6061.

Inohara, Chamaillard, McDonald C, and Nunez G (2005) NOD-LRR proteins: role in host-microbial interactions and inflammatory disease. Annu Rev Biochem 74: 355–383.

Ishiguro T, Naito M, Yamamoto T, Hasegawa G, Gejyo F, Mitsuyama M, Suzuki H, and Kodama T (2001) Role of macrophage scavenger receptors in response to *Listeria monocytogenes* infection in mice. Am J Pathol 158: 179–188.

Jacquet C, Doumith M, Gordon JI, Martin PM, Cossart P, and Lecuit M (2004) A molecular marker for evaluating the pathogenic potential of foodborne *Listeria monocytogenes*. J Infect Dis 189: 2094–2100.

Jonquieres R, Bierne H, Fiedler F, Gounon P, and Cossart P (1999) Interaction between the protein InlB of *Listeria monocytogenes* and lipoteichoic acid: a novel mechanism of protein association at the surface of gram-positive bacteria. Mol Microbiol 34: 902–914.

Jonquieres R, Pizarro-Cerda J, and Cossart P (2001) Synergy between the N- and C-terminal domains of InlB for efficient invasion of non-phagocytic cells by *Listeria monocytogenes*. Mol Microbiol 42: 955–965.

Joseph B, Przybilla K, Stühler C, Schauer K, Slaghuis J, Fuchs TM, and Goebel W (2006) Identification of *Listeria monocytogenes* genes contributing to intracellular replication by expression profiling and mutant screening. J Bacteriol 188: 556–568.

Kamisango K, Saiki I, Tanio Y, Okumura H, Araki Y, Sekikawa I, Azuma I, and Yamamura Y (1982) Structures and biological activities of peptidoglycans of *Listeria monocytogenes* and *Propionibacterium acnes*. J Biochem (Tokyo) 92: 23–33.

Kaya S, Araki Y, and Ito E (1985) Characterization of a novel linkage unit between ribitol teichoic acid and peptidoglycan in *Listeria monocytogenes* cell walls. Eur J Biochem 146: 517–522.

Khelef N, Lecuit M, Bierne H, and Cossart P (2006) Species specificity of the *Listeria monocytogenes* InlB protein. Cell Microbiol 8: 457–470.

Kloszewska M, Quintela JC, Allmaier G, de Pedro MA, Popowska M, and Markiewicz Z (2006) The fine structure of the cell wall murein of *Listeria monocytogenes*. FEMS Microbiol Lett. (unpublished).

Kobayashi KS, Chamaillard M, Ogura Y, Henegariu O, Inohara N, Nunez G, and Flavell RA (2005) Nod2-dependent regulation of innate and adaptive immunity in the intestinal tract. Science 307: 731–734.

Kolb-Maurer A, Pilgrim S, Kampgen E, McLellan AD, Brocker EB, Goebel W, and Gentschev I (2001) Antibodies against *Listerial* protein 60 act as an opsonin for phagocytosis of *Listeria monocytogenes* by human dendritic cells. Infect Immun 69: 3100–3109.

Kolb-Maurer A, Kammerer U, Maurer M, Gentschev I, Brocker EB, Rieckmann P, and Kampgen E (2003) Production of IL-12 and IL-18 in human dendritic cells upon infection by *Listeria monocytogenes*. FEMS Immunol Med Microbiol 35: 255–262.

Korsak D, Vollmer W, and Markiewicz Z (2005) *Listeria monocytogenes* EGD lacking penicillin-binding protein 5 (PBP5) produces a thicker cell wall. FEMS Microbiol Lett 251: 281–288.

Kuhn M, and Goebel W (1989) Identification of an extracellular protein of *Listeria monocytogenes* possibly involved in intracellular uptake by mammalian cells. Infect Immun 57: 55–61.

Lecuit M, Dramsi S, Gottardi C, Fedor-Chaiken M, Gumbiner B, and Cossart P (1999) A single amino acid in E-cadherin responsible for host specificity towards the human pathogen *Listeria monocytogenes*. EMBO J 18: 3956–3963.

Lecuit M, Vandormael-Pournin S, Lefort J, Huerre M, Gounon P, Dupuy C, Babinet C, and Cossart P (2001) A transgenic model for listeriosis: role of internalin in crossing the intestinal barrier. Science 292: 1722–1725.

Lecuit M, Nelson DM, Smith SD, Khun H, Huerre M, Vacher-Lavenu MC, Gordon JI, and Cossart P (2004) Targeting and crossing of the human maternofetal barrier by *Listeria monocytogenes*: role of internalin interaction with trophoblast E-cadherin. Proc Natl Acad Sci U S A 101: 6152–6157.

Lei XH, Fiedler F, Lan Z, and Kathariou S (2001) A novel serotype-specific gene cassette (*gltA-gltB*) is required for expression of teichoic acid-associated surface antigens in *Listeria monocytogenes* of serotype 4b. J Bacteriol 183: 1133–1139.

Lenz LL, Mohammadi S, Geissler A, and Portnoy DA (2003) SecA2-dependent secretion of autolytic enzymes promotes *Listeria monocytogenes* pathogenesis. Proc Natl Acad Sci U S A 100: 12432–12437.

Li Z, Dai J, Zheng H, Liu B, and Caudill M (2002) An integrated view of the roles and mechanisms of heat shock protein gp96-peptide complex in eliciting immune response. Front Biosci 7: d731–d751.

Lingnau A, Domann E, Hudel M, Bock M, Nichterlein T, Wehland J, and Chakraborty T (1995) Expression of the *Listeria monocytogenes* EGD *inlA* and *inlB* genes, whose products mediate bacterial entry into tissue culture cell lines, by PrfA-dependent and-independent mechanisms. Infect Immun 63: 3896–3903.

Machata S, Hain T, Rohde M, and Chakraborty T (2005) Simultaneous deficiency of both MurA and p60 proteins generates a rough phenotype in *Listeria monocytogenes*. J Bacteriol 187: 8385–8394.

Mandin P, Fsihi H, Dussurget O, Vergassola M, Milohanic E, Toledo-Arana A, Lasa I, Johansson J, and Cossart P (2005) VirR, a response regulator critical for *Listeria monocytogenes* virulence. Mol Microbiol 57: 1367–1380.

Marraffini LA, and Schneewind O (2005) Anchor structure of staphylococcal surface proteins. V. Anchor structure of the sortase B substrate IsdC. J Biol Chem 280: 16263–16271.

Mazmanian SK, Liu G, Ton-That H, and Schneewind O (1999) *Staphylococcus aureus* sortase, an enzyme that anchors surface proteins to the cell wall. Science 285: 760–763.

Mazmanian SK, Ton-That H, and Schneewind O (2001) Sortase-catalysed anchoring of surface proteins to the cell wall of *Staphylococcus aureus*. Mol Microbiol 40: 1049–1057.

Mazmanian SK, Ton-That H, Su K, and Schneewind O (2002) An iron-regulated sortase anchors a class of surface protein during *Staphylococcus aureus* pathogenesis. Proc Natl Acad Sci U S A 99: 2293–2298.

McLaughlan AM, and Foster SJ (1997) Characterisation of the peptidoglycan hydrolases of *Listeria monocytogenes* EGD. FEMS Microbiol Lett 152: 149–154.

McLaughlan AM, and Foster SJ (1998) Molecular characterization of an autolytic amidase of *Listeria monocytogenes* EGD. Microbiology 144 (Pt 5): 1359–1367.

Milohanic E, Jonquieres R, Cossart P, Berche P, and Gaillard JL (2001) The autolysin Ami contributes to the adhesion of *Listeria monocytogenes* to eukaryotic cells via its cell wall anchor. Mol Microbiol 39: 1212–1224.

Milohanic E, Jonquieres R, Glaser P, Dehoux P, Jacquet C, Berche P, Cossart P, and Gaillard JL (2004) Sequence and binding activity of the autolysin-adhesin Ami from epidemic *Listeria monocytogenes* 4b. Infect Immun 72: 4401–4409.

Murray PJ (2005) NOD proteins: an intracellular pathogen-recognition system or signal transduction modifiers? Curr Opin Immunol 17: 352–358.

Navarre WW, and Schneewind O (1999) Surface proteins of gram-positive bacteria and mechanisms of their targeting to the cell wall envelope. Microbiol Mol Biol Rev 63: 174–229.

Nelson KE, Fouts DE, Mongodin EF, Ravel J, DeBoy RT, Kolonay JF, Rasko DA, Angiuoli SV, Gill SR, Paulsen IT, Peterson J, White O, Nelson WC, Nierman W, Beanan MJ, Brinkac LM, Daugherty SC, Dodson RJ, Durkin AS, Madupu R, Haft DH, Selengut J, Van Aken S, Khouri H, Fedorova N, Forberger H, Tran B, Kathariou S, Wonderling LD, Uhlich GA, Bayles DO, Luchansky JB, and Fraser CM (2004) Whole genome comparisons of serotype 4b and 1/2a strains of the food-borne pathogen *Listeria monocytogenes* reveal new insights into the core genome components of this species. Nucleic Acids Res 32: 2386–2395.

Neuhaus FC, and Baddiley J (2003) A continuum of anionic charge: structures and functions of D-alanyl-teichoic acids in gram-positive bacteria. Microbiol Mol Biol Rev 67: 686–723.

Newton SM, Klebba PE, Raynaud C, Shao Y, Jiang X, Dubail I, Archer C, Frehel C, and Charbit A (2005) The *svpA-srtB* locus of *Listeria monocytogenes*: fur-mediated iron regulation and effect on virulence. Mol Microbiol 55: 927–940.

Opitz B, Puschel A, Beermann W, Hocke AC, Forster S, Schmeck B, van Laak V, Chakraborty T, Suttorp N, and Hippenstiel S (2006) *Listeria monocytogenes* Activated p38 MAPK and Induced IL-8 Secretion in a Nucleotide-Binding Oligomerization Domain 1-Dependent Manner in Endothelial Cells. J Immunol 176: 484–490.

Osaki M, Takamatsu D, Shimoji Y, and Sekizaki T (2002) Characterization of *Streptococcus suis* genes encoding proteins homologous to sortase of gram-positive bacteria. J Bacteriol 184: 971–982.

Paquet A, Jr, Raines KM, and Brownback PC (1986) Immunopotentiating activities of cell walls, peptidoglycans, and teichoic acids from two strains of *Listeria monocytogenes*. Infect Immun 54: 170–176.

Parida SK, Domann E, Rohde M, Muller S, Darji A, Hain T, Wehland J, and Chakraborty T (1998) Internalin B is essential for adhesion and mediates the invasion of *Listeria monocytogenes* into human endothelial cells. Mol Microbiol 28: 81–93.

Perry AM, Ton-That H, Mazmanian SK, and Schneewind O (2002) Anchoring of surface proteins to the cell wall of *Staphylococcus aureus*. III. Lipid II is an in vivo peptidoglycan substrate for sortase-catalyzed surface protein anchoring. J Biol Chem 277: 16241–16248.

Pierre J, Boisivon A, and Gutmann L (1990) Alteration of PBP 3 entails resistance to imipenem in *Listeria monocytogenes*. Antimicrob Agents Chemother 34: 1695–1698.

Pilgrim S, Kolb-Maurer A, Gentschev I, Goebel W, and Kuhn M (2003) Deletion of the gene encoding p60 in *Listeria monocytogenes* leads to abnormal cell division and loss of actin-based motility. Infect Immun 71: 3473–3484.

Popowska M (2004) Analysis of the peptidoglycan hydrolases of *Listeria monocytogenes*: multiple enzymes with multiple functions. Pol J Microbiol 53 Suppl: 29–34.

Popowska M, and Markiewicz Z (2006) Characterization of *Listeria monocytogenes* protein Lmo327 with murein hydrolase activity. Arch Microbiol 186: 69–86.

Promadej N, Fiedler F, Cossart P, Dramsi S, and Kathariou S (1999) Cell wall teichoic acid glycosylation in *Listeria monocytogenes* serotype 4b requires gtcA, a novel, serogroup-specific gene. J Bacteriol 181: 418–425.

Pron B, Boumaila C, Jaubert F, Sarnacki S, Monnet JP, Berche P, and Gaillard JL (1998) Comprehensive study of the intestinal stage of listeriosis in a rat ligated ileal loop system. Infect Immun 66: 747–755.

Pucciarelli MG, Calvo E, Sabet C, Bierne H, Cossart P, and Garcia-Del Portillo F (2005) Identification of substrates of the *Listeria monocytogenes* sortases A and B by a non-gel proteomic analysis. Proteomics 5: 4808–4817.

Raffelsbauer D, Bubert A, Engelbrecht F, Scheinpflug J, Simm A, Hess J, Kaufmann SH, and Goebel W (1998) The gene cluster inlC2DE of *Listeria monocytogenes* contains additional new internalin genes and is important for virulence in mice. Mol Gen Genet 260: 144–158.

Ramnath M, Rechinger KB, Jansch L, Hastings JW, Knochel S, and Gravesen A (2003) Development of a *Listeria monocytogenes* EGD-e partial proteome reference map and comparison with the protein profiles of food isolates. Appl Environ Microbiol 69: 3368–3376.

Ruhland GJ, Hellwig M, Wanner G, and Fiedler F (1993) Cell-surface location of *Listeria*-specific protein p60-detection of *Listeria* cells by indirect immunofluorescence. J Gen Microbiol 139: 609–616.

Sabet C, Lecuit M, Cabanes D, Cossart P, and Bierne H (2005) LPXTG protein InlJ, a newly identified internalin involved in *Listeria monocytogenes* virulence. Infect Immun 73: 6912–6922.

Saiki I, Kamisango K, Tanio Y, Okumura H, Yamamura Y, and Azuma I (1982) Adjuvant activity of purified peptidoglycan of *Listeria monocytogenes* in mice and guinea pigs. Infect Immun 38: 58–65.

Schaumburg J, Diekmann O, Hagendorff P, Bergmann S, Rohde M, Hammerschmidt S, Jansch L, Wehland J, and Karst U (2004) The cell wall subproteome of *Listeria monocytogenes*. Proteomics 4: 2991–3006.

Schleifer KH, and Kandler O (1972) Peptidoglycan types of bacterial cell walls and their taxonomic implications. Bacteriol Rev 36: 407–477.

Schubert WD, Gobel G, Diepholz M, Darji A, Kloer D, Hain T, Chakraborty T, Wehland J, Domann E, and Heinz DW (2001) Internalins from the human pathogen *Listeria monocytogenes* combine three distinct folds into a contiguous internalin domain. J Mol Biol 312: 783–794.

Schubert WD, Urbanke C, Ziehm T, Beier V, Machner MP, Domann E, Wehland J, Chakraborty T, and Heinz DW (2002) Structure of internalin, a major invasion protein of *Listeria monocytogenes*, in complex with its human receptor E-cadherin. Cell 111: 825–836.

Shen Y, Naujokas M, Park M, and Ireton K (2000) InlB-dependent internalization of *Listeria* is mediated by the Met receptor tyrosine kinase. Cell 103: 501–510.

Smith TJ, Blackman SA, and Foster SJ (2000) Autolysins of *Bacillus subtilis*: multiple enzymes with multiple functions. Microbiology 146 (Pt 2): 249–262.

Srivastava KK, and Siddique IH (1973) Quantitative chemical composition of peptidoglycan of *Listeria monocytogenes*. Infect Immun 7: 700–703.

Takeda K, and Akira S (2005) Toll-like receptors in innate immunity. Int Immunol 17: 1–14.

Ton-That H, Liu G, Mazmanian SK, Faull KF, and Schneewind O (1999) Purification and characterization of sortase, the transpeptidase that cleaves surface proteins of *Staphylococcus aureus* at the LPXTG motif. Proc Natl Acad Sci U S A 96: 12424–12429.

Ton-That H, Mazmanian SK, Faull KF, and Schneewind O (2000) Anchoring of surface proteins to the cell wall of *Staphylococcus aureus*. Sortase catalyzed in vitro transpeptidation reaction using LPXTG peptide and NH(2)-Gly(3) substrates. J Biol Chem 275: 9876–9881.

Ton-That H, Marraffini LA, and Schneewind O (2004) Protein sorting to the cell wall envelope of gram-positive bacteria. Biochim Biophys Acta 1694: 269–278.

Torres D, Barrier M, Bihl F, Quesniaux VJ, Maillet I, Akira S, Ryffel B, and Erard F (2004) Toll-like receptor 2 is required for optimal control of *Listeria monocytogenes* infection. Infect Immun 72: 2131–2139.

Trost M, Wehmhoner D, Karst U, Dieterich G, Wehland J, and Jansch L (2005) Comparative proteome analysis of secretory proteins from pathogenic and nonpathogenic *Listeria* species. Proteomics 5: 1544–1557.

Uchikawa K, Sekikawa I, and Azuma I (1986a) Structural studies on lipoteichoic acids from four *Listeria* strains. J Bacteriol 168: 115–122.

Uchikawa K, Sekikawa I, and Azuma I (1986b) Structural studies on teichoic acids in cell walls of several serotypes of *Listeria monocytogenes*. J Biochem (Tokyo) 99: 315–327.

Ullmann WW, and Cameron JA (1969) Immunochemistry of the cell walls of *Listeria monocytogenes*. J Bacteriol 98: 486–493.

Vicente MF, Berenguer J, de Pedro MA, Perez-Diaz JC, and Baquero F (1990a) Penicillin binding proteins in *Listeria monocytogenes*. Acta Microbiol Hung 37: 227–231.

Vicente MF, Perez-Daz JC, Baquero F, de Pedro M, and Berenguer J (1990b) Penicillin-binding protein 3 of *Listeria monocytogenes* as the primary lethal target for beta-lactams. Antimicrob Agents Chemother 34: 539–542.

Wietzerbin JB, Das C, Petit JF, Lederer E, Leyh-Bouille M, and Ghuysen JM (1974) Occurrence of D-alanyl-(D)-*meso*-diaminopimelic acid and *meso*-diaminopimelyl-*meso*-diaminopimelic acid interpeptide linkages in the peptidoglycan of Mycobacteria. Biochemistry 13: 3471–3476.

Wuenscher MD, Kohler S, Bubert A, Gerike U, and Goebel W (1993) The iap gene of *Listeria monocytogenes* is essential for cell viability, and its gene product, p60, has bacteriolytic activity. J Bacteriol 175: 3491–3501.

Zawadzka-Skomial J, Markiewicz Z, Nguyen-Disteche H, Devreese B, Frere JM, and Terrak M (2006) Characterization of the bifunctional glycosyl-transferase/acyl-transferase penicillin-binding protein 4 from *Listeria monocytogenes*. J Bacteriol 188: 1875–1881.

6
Environmental Reservoir and Transmission into the Mammalian Host

Haley F. Oliver, Martin Wiedmann, and Kathryn J. Boor
Department of Food Science, Cornell University, Ithaca, NY 14853, USA
e-mail: kjb4@cornell.edu

Abstract: The widespread presence of *Listeria monocytogenes* in various diverse environments, including those that are natural (i.e., nonagricultural), agricultural, and food-associated, suggests that these environments may serve as sources or reservoirs of *L. monocytogenes* that can be transmitted to various hosts, including humans. As the vast majority of human listeriosis infections are recognized to occur through consumption of contaminated foods, and as animal listeriosis infections also appear to be predominantly feed-borne, development of effective intervention strategies for reducing the incidence of listeriosis among susceptible human and animal populations requires elucidation of specific routes of *L. monocytogenes* transmission among different ecosystems and compartments within food and feed production systems. Current knowledge of *L. monocytogenes* ecology is presented to provide insight into the primary sources that appear to contribute to its introduction into human food-associated environments and foods as well as its transmission among various compartments in food and agricultural production systems.

6.1.   Introduction

Listeria species, including *L. monocytogenes*, are often described as ubiquitous in nature as they have been isolated from a diverse array of natural, man-made, agricultural, and food-associated environments (Farber and Peterkin 1991; Fenlon 1999; Gravani 1999; Kathariou 2002; Roberts and Wiedmann 2003). The vast majority of human listeriosis infections (99%) are foodborne (Mead et al. 1999), while animal listeriosis infections appear to be predominantly feed-borne. An emerging understanding of *L. monocytogenes* ecology has provided insight into the primary sources that appear to contribute to its introduction into human food-associated environments and foods, as well as its transmission among various compartments in food production systems. As farm ruminants represent the mammalian hosts most commonly affected by clinical listeriosis, ruminant animal agricultural systems are likely to serve as reservoirs or sources of *L. monocytogenes* that are transmitted into the human food chain. Therefore, for a complete understanding of *L. monocytogenes* transmission to human food and humans, it is also important to understand the transmission of *L. monocytogenes* into other mammals, and particularly farm ruminants. This chapter reviews our knowledge of the ecology and transmission of *L. monocytogenes* in natural, nonagricultural environments, agricultural environments, and food-associated environments and addresses the potential contributions of these environments as sources or reservoirs of *L. monocytogenes* that can be transmitted to mammalian and particularly human hosts. We propose that, given the ability of *L. monocytogenes* to survive for prolonged time periods in many different environments, as well as its wide distribution and prevalence in different environments, it is likely that selective pressures associated with different environments play an important role in the evolution of this pathogen. We further propose that humans likely represent an accidental host for this environmental pathogen and that the true importance of most virulence factors may lie in their role in enhancing *L. monocytogenes* survival in nonhuman host-associated environments (Kreft et al. 1999).

6.2.   Methods for Studying *L. monocytogenes* Transmission

Studies on transmission of any pathogen are critically dependent on our ability to reliably detect, and ideally quantitate, a given pathogen in different environments and hosts. Methodological capabilities for accurately recovering and characterizing the representative diversity of a pathogen present in a given environment determine the quality of the information obtained. Since limitations in detection and subtyping methods critically impact our ability to understand the ecology of *L. monocytogenes*, we will briefly review commonly used detection and subtyping methods for *L. monocytogenes*, including their relevant limitations. A number of recent reviews and book chapters provide more comprehensive coverage of *L. monocytogenes* detection and subtyping methods (Wiedmann 2002a,b; Windham et al. 2005; Sauders and Wiedmann submitted).

6.2.1.   Detection Methods for L. monocytogenes

Despite their wide distribution in nature, *L. monocytogenes* and other *Listeria* spp. usually occur in small numbers, within the context of large numbers of other microorganisms, in most natural habitats. Therefore, detection methods for *Listeria* spp. typically include a selective enrichment step to allow amplification of the small numbers of *Listeria* spp. initially present, followed by plating on selective and differential media to enable their detection. This strategy is prone to providing false negative results, particularly if the *Listeria* spp. present in the sample are injured prior to exposure to the selective medium. On the other hand, use of a nonselective pre-enrichment step may allow other microorganisms to overgrow *Listeria* spp. in a given sample, thus also yielding false negative results. While these issues have led to the development of a variety of different enrichment media and procedures for *Listeria* spp., in general, a single enrichment procedure is unlikely to detect all *L. monocytogenes* that may be present in a given set of samples (Sauders and Wiedmann in press). For example, *L. innocua* has been shown to outcompete *L. monocytogenes* during some enrichment procedures (MacDonald and Sutherland 1994), and different enrichment media appear to favor recovery of different bacterial subtypes from the same sample (Ryser et al. 1996; Donnelly 2002). As a consequence, the use of a combination of different enrichment and plating procedures in parallel will provide the most sensitive detection of *Listeria* spp.; however, this approach is usually cost-prohibitive and therefore not practical. Since many environments can contain multiple *Listeria* species and/or multiple *Listeria* strains (Pritchard et al. 1995; Ryser et al. 1996, 1997) and because *L. monocytogenes* can be overgrown by other *Listeria* spp. during enrichment, *L. monocytogenes* prevalences reported for different environments likely underestimate the true prevalence and diversity of *L. monocytogenes* in a given sample.

Quantitative data on the presence of *Listeria* spp. and *L. monocytogenes* are important for understanding the ecology and transmission of *Listeria*. Most probable number (MPN) methods are generally used for quantification since *Listeria* populations in most environments are usually $<100\,CFU\,g^{-1}$ (Sauders and Wiedmann in press). A paucity of quantitative data on *L. monocytogenes* loads in different environments (Bernagozzi et al. 1994) other than in food samples (Yu and Fung 1993) exists due to the labor- and cost-intensive nature of MPN methods.

6.2.2. Subtyping Methods for L. monocytogenes

Application of subtyping methods is critical to our ability to understand the ecology and transmission of *L. monocytogenes*. While serotyping is commonly used to characterize *L. monocytogenes*, it provides limited discriminatory power as only 13 *L. monocytogenes* serotypes can be differentiated with three serotypes (1/2a, 1/2b, and 4b) representing the vast majority of human listeriosis isolates. Multilocus enzyme electrophoresis (MLEE) and phage typing provided initial insight into the population genetics and transmission of *L. monocytogenes*. Molecular subtyping methods, including ribotyping (Wiedmann et al. 1996, 1997), pulsed-field gel electrophoresis (PFGE) (Lozniewski et al. 2001; Vela et al. 2001), and, more recently, multilocus sequence-based typing (MLST) (Cai et al. 2002; Salcedo et al. 2003), have provided recent advances in our understanding of *L. monocytogenes* ecology and transmission and have been used in many studies on *L. monocytogenes* ecology reported since 1995. While all three molecular methods provide discriminatory power, PFGE was shown to be more discriminatory for *L. monocytogenes* than MLST or ribotyping. MLST and ribotyping have provided important and relevant subtyping information, however, including identification of epidemic clones and virulence-attenuated subtypes (Gray et al. 2004; Nightingale et al. 2005a,b).

Characterization of *L. monocytogenes* isolates from a variety of different hosts and environments by a variety of different subtyping methods, including initial MLEE work by Piffaretti et al. (1989), has also shown that strains comprising the species *L. monocytogenes* represent at least three distinct genetic lineages. While different nomenclatures have been used to designate these *L. monocytogenes* lineages (Kathariou 2002), the main lineages described in different studies appear to be identical, as supported by consistent grouping of specific *L. monocytogenes* serotypes into lineages (Piffaretti et al. 1989; Nadon et al. 2001). Based on the lineage designations used by most groups (Wiedmann et al. 1997; Ward et al. 2004), lineage I predominantly includes serotypes 1/2b, 3b, 3c, and 4b strains, while lineage II primarily includes serotypes 1/2a, 1/2c, and 3a (Nadon et al. 2001). Interestingly, previous reports have shown that lineage I strains are significantly overrepresented among human clinical listeriosis cases as compared to their prevalence among animal listeriosis cases and contaminated foods (Nadon et al. 2001; Norton et al. 2001; Gray et al. 2004). On the other hand, lineage II strains show a significantly higher prevalence among food isolates and animal clinical cases than among human listeriosis cases (Jeffers et al. 2001; Gray et al. 2004). In addition, lineage I isolates appear to have significantly greater pathogenic potential, as determined by their ability to spread to neighboring host cells in a cell culture plaque assay, when compared to lineage II isolates (Norton et al. 2001; Gray et al. 2004). Lineage III predominantly includes serotypes 4a and 4c, as well as some serotype 4b strains that are distinct from those grouped into lineage I (Nadon et al. 2001). Strains classified in lineage III appear to be associated with isolation from animals and are occasionally isolated from human listeriosis cases with clinical disease, but are rarely isolated from foods (Jeffers et al. 2001; Gray et al. 2004). Increasing evidence thus exists that *L. monocytogenes* strains represent multiple lineages that appear to differ in their abilities to be transmitted to humans, as also supported by recent subtype-specific mathematical modeling data, which indicate that the likelihood of human disease caused by *L. monocytogenes* classified into different lineages can differ by more than 2 log (Chen et al. 2006).

6.3. *Listeria monocytogenes* in Natural and Other Nonagricultural Environments

While most studies on the presence of *L. monocytogenes* in different environments have focused on food-associated and farm environments (Fenlon et al. 1995; Nightingale et al. 2004), multiple studies have reported that *L. monocytogenes* are common in natural and other nonagricultural environments, and can also survive for extended time periods in soil and water. Recent subtyping studies also indicate that at least some of the subtypes found in natural and other nonagricultural environments are also found among human listeriosis cases, indicating that these environments may represent a source of human pathogenic *L. monocytogenes* subtypes.

6.3.1.   Listeria monocytogenes *Prevalence and Load in Natural and Other Nonagricultural Environments*

Initial studies by Welshimer and Donker-Voet (1971), which investigated the presence of *L. monocytogenes* in soil and plant materials from agricultural and nonagricultural environments, reported that six of seven nonagricultural sites were positive for *L. monocytogenes* during the spring but not the fall. While this study has to be interpreted carefully, since all recognized *Listeria* spp. were classified as *L. monocytogenes* at that time, it provided initial evidence for the presence of *Listeria* spp. in natural environments. Later, Weis and Seeliger (1975) found *Listeria* spp. from plant samples collected from cornfields (9.7% of samples were positive), grain fields (13.3%), cultivated fields (12.5%), uncultivated fields (44%), meadows and pastures (15.5%), forests (21.3%), and wildlife feeding areas (23.1%) in southern Germany. While the original paper by Weis and Seeliger reported these numbers as *L. monocytogenes* prevalence, only 37 of 103 *Listeria* isolates elicited disease consistent with listeriosis in their mouse bioassay, suggesting that as many as 64% of their isolates may have been *Listeria* spp. other than *L. monocytogenes* or *L. ivanovii*. Fenlon et al. (1996) did not find any *L. monocytogenes* in the soil samples associated with vegetable crops that they examined, but they isolated *L. monocytogenes* from the soil collected from fields where cattle- or sheep-fed silage diets had been kept, indicating the importance of animals as sources of *L. monocytogenes* in soil and on plants. One of the most comprehensive, recent studies on *L. monocytogenes* and *Listeria* spp. prevalences in different environments was conducted by Sauders (2005) who tested approximately 900 samples from each natural and urban environments (e.g., soil, water, and plant materials) in New York State over a 2-year period. In this study, *L. monocytogenes* prevalences were significantly higher in urban environments (7.5%) as compared to natural environments (1.4%). Additional recent studies have further confirmed that *L. monocytogenes* can be found in a number of different natural and nonagricultural environments, including surface waters (Frances et al. 1991; Arvanitidou et al. 1997), estuarine environments (Colburn et al. 1990), and sewage (Watkins and Sleath 1981; Geuenich et al. 1985; al-Ghazali MR and al-Azawi 1988a,b; MacGowan et al. 1994; De Luca et al. 1998; Garrec et al. 2003).

Overall, most studies on *Listeria* in nonfarm-associated natural environments indicate that *L. monocytogenes* is found at a lower prevalence than other *Listeria* spp. (MacGowan et al. 1994; Sauders 2005). Interestingly, MacGowan et al. (1994) found that *L. seeligeri* was the *Listeria* spp. most frequently isolated from soils, while Sauders (2005) found that *L. seeligeri* was the most common

Listeria spp. isolated from both urban and natural environments. This is particularly intriguing since *L. seeligeri* while not considered a mammalian pathogen, does contain the *Listeria* virulence gene cluster and is hemolytic, which suggests a potential role of at least some *Listeria* virulence genes for survival in selected environments.

6.3.2.   Listeria monocytogenes *Growth and Survival in Natural and Other Nonagricultural Environments*

Some of the first efforts to characterize *L. monocytogenes* as a naturally occurring, saprophytic organism took place in the late 1950s. H.J. Welshimer conducted a study on the survival of *L. monocytogenes* and concluded that *L. monocytogenes* could survive for at least 295 days in certain types of soil under defined conditions (Welshimer 1960). Another study (Botzler et al. 1974) also showed that *L. monocytogenes* was able to survive, and in some instances, multiply, in nonsterilized (i.e., natural) and sterilized soil and water at ambient winter temperatures ranging from -15 to $+18°C$. In this study, sterile and natural soil both supported *L. monocytogenes* survival and growth. For example, *L. monocytogenes* inoculated into a sterile soil suspension at approximately $10^5\,CFU\,ml^{-1}$ increased up to $2.14 \times 10^7\,CFU\,ml^{-1}$ over a 154-day period. These data provide evidence of the ability of *L. monocytogenes* to survive and multiply in different niches in natural and other nonagricultural environments.

6.3.3.   *Subtype Analysis of* L. monocytogenes *Found in Natural and Nonagricultural Environments*

Early serotype analysis by Weis and Seeliger (1975) of *L. monocytogenes* isolated from natural and agricultural fields found that serotypes 1/2b and 4b were the two most prevalent serotypes isolated from soil and plant samples, which provided initial evidence that human pathogenic *L. monocytogenes* may be present in diverse environments, since serotypes 1/2b and 4b are commonly associated with human disease (McLauchlin 1990). Ribotype analysis of 80 *L. monocytogenes* isolates collected from urban and natural environments by Sauders et al. (2006) also found that a number of ribotypes identified among these isolates had previously been linked to human listeriosis cases, including outbreaks. Specifically, a number of isolates from urban sites and a single isolate from a natural environment were ribotype DUP-1038B, which represents a subtype classified into one of the three *L. monocytogenes* epidemic clones (ECII) (Kathariou 2002) that has been associated with multiple human listeriosis outbreaks (Jeffers et al. 2001; Sauders et al. 2003). To illustrate, DUP-1038B was the predominant ribotype isolated over more than a year from multiple sites in a single urban environment, indicating the persistence of this subtype in an urban environment. Overall, these data provide evidence that urban environments represent sources of human pathogenic *L. monocytogenes* strains.

Interestingly, the vast majority of isolates (>90%) from natural environments were classified as *L. monocytogenes* lineage II. The lineage II classification is significantly less common among human isolates than classification into lineage I (Sauders et al. 2006), and strains in lineage II appear less likely to cause human disease as compared to those in lineage I (Chen et al. 2006). The ribotype most frequently isolated from natural environments (DUP-1039C) is also commonly found in farm environments and animal listeriosis cases, as well as infrequently among human cases, supporting the hypothesis that the natural environment represents a source of animal and human pathogenic *L. monocytogenes*, even though many of the strains found in natural environments may be less virulent for humans than strains found in other environments (e.g., the urban environment). Furthermore, the presence in natural environments of a ribotype that is commonly found in farm environments could also indicate that farms, farm environments, and farm animals may represent a source of *L. monocytogenes* introduction into natural environments, e.g., via runoff from farms or animal movement.

6.4. *Listeria monocytogenes* in Agricultural Environments

Listeriosis was first observed in animals (i.e., rabbits) in 1926 (Murray et al. 1926) and has since been reported in a number of domesticated and wild animals (Wesley 1999; Hayashidani et al. 2002). Most of the reported animal listeriosis cases have occurred in farm ruminants, including cattle, goats, and sheep, therefore, most of the available information on *L. monocytogenes* in agricultural environments focuses on the presence and ecology of *L. monocytogenes* in ruminants and on ruminant farms. Overall, *L. monocytogenes* prevalence in ruminants and on ruminant farms varies, but appears to be highest in animals fed silage (fermented plant material, such as grass, hay, or chopped field corn) (Nightingale et al. 2004). Prevalence in farm environments and in fecal material of silage-fed animals appears to, on average, exceed 20% (Nightingale et al. 2004) and often includes human disease–associated *L. monocytogenes* subtypes. Farm environments and farm animals may thus be an important source, and potentially a reservoir, of human pathogenic *L. monocytogenes*.

6.4.1. Listeria monocytogenes *Prevalence and Load in Agricultural Environments*

A number of studies (Welshimer 1968; Weis and Seeliger 1975; Fenlon 1985; Wesley 1999; Nightingale et al. 2004) have reported that *L. monocytogenes* is commonly present throughout the agricultural environment, particularly in environments associated with ruminants, including farm soil, vegetation, and water, as well as in animal feeds (especially silage), in fecal material, and on animal hides and external surfaces. The presence of *L. monocytogenes* in silage and further dispersal through fecal shedding in ruminant-associated agricultural environments likely have the greatest impacts on transmission of *L. monocytogenes,* both within animal populations as well as from animal populations and farm environments to humans. Fecal shedding is particularly likely to contribute to environmental dispersal of *L. monocytogenes,* i.e., onto plant materials and fields that may be a source of animal feed or human food (e.g., vegetables). A number of studies have shown that the prevalence of fecal shedding in farm ruminants can range from a few percent of total animals (Unnerstad et al. 2006; Sammarco et al. 2005) to more than 50% (Skovgaard and Morgen 1988), with a higher prevalence of fecal shedding in silage-fed animals (Husu 1990). For example, a recent large study on *L. monocytogenes* prevalence and ecology in farm animals and farm environments found an average fecal prevalence of 20.2% among cattle (Nightingale et al. 2004). While fecal shedding can occur in clinical listeriosis cases, most cattle with fecal samples that test positive for *L. monocytogenes* do not show listeriosis symptoms (Gronstol 1979; Loken et al. 1982). On the other hand, small ruminants (e.g., sheep and goats) generally show a lower *L. monocytogenes* prevalence in fecal samples, most likely since silage feeding, and hence exposure to *L. monocytogenes,* is less common on small ruminant farms. Quantitative data on *L. monocytogenes* levels in ruminant fecal samples are limited; however, Fenlon et al. (1996) reported *L. monocytogenes* levels as high as $5.0 \times 10^2 \, \text{CFU} \, \text{g}^{-1}$ among cattle fed silage, while fecal levels among grazing cattle were $0.4 \, \text{CFU} \, \text{g}^{-1}$. Oral exposure to *L. monocytogenes* appears to be a critical risk factor for fecal shedding of this pathogen, but other external factors such as stress, e.g., climate and feed changes, transport, and changes in immunological state, such as pregnancy, may also enhance the likelihood of fecal shedding in ruminants (Gronstol 1979; Gronstol and Overas 1980; Loken et al. 1982; Fenlon et al. 1996).

Similar to human listeriosis, animal listeriosis appears to be predominantly a feed-borne disease, with consumption of silage, and particularly improperly fermented (and *L. monocytogenes* contaminated) silage, as a major risk factor for clinical listeriosis in ruminants (Nightingale et al. 2005) and for *L. monocytogenes* fecal shedding in ruminants (Nightingale et al. 2004) as well as the presence of *L. monocytogenes* in raw milk (Sanaa et al. 1993). Silage is commonly used for feed in modern ruminant production, and particularly for dairy cattle feed, due to its year-round availability. While properly fermented silage has a pH of ≤ 4.5 which helps to inhibit growth of spoilage microorganisms and pathogens, including *L. monocytogenes,* improperly fermented silage often has an elevated pH (e.g., >5.5) which allows for the growth of spoilage microorganisms and pathogens (Ryser et al. 1997). Since crops used for silage may be contaminated with *L. monocytogenes* prior to harvest through a variety of pathways, including fecal deposition by wild (Fenlon 1985) or farm animals, contaminated soil (Fenlon et al. 1996), or deposition of sewage sludge and manure, *L. monocytogenes* is commonly found in poorly fermented silage. As silage is an important source of *L. monocytogenes* infection in ruminants, a number of studies on *L. monocytogenes* prevalence and loads in silage have been conducted. Overall, *L. monocytogenes* has been isolated from up to 44% of silage samples tested (Fenlon 1985) with widely ranging bacterial loads (Fenlon 1986; Wiedmann et al. 1994) as high as $1.0 \times 10^8 \, \mathrm{CFU\,g^{-1}}$ reported (Wiedmann et al. 1996). Considering the high *L. monocytogenes* prevalence and densities that can be found on ruminant farms (Nightingale et al. 2004), these environments represent a likely point of introduction for *L. monocytogenes* into the human food chain through a variety of pathways, including the use of contaminated manure for fertilization of human food crops, consumption of animal products lacking a listeriocidal heat treatment (e.g., raw milk), and transmission of the organism via fomites into food processing environments, where *L. monocytogenes* may subsequently persist for extended time periods, thus enabling recontamination of processed foods.

In addition to ruminant species, *L. monocytogenes* also can be isolated from a number of nonruminant species and nonruminant agricultural environments. For example, *L. monocytogenes* has been isolated from the feces of wild birds (Fenlon 1985), horses (Weber et al. 1995; Gudmundsdottir et al. 2004), swine (Hayashidani et al. 2002; Yokoyama et al. 2005), poultry (Weber et al. 1995), and other domestic animals (Weber et al. 1995), as well as from eviscerated farmed fish (Miettinen and Wirtanen 2005). While *L. monocytogenes* in ruminants and on ruminant farms are more likely to contribute directly to human disease, i.e., through human consumption of raw milk (Ryser 1999), the presence of *L. monocytogenes* in other food animals is more likely to contribute indirectly to food contamination and human disease, e.g., by facilitating introduction of this pathogen into food processing plants or onto vegetables through contaminated manure (e.g., Fenlon et al. 1996; Rorvik et al. 2003). One example of food products that appear to become contaminated both directly and indirectly is cold-smoked fish products. Production of cold-smoked fish products does not include a listeriocidal heat treatment. Matching *L. monocytogenes* subtypes have been found occasionally in both raw fish and cold-smoked products produced from the contaminated raw materials, supporting a direct route of transmission (Markkula et al. 2005); however, in most cases, *L. monocytogenes* contamination of these products appears to occur from the processing plant environment rather than from the raw material (Norton et al. 2001; Hoffman et al. 2003).

In addition to animal-associated agricultural environments, *L. monocytogenes* has also been isolated from a number of plant-based agricultural systems (Weis and Seeliger 1975; Beuchat 1996; Fenlon et al. 1996). Reported prevalences of *L. monocytogenes* in raw vegetables have ranged from 1.1 to 85.7% with an average prevalence of 11.4% (Beuchat 1996). Raw vegetables have been implicated as sources in multiple human listeriosis outbreaks (Schlech et al. 1983; Ho et al. 1986; Allerberger and Guggenbichler 1989). Vegetables may become increasingly important in human listeriosis transmission since current trends in food consumption patterns reflect increasing consumption of raw and ready-to-eat (RTE) vegetables.

6.4.2. Listeria monocytogenes—*Growth and Survival in Agricultural Environments*

In addition to field studies that indicate survival and persistence of *L. monocytogenes* in agricultural environments over time periods up to 6 years (Fenlon 1999), a number of laboratory studies have also provided evidence that *Listeria* spp. survive in animal feces (Dijkstra 1971) and agricultural soil (Fenlon 1999), including manure-amended soil (Jiang et al. 2004), for prolonged time periods. For example, *Listeria* spp. have been shown to survive in bovine feces from 182 to 2,190 days (Fenlon 1999), for several weeks in soil (Jiang et al. 2004; Nicholson et al. 2005), and for >56 days (Fenlon 1999) in sewage sludge cake sprayed onto fields.

6.4.3. *Subtype Analysis of* L. monocytogenes *Found in Agricultural Environments*

Serotyping data provided initial evidence that *L. monocytogenes* serotypes associated with human disease (i.e., serotypes 1/2a, 1/2b, and 4b) are present in agricultural environments (Dijkstra 1978; Borucki et al. 2005), ruminant fecal samples (Ralovich et al. 1986; Skovgaard and Morgen 1988), and raw milk collected on farms (Fenlon et al. 1995). Molecular subtyping studies further support the presence of human disease–associated *L. monocytogenes* strains in agricultural environments and on farms. For example, in one study, 23% of *L. monocytogenes* isolates from human sporadic cases were found to have identical or similar PFGE patterns to *L. monocytogenes* isolates from farm environments (Borucki et al. 2004). Nightingale et al. (2004) found that 25 of the 35 *Eco*RI ribotypes with greater than five occurrences among farm environments and ruminant fecal samples had also been isolated previously from human listeriosis cases (Sauders et al. 2006). This study also found that all three ribotypes linked to multiple human listeriosis outbreaks, which represent *L. monocytogenes* epidemic clones, were each found on multiple farms. Similarly, isolates with a PFGE type identical to the strain responsible for the 1985 listeriosis outbreak in Los Angeles (Linnan et al. 1988) have also been isolated from a dairy farm by Borucki et al. (2004). Subtyping data thus clearly support that *L. monocytogenes* subtypes linked to sporadic human listeriosis cases and to human listeriosis outbreaks are commonly found in agricultural environments and on farms.

6.5. *Listeria monocytogenes* in Food-Associated Environments and Foods

Since the vast majority of human listeriosis cases are food-borne, a body of knowledge on the presence of *L. monocytogenes* in food-associated environments and foods has been published. Since a comprehensive review of these studies is beyond the scope of this chapter, only a brief summary is provided here. Detailed information on food-associated environments and outbreaks of listeriosis can be obtained from a number of review articles (Kathariou 2002; Tompkin 2002) and book chapters (e.g., Ryser 1999). *L. monocytogenes,* including subtypes associated with human listeriosis cases and outbreaks, are not uncommon in natural, nonagricultural, and agricultural environments, including raw food commodities. Therefore, an understanding of *L. monocytogenes* ecology and survival in food-associated environments is critical for elucidating the transmission of human listeriosis and reducing human infections.

6.5.1. Listeria monocytogenes—*Prevalence and Loads in Foods and Food-Associated Environments*

A relatively small number of human listeriosis cases have been associated with food products contaminated during primary production in agricultural environments, such as raw vegetables, raw milk, and raw milk dairy products, since heating regimes typically applied during food cooking or commercial processing effectively inactivate *L. monocytogenes*. Instead, most human listeriosis cases are caused by the consumption of RTE meats and dairy food products that contain *L. monocytogenes* as a consequence of contamination from environmental sources in processing plants and other food-associated environments (e.g., retail food businesses) (Tompkin 2002; Reij and Den Aantrekker 2004) and that are not thoroughly reheated immediately prior to consumption. Foods at particular risk for transmission of this pathogen include RTE foods that are not aseptically packaged after processing and that require refrigerated storage. *L. monocytogenes* prevalences in these products have been reported at 5% and above (Farber and Peterkin 1991, 1999). Since the infectious dose of *L. monocytogenes* appears to be high (10^6) (Vazquez-Boland et al. 2001) and initial contamination of RTE foods typically occurs at low levels, RTE foods with the highest likelihood of transmitting listeriosis are those that support the growth of this pathogen and that are stored long enough to allow bacterial numbers to increase to high levels. Foods that have been implicated as sources of human infections include deli meats (Gottlieb et al. 2006), coleslaw (Ryser 1999), cheeses (Goulet et al. 1995), hot dogs (CDC 1999), and smoked fish (Ericsson et al. 1997). Recent *L. monocytogenes* risk assessments (FDA/USDA 2003; FAO/WHO 2004) provide comprehensive information on food products most commonly linked to human listeriosis cases as well as *L. monocytogenes* contamination prevalences and levels in these foods. In the United States, the foods that appear to be most commonly responsible for human listeriosis cases include RTE deli meats (average *L. monocytogenes* prevalence of 1.9%) followed by hot dogs (4.8%) (FDA/USDA 2003). *L. monocytogenes* is prevalent in smoked seafood (12.0%), fruit (11.8%), preserved fish (9.8%), raw seafood (7%), and pate/meat spreads (6.5%) (FDA/USDA 2003). In general, RTE foods that are handled extensively after heat treatment, such as deli salads that are prepared in retail environments, show the highest *L. monocytogenes* prevalences. Foods produced under poor hygienic conditions also can have very high *L. monocyto-genes* prevalences. For example, Van Coillie et al. (2004) found that prepared minced meat collected in Belgian markets showed *L. monocytogenes* prevalences of 94.7%. In addition, specific food processing plants may manufacture products with extremely high prevalences of *L. monocytogenes*. For example, Fonnesbech Vogel et al. (2001) reported that cold-smoked fish from one processing plant in Denmark had a *L. monocytogenes* prevalence of 85% during their 1-year study.

Due to the importance of the food processing environment as a source of post-processing contamination of RTE foods with *L. monocytogenes*, a variety of studies have evaluated the prevalence, transmission, and ecology of *L. monocy-togenes* in food processing plants, including the presence of this pathogen on worker gloves and aprons, as well as in free-standing water, aerosolized dust particles (De Roin et al. 2003), walls (Chasseignaux et al. 2001), floors and drains (Rorvik et al. 1997; Chasseignaux et al. 2001; Norton et al. 2001), and on processing equipment (Eklund et al. 1995; Rorvik et al. 1995; Autio et al. 1999; Norton et al. 2001; Hoffman et al. 2003; Thimothe et al. 2004; Markkula et al. 2005; Hu et al. 2006). Reported *L. monocytogenes* prevalences in food processing environments have varied tremendously among studies. For example, some studies have shown *L. monocytogenes* prevalences in food processing plant drains to be close to 100% (Gravani 1999), while others have found very low prevalences in drains (<1%) (Autio et al. 1999). *L. monocytogenes* prevalences in processing plant environments are greatly affected by the charac-teristics of a given processing plant (e.g., sanitary practices, age and design of the facility, processing run times; Norton et al. 2001; Thimothe et al. 2004) and also by the types of samples tested and the time of sample collection (e.g., sampling of processing equipment prior to production initiation should yield lower prevalences than sampling of drains at the end of a production day). Further, *L. monocytogenes* prevalences and contamination patterns in processing plant environments can vary considerably from week to week and even daily (Hu et al. 2006).

Since the mid-1990s, use of molecular subtyping techniques for character-izing *L. monocytogenes* has enhanced our understanding of the ecology and transmission of *L. monocytogenes* in food processing plants. Most importantly, subtyping studies (Lappi et al. 2004; Thimothe et al. 2004) have shown that one or more specific *L. monocytogenes* subtypes can persist in a given processing plant from a few months to up to 10 years (Norton et al. 2001; Thimothe et al. 2004). Persistent *L. monocytogenes* contamination has been identified in a variety of food processing environments, including in plants that produce milk (Kells and Gilmour 2004), cheese (Kabuki et al. 2004), RTE meat (Samelis and Metaxopoulous 1999), pork (Chasseignaux et al. 2002) and poultry (Lawrence and Gilmour 1995; Chasseignaux et al. 2002) products, RTE crawfish (Lappi et al. 2004), and smoked seafoods (Hoffman et al. 2003). Persistent isolates in plants often appear to be "plant-specific." For example, different plants in close geographical proximity can host distinct persistent strains (Norton et al. 2001), suggesting that specific strains can establish themselves as resident microflora in a given processing facility. Importantly, subtyping studies have also shown that subtypes persisting in a given processing environment are often also isolated from finished products, indicating that *L. monocytogenes* persistence in processing plants can be a major factor contributing to finished product contamination (Lappi et al. 2004; Thimothe et al. 2004), particularly if the persistent strain is also associated with food contact surfaces.

While *L. monocytogenes* transmission and persistence in food processing plants have been examined in numerous studies, the primary sources responsible for introduction of *L. monocytogenes* into the processing environment have not been clearly identified. Potential sources include contaminated raw materials, employee's shoes and attire as well as equipment introduced into the plant, including equipment tires (Rocourt and Cossart 1997). Fecal shedding by healthy human carriers has also been proposed as a potential source for introduction of *L. monocytogenes* into food processing plants. Early studies in high-risk populations (e.g., household contacts of listeriosis patients; Schuchat et al. 1993) suggested the potential for a high prevalence of fecal shedders among humans. However, more recent studies on broader human populations indicate that the prevalence of *L. monocytogenes* shedding is generally low (<0.12%, Sauders et al. 2005 and <0.17% Schlech et al. 2005) and usually of short duration (Grif et al. 2003), suggesting that human fecal shedding is likely to play a limited role as a source of *L. monocytogenes* introduction into food processing environments (Ivanek et al. 2006).

The majority of reports on *L. monocytogenes* ecology and transmission have focused on food processing environments, while less is known about *L. monocytogenes* prevalence and transmission in other food-related environments, including the retail environment and consumer homes. A study by Gombas et al. (2003) found that luncheon meats, deli salads, and seafood salads packaged at retail were 6.8, 2.6, and 5 times more likely to be contaminated with *L. monocytogenes* as compared to manufacturer-packaged equivalents, suggesting a considerable risk for retail *L. monocytogenes* contamination of at least some types of RTE foods. Further, a molecular subtyping study by Sauders et al. (Sauders et al. 2004) showed that a number of different *L. monocytogenes* strains appear to persist in different retail environments, consistent with the well-established ability of this pathogen to establish persistent contamination in food processing environments. Not surprisingly, *L. monocytogenes* has also been isolated from domestic household environments, including food preparation–associated surfaces such as kitchen sinks, dish-cloths, and washing-up brushes (Beumer et al. 1996), indicating consumer households as potential contamination sources. The contributions of points after primary processing, including in retail and consumer or commercial kitchens, to *L. monocytogenes* contamination of RTE foods, and hence to human disease incidence, remain to be elucidated. If time intervals are short between contamination and consumption of food products in food service operations and homes, it is less likely that *L. monocytogenes* will be able to grow to levels associated with human disease, particularly if initial contamination levels are low.

6.5.2.   Listeria monocytogenes–*Growth and Survival in Foods and Food-Associated Environments*

While *L. monocytogenes* can survive and grow under a variety of environmental conditions, including high salt concentrations and low pH, one of the most important characteristics contributing to human exposure is its ability to grow at refrigeration temperatures (Gray and Killinger 1966), which can enable it to multiply to high numbers in RTE foods that support its growth. Numerous studies have been conducted to characterize the ability of *L. monocytogenes* to grow in different foods held under different temperatures. *L. monocytogenes* has been found to grow, albeit slowly, even at refrigeration temperatures close to 0°C, with increasing growth rates as storage temperature increases. Many RTE foods that inherently permit growth of *L. monocytogenes* can allow growth of this pathogen to high numbers (e.g., up to $2.5 \times 10^6\,\mathrm{CFU\,g^{-1}}$ in corned beef and up to $1.8 \times 10^7\,\mathrm{CFU\,g^{-1}}$ in ham; Sim et al. 2002). A comprehensive review and summary of *L. monocytogenes* growth and survival characteristics in a variety of foods can be found in *Microorganisms in Foods 5* (ICMSF 1996).

6.5.3.   *Subtype Analysis of* L. monocytogenes *Found in Foods and Food-Associated Environments*

While serotyping studies indicated that a number of isolates from RTE foods represented serotypes associated with human disease (i.e., serotypes 1/2a, 1/2b, and 4b) (Ryser 1999), a number of food isolates also represented serotypes rarely associated with human infections; in particular, serotype 1/2c (Farber and Peterkin 1991). The observations that serotypes 1/2b and 4b were often overrepresented among human clinical isolates as compared to their prevalence among food-associated isolates, while serotypes 1/2a, and particularly 1/2c, were generally overrepresented among food isolates as compared to their prevalence among human clinical isolates, provided initial evidence that *L. monocytogenes* strains and serotypes differ in their abilities to cause human disease. Molecular subtyping studies further supported these observations by showing that *L. monocytogenes* strains grouped into lineage I (which includes serotypes 1/2b and 4b) were more common among human clinical isolates as compared to food isolates, while strains grouped into lineage II (which includes serotypes 1/2a and 1/2c) were more common among food isolates than human clinical isolates (Wiedmann et al. 1997; Jeffers et al. 2001). A study of almost 1,000 *L. monocytogenes* isolates from human clinical cases and foods showed that a number of specific ribotypes within the different *L. monocytogenes* lineages differed significantly in their prevalences among food and human isolates (Gray et al. 2004). Three ribotypes that were overrepresented among isolates from human listeriosis patients represented subtypes previously associated with multiple human listeriosis cases (i.e., epidemic clones), and also showed significantly higher ability to spread from cell to cell in a tissue culture plaque assay, providing phenotypic data supporting enhanced mammalian virulence of these subtypes and epidemic clones (Gray et al. 2004). Conversely, a number of the specific subtypes that were more common among food isolates in both the United States and France also showed reduced invasion efficiencies for human intestinal epithelial Caco-2 cells. Interestingly, the reduced invasion phenotype was found to be caused by one of several possible mutations leading to premature stop codons in *inlA*, which encodes for internalin A, a listerial surface molecule critical for invasion of human intestinal epithelial cells, thus providing a clear biological explanation for attenuated human virulence in these strains. Importantly, strains with premature stop codon mutations in *inlA* appear to represent about 30% of food isolates as established by independent studies in France (Jacquet et al. 2004) and the USA (Gray et al. 2004; Nightingale et al. 2005) supporting that a number of food isolates show reduced virulence. In addition, a smaller proportion of food isolates appears to show attenuated human virulence due to mutations in other virulence genes, including *hly* and *prfA* (Roberts et al. 2005; Roche et al. 2005). The combination of molecular subtyping with phenotypic characterization has thus provided substantial evidence for virulence differences among *L. monocytogenes* subtypes. These experimental observations have also been supported by mathematical modeling data that indicate greater than 5 log differences in the likelihood of different *L. monocytogenes* subtypes to cause human disease (Chen et al. 2006).

Subtyping studies on *L. monocytogenes* isolated from food processing and retail environments showed that plant- or retail-specific *L. monocytogenes* subtypes can persist in these environments and also showed that the subtypes found in these environments represent both subtypes rarely associated with human disease, as well as those commonly associated with human disease, including subtypes that have caused multiple human listeriosis outbreaks (i.e., epidemic clones, Sauders et al. 2004). The observation that human disease–associated *L. monocytogenes* can persist in the environment without apparent loss of human virulence is further supported by the observation that a human listeriosis outbreak in 2000 was linked to contaminated RTE deli turkey produced in a processing plant in which the causative strain appears to have persisted for more than 10 years (Olsen et al. 2005). The strain responsible for the outbreak in 2000 also caused a single human listeriosis case in 1989 that was linked to consumption of contaminated hot dogs produced (and contaminated) in the same plant (Olsen et al. 2005).

6.6. Transmission into the Mammalian Host

Transmission of *L. monocytogenes* into mammalian hosts and development of infection are dependent on various host factors, pathogen-related factors, environmental factors, as well as interactions among these factors (Roberts and Wiedmann 2003). Critical factors for transmission of *L. monocytogenes* include (1) presence of sufficiently high numbers of the pathogen in food or feed, (2) presence of *L. monocytogenes* strains of sufficient virulence in the food, as well as (3) exposure of a sufficiently susceptible mammalian host. The interplay among these factors is also critical for development of listeriosis. For example, even a highly virulent *L. monocytogenes* strain present at high levels in a food is unlikely to cause disease if the exposed host is young and highly immuno-competent. At the other extreme, even a virulence-attenuated *L. monocytogenes* strain (e.g., a strain with a premature stop codon in *inlA*) can cause human disease, even if present in foods at fairly low levels, if the exposed host is severely immunocompromised. Transmission of *L. monocytogenes* appears to differ between human and nonhuman mammalian hosts (specifically in different ruminants); therefore, key aspects of *L. monocytogenes* transmission into these host populations is discussed in the following sections.

6.6.1. *Transmission of* L. monocytogenes *to Nonhuman Mammals*

While *L. monocytogenes* can cause disease in various nonprimate mammalian hosts, this section focuses predominantly on transmission in ruminants (e.g., cattle), since very little is known about natural transmission in other nonprimate mammalian hosts. This selected focus is not intended to imply that other mammalian hosts do not have important roles in the overall ecology and trans-mission of *L. monocytogenes*.

Transmission of *L. monocytogenes* in silage-fed ruminant hosts and the ecology of *L. monocytogenes* in ruminant hosts and ruminant farm environ-ments (specifically, those feeding silage) appear to be characterized by a high prevalence of this pathogen in the environment (Nightingale et al. 2004), including high levels (up to 100%) of fecal shedding among cattle as well as potentially high *L. monocytogenes* loads in silage (up to $1 \times 10^8 \, \text{CFU} \, \text{g}^{-1}$ silage; Wiedmann et al. 1996). Preliminary analyses suggest intrahost ampli-fication of ingested *L. monocytogenes* in cattle that show fecal shedding (Nightingale et al. 2004), even though fecal shedding generally appears to be short; further studies, including mathematical modeling of transmission, are needed to further confirm this hypothesis. The data available to date suggest that silage-fed ruminants and the associated farm environment can maintain high *L. monocytogenes* densities, most likely due to a combination of multiple factors, including *L. monocytogenes* growth in poorly fermented silage as well as fecal shedding by animals. *L. monocytogenes* that are fecally shed are dispersed into the environment, e.g., onto plant material that may be used for subsequent silage production, thus effectively maintaining an infectious cycle. Interestingly, clinical disease in ruminant populations that are fed silage year-round appears to be uncommon and is generally limited to a single or few animals in a herd if disease occurs. In these ruminant populations, disease cases often appear to be linked to either consumption of silage contaminated with extremely high levels of *L. monocytogenes* or immunosuppression of cows or both (Wesley 1999; Roberts and Wiedmann 2003). The lack of frequent signs of overt disease despite the presence of high pathogen numbers in the environment may possibly reflect herd immunity against *L. monocytogenes* due to frequent exposure to the pathogen through consumption of contaminated silage. In the silage-fed large ruminant ecosystem, the host and the pathogen may have established an equilibrium that allows high pathogen population densities with limited animal disease. The importance of constant or at least frequent *L. monocytogenes* exposure, and consequent immunity, is highlighted by the fact that small ruminant populations (e.g., sheep and goats) that are fed silage only seasonally (i.e., in the winter) show higher prevalences of listeriosis with more severe disease outcomes following exposure to contaminated silage, possibly due to reduced immunity after extended time periods without silage feeding and thus without the *L. monocytogenes* exposure needed to maintain or build anti-listerial immunity.

Interestingly, preliminary data indicate that the majority of farm environment and ruminant-associated *L. monocytogenes* isolates examined to date are fully able to invade human intestinal epithelial cells (i.e., they do not carry *inlA* premature stop codon mutations that are responsible for virulence attenuation in a proportion of human food-associated *L. monocytogenes* strains; Nightingale et al. 2005). It is thus tempting to speculate that silage-fed ruminants and the associated agricultural environments represent an important, but unlikely sole, reservoir for human virulent *L. monocytogenes* strains. On-farm sources appear to be rarely linked directly to food contamination and human disease, as most RTE foods appear to be contaminated in the processing plant environment rather than from the farm environment; therefore, the importance of ruminant farms and agricultural environments as direct or indirect sources of human *L. monocytogenes* infections remain to be elucidated.

6.6.2. *Transmission of* L. monocytogenes *into the Human Host*

Transmission of *L. monocytogenes* into the human host is almost exclusively food-borne (Mead et al. 1999), and RTE foods (i.e., foods that do not undergo an additional cooking step before consumption) that allow growth of *L. monocytogenes* during storage are most commonly implicated as vehicles of human infections. As described in Sect. 6.5.1, specific food categories most commonly associated with human listeriosis cases include RTE deli meats and hot dogs, deli salads, soft cheeses, raw milk and raw milk dairy products, pate, smoked seafoods, and vegetables (Ryser 1999; FDA/USDA 2003).

Unlike many other food-borne pathogens, such as *Salmonella* and enterohemorrhagic *Escherichia coli*, human infections with *L. monocytogenes* usually require a high pathogen dose (Vazquez-Boland et al. 2001; FDA/USDA 2003). In addition, human hosts that present clinical symptoms after food-borne exposure to *L. monocytogenes* usually are fetuses or severely immunocompromised individuals. Since human food-borne exposure to *L. monocytogenes*, even at high doses, is not uncommon and usually does not result in human disease, it is tempting to speculate that the majority of the human population has some immunity against this pathogen. Considering the rarity of human infections as well as the apparent short duration and low prevalence of human fecal shedding, it appears that *L. monocytogenes* represents an opportunistic human pathogen and that human infections are likely to contribute little if anything to the ecological success or dispersal of *L. monocytogenes*.

6.7. Overall *L. monocytogenes* Transmission and Conclusions

L. monocytogenes is a widely distributed, if not ubiquitous, bacterial pathogen. While the importance of feed- and food-borne transmission to ruminant and human hosts has been well defined, its routes of transmission among different ecosystems and compartments within food production systems appear complex and remain to be clearly elucidated. Despite the fact that human infections with *L. monocytogenes* appear rare, particularly given the frequent prevalence and occasional high load of *L. monocytogenes* in many different environments, including in human foods and animal feeds, it is tempting to propose an anthropocentric transmission pathway for *L. monocytogenes* from the general environment through animal populations to food processing environments and foods to humans. While subtyping studies have clearly shown that human disease–associated *L. monocytogenes* strains, including epidemic clones, can be found in many environments, including natural, urban, and farm environments, directionality of transfer and transmission is difficult to establish. Consequently, future work remains to identify and characterize *L. monocytogenes* hosts, reservoirs, and transmission pathways, with consideration given to the possibility that the true natural host(s) of *L. monocytogenes* could be currently unidentified mammalian or even nonmammalian species. The ecological success of *L. monocytogenes* as a globally distributed microorganism may lie in its ability to survive in a large number of hosts as well as in non-host-associated environments, with the ability to establish high population densities in some host-associated ecosystems.

A number of distinct *L. monocytogenes* phylogenetic lineages and clonal groups have been identified and classified based on differences in abilities to cause human disease. Key groups important to the overall picture of *L. monocytogenes* ecology and transmission include (1) virulence-attenuated strains (such as those characterized by premature stop codons in *inlA* (Nightingale et al. 2005) or by mutations in other virulence genes (Roberts et al. 2005)), (2) epidemic clones, which appear to show increased human virulence as compared to other strains, and (3) lineage III strains that appear to be associated with animal hosts (Jeffers et al. 2001) and that have limited ability to survive or multiply in non-host-associated environments (Gray et al. 2004; Roberts et al. 2006). Evolution of *L. monocytogenes* strains and lineages likely represents adaptation of specific strains to different niches, including many that may remain to be defined (e.g., in alternate host species). An improved understanding of the evolution of different *L. monocytogenes* ecotypes will thus provide an opportunity to better understand the ecology and transmission of *L. monocytogenes*, including its reservoirs and hosts.

References

al-Ghazali MR, al-Azawi SK (1988a) Storage effects of sewage sludge cake on the survival of *Listeria monocytogenes*. J Appl Bacteriol 65:209–213

al-Ghazali MR, al-Azawi SK (1988b) Effects of sewage treatment on the removal of *Listeria monocytogenes*. J Appl Bacteriol 65:203–208

Allerberger F, Guggenbichler JP (1989) Listeriosis in Austria—report of an outbreak in 1986. Acta Microbiol Hung 36:149–152

Arvanitidou M, Papa A, Constantinidis TC, Danielides V, Katsouyannopoulos V (1997) The occurrence of *Listeria* spp. and *Salmonella* spp. in surface waters. Microbiol Res 152:395–397

Autio T, Hielm S, Miettinen M, Sjoberg AM, Aarnisalo K, Bjorkroth J, Mattila-Sandholm T, Korkeala H (1999) Sources of *Listeria monocytogenes* contamination in a cold-smoked rainbow trout processing plant detected by pulsed-field gel electrophoresis typing. Appl Environ Microbiol 65:150–155

Bernagozzi M, Bianucci F, Sacchetti R, Bisbini P (1994) Study of the prevalence of *Listeria* spp. in surface water. Zentralbl Hyg Umweltmed 196:237–244

Beuchat LR (1996) *Listeria monocytogenes* incidence on vegetables. Food Control 7:223–228

Beumer RR, te Giffel MC, Spoorenberg E, Rombouts FM (1996) *Listeria* species in domestic environments. Epidemiol Infect 117:437–442

Borucki MK, Reynolds J, Gay CC, McElwain KL, Kim SH, Knowles DP, Hu J (2004) Dairy farm reservoir of *Listeria monocytogenes* sporadic and epidemic strains. J Food Prot 67:2496–2499

Borucki MK, Gay CC, Reynolds J, McElwain KL, Kim SH, Call DR, Knowles DP (2005) Genetic diversity of *Listeria monocytogenes* strains from a high-prevalence dairy farm. Appl Environ Microbiol 71:5893–5899

Botzler RG, Cowan AB, Wetzler TF (1974) Survival of *Listeria monocytogenes* in soil and water. J Wildl Dis 10:204–212

Cai S, Kabuki DY, Kuaye AY, Cargioli TG, Chung MS, Nielsen R, Wiedmann M (2002) Rational design of DNA sequence-based strategies for subtyping *Listeria monocytogenes*. J Clin Microbiol 40:3319–3325

CDC (1999) Update: multistate outbreak of listeriosis—United States. MMWR 47: 1117–1118

Chasseignaux E, Toquin MT, Ragimbeau C, Salvat G, Colin P, Ermel G (2001) Molecular epidemiology of *Listeria monocytogenes* isolates collected from the environment, raw meat and raw products in two poultry- and pork-processing plants. J Appl Microbiol 91:888–899

Chasseignaux E, Gerault P, Toquin MT, Salvat G, Colin P, Ermel G (2002) Ecology of *Listeria monocytogenes* in the environment of raw poultry meat and raw pork meat processing plants. FEMS Microbiol Lett 210:271–275

Chen Y, Ross WH, Gray MJ, Wiedmann M, Whiting RC, Scott VN (2006) Attributing risk to *Listeria monocytogenes* subgroups: dose response in relation to genetic lineages. J Food Prot 69:335–344

Colburn KG, Kaysner CA, Abeyta CJ, Wekell MM (1990) *Listeria* species in a California coast estuarine environment. Appl Environ Microbiol 56:2007–2011

De Luca G ZF, Fateh-Moghadm P, Stampi S (1998) Occurrence of *Listeria monocytogenes* in sewage sludge. Zentralbl Hyg Umweltmed 201:269–277

De Roin MA, Foong SC, Dixon PM, Dickson JS (2003) Survival and recovery of *Listeria monocytogenes* on ready-to-eat meats inoculated with a desiccated and nutritionally depleted dustlike vector. J Food Prot 66:962–969

Dijkstra RG (1971) Investigations on the survival times of *Listeria* bacteria in suspensions of brain tissue, silage and faeces and in milk. Zentralbl Bakteriol [Orig] 216:92–95

Dijkstra RG (1978) Incidence of *Listeria monocytogenes* in the intestinal contents of broilers on different farms. Tijdschr Diergeneeskd 103:229–231

Donnelly CW (2002) Detection and isolation of *Listeria monocytogenes* from food samples: implications of sublethal injury. J AOAC Int 85:495–500

Eklund M, Poysky F, Paranjpye R, Lashbrook L, Peterson M, Pelroy G (1995) Incidence and sources of *Listeria monocytogenes* in cold smoking fishery products and processing plants. J Food Prot 58:502–508

Ericsson H, Eklow A, Danielsson-Tham M, Loncarevic S, Mentzing L, Persson I, Unnerstad H, Tham W (1997) An outbreak of listeriosis suspected to have been caused by rainbow trout. J Clin Microbiol 35:2904–2907

FAO/WHO (2004) Microbiological risk assessment series 5: risk assessment of *Listeria monocytogenes* in ready-to-eat foods. Available at: http://www.fao.org/documents

Farber JM, Peterkin PI (1991) *Listeria monocytogenes*, a food-borne pathogen. Microbiol Rev 55:476–511

Farber JM, Peterkin PI (1999) Incidence and behavior of *Listeria monocytogenes* in meat products. In: Ryser ET, Marth EH (eds) *Listeria*, listeriosis, and food safety. 2nd edn. Marcel Decker, New York, pp. 505–564

FDA/USDA (2003) Quanitative assessment of relative risk to public health from foodborne *Listeria monocytogenes* among selected categories of ready-to-eat foods. Available at: http://wwwfoodsafetygov/~dms/LMr2-tochtml

Fenlon DR (1985) Wild birds and silage as reservoirs of *Listeria* in the agricultural environment. J Appl Bacteriol 59:537–543

Fenlon DR (1986) Rapid quantitative assessment of the distribution of *Listeria* in silage implicated in a suspected outbreak of listeriosis in calves. Vet Rec 118: 240–242

Fenlon DR (1999) *Listeria monocytogenes* in the natural environment. In: Ryser ET, Marth EH (eds) *Listeria*, listeriosis, and food safety. 2nd edn. Marcel Decker, New York, pp. 21–37

Fenlon DR, Stewart T, Donachie W (1995) The incidence, numbers and types of *Listeria monocytogenes* isolated from farm bulk tank milks. Lett Appl Microbiol 20:57–60

Fenlon DR, Wilson J, Donachie W (1996) The incidence and level of *Listeria monocytogenes* contamination of food sources at primary production and initial processing. J Appl Bacteriol 81:641–650

Fonnesbech Vogel B, Huss HH, Ojeniyi B, Ahrens P, Gram L (2001) Elucidation of *Listeria monocytogenes* contamination routes in cold-smoked salmon processing plants detected by DNA-based typing methods. Appl Environ Microbiol 67: 2586–2595

Frances N, Hornby H, Hunter PR (1991) The isolation of *Listeria* species from fresh-water sites in Cheshire and North Wales. Epidemiol Infect 107:235–238

Garrec N, Picard-Bonnaud F, Pourcher AM (2003) Occurrence of *Listeria* sp. and *L. monocytogenes* in sewage sludge used for land application: effect of dewatering, liming and storage in tank on survival of *Listeria* species. FEMS Immunol Med Microbiol 35:275–283

Geuenich HH, Muller HE, Schretten-Brunner A, Seeliger HP (1985) The occurrence of different *Listeria* species in municipal waste water. Zentralbl Bakteriol Mikrobiol Hyg [B] 181:563–565

Gombas DE, Chen Y, Clavero RS, Scott VN (2003) Survey of *Listeria monocytogenes* in ready-to-eat foods. J Food Prot 66:559–569

Gottlieb SL, Newbern EC, Griffin PM, Graves LM, Hoekstra RM, Baker NL, Hunter SB, Holt KG, Ramsey F, Head M, Levine P, Johnson G, Schoonmaker-Bopp D, Reddy V, Kornstein L, Gerwel M, Nsubuga J, Edwards L, Stonecipher S, Hurd S, Austin D, Jefferson MA, Young SD, Hise K, Chernak ED, Sobel J, Group LOW (2006) Multistate outbreak of listeriosis linked to turkey deli meat and subsequent changes in US regulatory policy. Clin Infect Dis 42:29–36

Goulet V, Jacquet C, Vaillant V, Rebiere I, Mouret E, Lorente C, Maillot E, Stainer F, Rocourt J (1995) Listeriosis from consumption of raw-milk cheese. Lancet 345:1581–1582

Gravani R (1999) Incidence and control of *Listeria* in food-processing facilities. In: Ryser ET, Marth EH (eds) *Listeria,* listeriosis, and food safety. 2nd edn. Marcel Decker, New York, pp. 657–700

Gray MJ, Zadoks RN, Fortes ED, Dogan B, Cai S, Chen Y, Scott VN, Gombas DE, Boor KJ, Wiedmann M (2004) *Listeria monocytogenes* isolates from foods and humans form distinct but overlapping populations. Appl Environ Microbiol 70: 5833–5841

Gray ML, Killinger AH (1966) *Listeria monocytogenes* and listeric infections. Bacteriol Rev 30:309–382

Grif K, Patscheider G, Dierich MP, Allerberger F (2003) Incidence of fecal carriage of *Listeria monocytogenes* in three healthy volunteers: a one-year prospective stool survey. Eur J Clin Microbiol Infect Dis 22:16–20

Gronstol H (1979) Listeriosis in sheep. *Listeria monocytogenes* excretion and immunological state in healthy sheep. Acta Vet Scand 20:168–179

Gronstol H, Overas J (1980) Listeriosis in sheep. Eperythrozoon ovis infection used as a model to study predisposing factors. Acta Vet Scand 21:523–532

Gudmundsdottir K, Svansson V, Aalbaek B, Gunnarsson E, Sigurdarson S (2004) *Listeria monocytogenes* in horses in Iceland. Vet Rec 155:456–459

Hayashidani H, Kanzaki N, Kaneko Y, Okatani A, Taniguchi T, Kaneko K, Ogawa M (2002) Occurrence of yersiniosis and listeriosis in wild boars in Japan. J Wildl Dis 38:202–205

Ho JL, Shands KN, Friedland G, Eckind P, Fraser DW (1986) An outbreak of type 4b *Listeria monocytogenes* infection involving patients from eight Boston hospitals. Arch Intern Med 146:520–524

Hoffman AD, Gall KL, Norton DM, Wiedmann M (2003) *Listeria monocytogenes* contamination patterns for the smoked fish processing environment and for raw fish. J Food Prot 66:52–60

Hu Y, Gall K, Ho A, Ivanek R, Grohn YT, Wiedmann M (2006) Daily variability of *Listeria* contamination patterns in a cold-smoked salmon processing operations. J Food Prot 69:2123–2133

Husu JR (1990) Epidemiological studies on the occurrence of *Listeria monocytogenes* in the feces of dairy cattle. Zentralbl Veterinarmed B 37:276–282

ICMSF (1996) *Listeria monocytogenes*. In: Roberts TA, Baird-Parker AC, Tompkin RB (eds) Microorganisms in foods 5. 1st edn. Blackie Academic & Professional, London, pp. 141–182

Ivanek R, Grohn YT, Wiedmann M (2006) *Listeria monocytogenes* in multiple habitats and host populations: review of available data for mathematical modeling. Foodborne Pathog Dis 3:319–336

Jacquet C, Doumith M, Gordon JI, Martin PM, Cossart P, Lecuit M (2004) A molecular marker for evaluating the pathogenic potential of foodborne *Listeria monocytogenes*. J Infect Dis 189:2094–2100

Jeffers GT, Bruce JL, McDonough PL, Scarlett J, Boor KJ, Wiedmann M (2001) Comparative genetic characterization of *Listeria monocytogenes* isolates from human and animal listeriosis cases. Microbiology 147:1095–1104

Jiang X, Islam M, Morgan J, Doyle MP (2004) Fate of *Listeria monocytogenes* in bovine manure-amended soil. J Food Prot 67:1676–1681

Kabuki DY, Kuaye AY, Wiedmann M, Boor KJ (2004) Molecular subtyping and tracking of *Listeria monocytogenes* in latin-style fresh-cheese processing plants. J Dairy Sci 87:2803–2812

Kathariou S (2002) *Listeria monocytogenes* virulence and pathogenicity, a food safety perspective. J Food Prot 65:1811–1829

Kells J, Gilmour A (2004) Incidence of *Listeria monocytogenes* in two milk processing environments, and assessment of *Listeria monocytogenes* blood agar for isolation. Int J Food Microbiol 91:167–174

Kreft J, Vazquez-Boland JA, Ng E, Goebel W (1999) Virulence gene clusters and putative pathogenicity islands in Listeriae. In: Kaper JB, Hacker J (eds) Pathogenicity islands and other mobile virulence elements. 1st edn. ASM Press, Washington, DC, pp. 219–232

Lappi VR, Thimothe J, Nightingale KK, Gall K, Scott VN, Wiedmann M (2004) Longitudinal studies on *Listeria* in smoked fish plants: impact of intervention strategies on contamination patterns. J Food Prot 67:2500–2514

Lappi VR, Thimothe J, Walker J, Bell J, Gall K, Moody MW, Wiedmann M (2004) Impact of intervention strategies on *Listeria* contamination patterns in crawfish processing plants: a longitudinal study. J Food Prot 67:1163–1169

Lawrence LM, Gilmour A (1995) Characterization of *Listeria monocytogenes* isolated from poultry products and from the poultry-processing environment by random amplification of polymorphic DNA and multilocus enzyme electrophoresis. Appl Environ Microbiol 61:2139–2144

Linnan M, Mascola L, Lou X, Goulet V, May S, Salminen C, Hird D, Yonkura M, Hayes PS, Weaver RE, Audurier A, Plikaytis BD, Fannin S, Kleks A, Broome CV (1988) Epidemic listeriosis associated with Mexican-style cheese. N Engl J Med 319:823–828

Loken T, Aspoy E, Gronstol H (1982) *Listeria monocytogenes* excretion and humoral immunity in goats in a herd with outbreaks of listeriosis and in a healthy herd. Acta Vet Scand 23:392–399

Lozniewski A, Humbert A, Corsaro D, Schwartzbrod J, Weber M, Le Faou A (2001) Comparison of sludge and clinical isolates of *Listeria monocytogenes*. Lett Appl Microbiol 32:336–339

MacDonald F, Sutherland AD (1994) Important differences between the generation times of *Listeria monocytogenes* and *List. innocua* in two *Listeria* enrichment broths. J Dairy Res 61:433–436

MacGowan AP, Bowker K, McLauchlin J, Bennett PM, Reeves DS (1994) The occurrence and seasonal changes in the isolation of *Listeria* spp. in shop bought food stuffs, human faeces, sewage and soil from urban sources. Int J Food Microbiol 21:325–334

Markkula A, Autio T, Lunden J, Korkeala H (2005) Raw and processed fish show identical *Listeria monocytogenes* genotypes with pulsed-field gel electrophoresis. J Food Prot 68:1228–1231

McLauchlin J (1990) Distribution of serovars of *Listeria monocytogenes* isolated from different categories of patients with listeriosis. Eur J Clin Microbiol Infect Dis 9:210–213

Mead PS, Slutsker L, Dietz V, McCaig LF, Bresee JS, Shapiro C, Griffin PM, Tauxe RV (1999) Food-related illness and death in the United States. Emerg Infect Dis 5:607–625

Miettinen H, Wirtanen G (2005) Prevalence and location of *Listeria monocytogenes* in farmed rainbow trout. Int J Food Microbiol 104:135–143

Murray EGD, Webb RA, Swann MBR (1926) A disease of rabbits characterized by a large mononuclear leucocytosis, caused by a hitherto undescribed bacillus *Bacterium monocytogenes* (n.sp.). J Pathol Bacteriol 29:407–439

Nadon CA, Woodward DL, Young C, Rodgers FG, Wiedmann M (2001) Correlations between molecular subtyping and serotyping of *Listeria monocytogenes*. J Clin Microbiol 39:2704–2707

Nicholson FA, Groves SJ, Chambers BJ (2005) Pathogen survival during livestock manure storage and following land application. Bioresour Technol 96:135–143

Nightingale KK, Schukken YH, Nightingale CR, Fortes ED, Ho AJ, Her Z, Grohn YT, McDonough PL, Wiedmann M (2004) Ecology and transmission of *Listeria monocytogenes* infecting ruminants and in the farm environment. Appl Environ Microbiol 70:4458–4467

Nightingale KK, Fortes ED, Ho AJ, Schukken YH, Grohn YT, Wiedmann M (2005a) Evaluation of farm management practices as risk factors for clinical listeriosis and fecal shedding of *Listeria monocytogenes* in ruminants. J Am Vet Med Assoc 227:1808–1814

Nightingale KK, Windham K, Martin KE, Yeung M, Wiedmann M (2005b) Select *Listeria monocytogenes* subtypes commonly found in foods carry distinct nonsense mutations in *inlA*, leading to expression of truncated and secreted internalin A, and are associated with a reduced invasion phenotype for human intestinal epithelial cells. Appl Environ Microbiol 71:8764–8772

Nightingale KK, Windham K, Wiedmann M (2005) Evolution and molecular phylogeny of *Listeria monocytogenes* isolated from human and animal listeriosis cases and foods. J Bacteriol 187:5537–5551

Norton DM, McCamey MA, Gall KL, Scarlett JM, Boor KJ, Wiedmann M (2001) Molecular studies on the ecology of *Listeria monocytogenes* in the smoked fish processing industry. Appl Environ Microbiol 67:198–205

Norton DM, Scarlett JM, Horton K, Sue D, Thimothe J, Boor KJ, Wiedmann M (2001) Characterization and pathogenic potential of *Listeria monocytogenes* isolates from the smoked fish industry. Appl Environ Microbiol 67:646–653

Olsen SJ, Patrick M, Hunter SB, Reddy V, Kornstein L, MacKenzie WR, Lane K, Bidol S, Stoltman GA, Frye DM, Lee I, Hurd S, Jones TF, LaPorte TN, Dewitt W, Graves L, Wiedmann M, Schoonmaker-Bopp DJ, Huang AJ, Vincent C, Bugenhagen A, Corby J, Carloni ER, Holcomb ME, Woron RF, Zansky SM, Dowdle G, Smith F, Ahrabi-Fard S, Ong AR, Tucker N, Hynes NA, Mead P (2005) Multistate outbreak of *Listeria monocytogenes* infection linked to delicatessen turkey meat. Clin Infect Dis 40:962–967

Piffaretti JC, Kressebuch H, Aeschbacher M, Bille J, Bannerman E, Musser JM, Selander RK, Rocourt J (1989) Genetic characterization of clones of the bacterium *Listeria monocytogenes* causing epidemic disease. Proc Natl Acad Sci USA 86:3818–3822

Pritchard TJ, Flanders KJ, Donnelly CW (1995) Comparison of the incidence of *Listeria* on equipment versus environmental sites within dairy processing plants. Int J Food Microbiol 26:375–384

Ralovich B, Audurier A, Hajtos I, Berkessy E, Pitron-Szemeredi M (1986) Comparison of *Listeria* serotypes and phage types isolated from sheep, other animals and humans. Acta Microbiol Hung 33:9–17

Reij MW, Den Aantrekker ED (2004) Recontamination as a source of pathogens in processed foods. Int J Food Microbiol 91:1–11

Roberts AJ, Wiedmann M (2003) Pathogen, host and environmental factors contributing to the pathogenesis of listeriosis. Cell Mol Life Sci 60:904–918

Roberts A, Chan Y, Wiedmann M (2005) Definition of genetically distinct attenuation mechanisms in naturally virulence-attenuated *Listeria monocytogenes* by comparative cell culture and molecular characterization. Appl Environ Microbiol 71:3900–3910

Roberts A, Nightingale K, Jeffers G, Fortes E, Kongo JM, Wiedmann M (2006) Genetic and phenotypic characterization of *Listeria monocytogenes* lineage III. Microbiology 152:685–693

Roche SM, Gracieux P, Milohanic E, Albert I, Virlogeux-Payant I, Temoin S, Grepinet O, Kerouanton A, Jacquet C, Cossart P, Velge P (2005) Investigation of specific substitutions in virulence genes characterizing phenotypic groups of low-virulence field strains of *Listeria monocytogenes*. Appl Environ Microbiol 71:6039–6048

Rocourt J, Cossart P (1997) *Listeria monocytogenes*. In: Doyle MP, Beuchat LR, Montville TJ (eds) Food microbiology, fundamentals and frontiers. 1st edn. ASM Press, Washington, DC, pp. 337–352

Rorvik LM, Caugant DA, Yndestad M (1995) Contamination pattern of *Listeria monocytogenes* and other *Listeria* spp. in a salmon slaughterhouse and smoked salmon processing plant. Int J Food Microbiol 25:19–27

Rorvik LM, Skjerve E, Knudsen BR, Yndestad M (1997) Risk factors for contamination of smoked salmon with*Listeria monocytogenes* during processing. Int J Food Microbiol 37:215–219

Rorvik LM, Aase B, Alvestad T, Caugant DA (2003) Molecular epidemiological survey of *Listeria monocytogenes* in broilers and poultry products. J Appl Microbiol 94:633–640

Ryser E (1999) Listeriosis in animals. In: Ryser ET, Marth EH (eds) *Listeria,* listeriosis, and food safety. 2nd edn. Marcel Decker, New York, pp. 229–358

Ryser E (1999) Foodborne listeriosis. In: Ryser ET, Marth EH (eds) *Listeria,* listeriosis, and food safety. 2nd edn. Marcel Decker, New York, pp. 299–358

Ryser ET, Arimi SM, Bunduki MM, Donnelly CW (1996) Recovery of different *Listeria* ribotypes from naturally contaminated, raw refrigerated meat and poultry products with two primary enrichment media. Appl Environ Microbiol 62:1781–1787

Ryser ET, Arimi SM, Donnelly CW (1997) Effects of pH on distribution of *Listeria* ribotypes in corn, hay, and grass silage. Appl Environ Microbiol 63:3695–3697

Salcedo C, Arreaza L, Alcala B, de la Fuente L, Vazquez JA (2003) Development of a multilocus sequence typing method for analysis of *Listeria monocytogenes* clones. J Clin Microbiol 41:757–762

Samelis J, Metaxopoulous J (1999) Incidence and principal sources of *Listeria* spp. and *L. monocytogenes* contamination in processed meats and a meat processing plant. Food Microbiology 16:465–477

Sammarco ML, Ripabelli G, Fanelli I, Grasso GM (2005) Prevalence of *Listeria* spp. in dairy farm and evaluation of antibiotic-resistance of isolates. Ann Ig 17:175–183

Sanaa M, Poutrel B, Menard JL, Serieys F (1993) Risk factors associated with contamination of raw milk by *Listeria monocytogenes* in dairy farms. J Dairy Sci 76: 2891–2898

Sauders BD (2005) Molecular epidemiology, diversity, distribution, and ecology of *Listeria*. PhD Thesis, Cornell University, Ithaca, NY

Sauders BD, Wiedmann M (in press) Ecology of *Listeria* species and *L. monocytogenes* in the natural environment. In: Ryser ET, Marth EH (eds) *Listeria, listeriosis, and food safety*. 3rd edn. Marcel Decker, New York

Sauders BD, Fortes ED, Morse DL, Dumas N, Kiehlbauch JA, Schukken Y, Hibbs JR, Wiedmann M (2003) Molecular subtyping to detect human listeriosis clusters. Emerg Infect Dis 9:672–680

Sauders BD, Mangione K, Vincent C, Schermerhorn J, Farchione CM, Dumas NB, Bopp D, Kornstein L, Fortes ED, Windham K, Wiedmann M (2004) Distribution of *Listeria monocytogenes* molecular subtypes among human and food isolates from New York State shows persistence of human disease-associated *Listeria monocytogenes* strains in retail environments. J Food Prot 67:1417–1428

Sauders BD, Pettit D, Currie B, Suits P, Evans A, Stellrecht K, Dryja DM, Slate D, Wiedmann M (2005) Low prevalence of *Listeria monocytogenes* in human stool. J Food Prot 68:178–181

Sauders BD, Durak MZ, Fortes E, Windham K, Schukken Y, Lembo AJ, Jr, Akey B, Nightingale KK, Wiedmann M (2006) Molecular characterization of *Listeria monocytogenes* from natural and urban environments. J Food Prot 69:93–105

Schlech III WF, Lavigne PM, Bortolussi RA, Allen AC, Haldane EV, Wort AJ, Hightower AW, Johnson SE, King SH, Nicholls ES, Broome CV (1983) Epidemic listeriosis—evidence for transmission by food. N Engl J Med 308:203–206

Schlech III WF, Schlech IV WF, Haldane H, Mailman TL, Warhuus M, Crouse N, Haldane DJM (2005) Does sporadic *Listeria* gastroenteritis exist? A 2-year population-based survey in Nova Scotia, Canada. Clin Infect Dis 41:778–784

Schuchat A, Deaver K, Hayes PS, Graves L, Mascola L, Wenger JD (1993) Gastrointestinal carriage of *Listeria monocytogenes* in household contacts of patients with listeriosis. J Infect Dis 167:1261–1262

Sim J, Hood D, Finnie L, Wilson M, Graham C, Brett M, Hudson JA (2002) Series of incidents of *Listeria monocytogenes* non-invasive febrile gastroenteritis involving ready-to-eat meats. Lett Appl Microbiol 35:409–413

Skovgaard N, Morgen CA (1988) Detection of *Listeria* spp. in faeces from animals, in feeds, and in raw foods of animal origin. Int J Food Microbiol 6:229–242

Thimothe J, Nightingale KK, Gall K, Scott VN, Wiedmann M (2004) Tracking of *Listeria monocytogenes* in smoked fish processing plants. J Food Prot 67:328–341

Tompkin RB (2002) Control of *Listeria monocytogenes* in the food-processing environment. J Food Prot 65:709–725

Unnerstad H, Romell A, Ericsson H, Danielsson-Tham ML, Tham W (2000) *Listeria monocytogenes* in faeces from clinically healthy dairy cows in Sweden. Acta Vet Scand 41:167–171

Van Coillie E, Werbrouck H, Heyndrickx M, Herman L, Rijpens N (2004) Prevalence and typing of *Listeria monocytogenes* in ready-to-eat food products on the Belgian market. J Food Prot 67:2480–2487

Vazquez-Boland JA, Kuhn M, Berche P, Chakraborty T, Dominguez-Bernal G, Goebel W, Gonzalez-Zorn B, Wehland J, Kreft J (2001) *Listeria* pathogenesis and molecular virulence determinants. Clin Microbiol Rev 14:584–640

Vela AI, Fernandez-Garayzabal JF, Vazquez JA, Latre MV, Blanco MM, Moreno MA, de La Fuente L, Marco J, Franco C, Cepeda A, Rodriguez Moure AA, Suarez G, Dominguez L (2001) Molecular typing by pulsed-field gel electrophoresis of Spanish animal and human *Listeria monocytogenes* isolates. Appl Environ Microbiol 67:5840–5843

Ward TJ, Gorski L, Borucki MK, Mandrell RE, Hutchins J, Pupedis K (2004) Intraspecific phylogeny and lineage group identification based on the prfA virulence gene cluster of *Listeria monocytogenes*. J Bacteriol 186:4994–5002

Watkins J, Sleath KP (1981) Isolation and enumeration of *Listeria monocytogenes* from sewage, sewage sludge and river water. J Appl Bacteriol 50:1–9

Weber A, Potel J, Schafer-Schmidt R, Prell A, Datzmann C (1995) Studies on the occurrence of *Listeria monocytogenes* in fecal samples of domestic and companion animals. Zentralbl Hyg Umweltmed 198:117–123

Weis J, Seeliger HP (1975) Incidence of *Listeria monocytogenes* in nature. Appl Microbiol 30:29–32

Welshimer HJ (1960) Survival of *Listeria monocytogenes* in soil. J Bacteriol 80:316–320

Welshimer HJ (1968) Isolation of *Listeria monocytogenes* from vegetation. J Bacteriol 95:300–303

Welshimer HJ, Donker-Voet J (1971) *Listeria monocytogenes* in nature. Appl Microbiol 21:516–519

Wesley IV (1999) Listeriosis in Animals. In: Ryser ET, Marth EH (eds) *Listeria,* listeriosis, and food safety. 2nd edn. Marcel Decker, New York, pp. 39–73

Wiedmann M (2002a) Molecular subtyping methods for *Listeria monocytogenes*. J AOAC Int 85:524–531

Wiedmann M (2002b) Subtyping technologies for bacterial foodborne pathogens. Nutr Rev 60:201–208

Wiedmann M, Czajka J, Bsat N, Bodis M, Smith MC, Divers TJ, Batt CA (1994) Diagnosis and epidemiological association of *Listeria monocytogenes* strains in two outbreaks of listerial encephalitis in small ruminants. J Clin Microbiol 32:991–996

Wiedmann M, Bruce JL, Knorr R, Bodis M, Cole EM, McDowell CI, McDonough PL, Batt CA (1996) Ribotype diversity of *Listeria monocytogenes* strains associated with outbreaks of listeriosis in ruminants. J Clin Microbiol 34:1086–1090

Wiedmann M, Arvik T, Bruce JL, Neubauer J, del Piero F, Smith MC, Hurley J, Mohammed HO, Batt CA (1997) Investigation of a listeriosis epizootic in sheep in New York state. Am J Vet Res 58:733–737

Wiedmann M, Bruce J, Keating C, Johnson A, McDonough P, Batt C (1997) Ribotypes and virulence gene polymorphisms suggest three distinct *Listeria monocytogenes* lineages with differences in pathogenic potential. Infect Immun 65:2707–2716

Windham K, Nightingale K, Wiedmann M (2005) Molecular evolution and diversity of foodborne pathogens. In: Shetty K, Pometto A, Paliyath G (eds) Food biotechnology. CRC Press, Boca Raton, pp. 1259–1291

Yokoyama E, Saitoh T, Maruyama S, Katsube Y (2005) The marked increase of *Listeria monocytogenes* isolation from contents of swine cecum. Comp Immunol Microbiol Infect Dis 28:259–268

Yu LS, Fung DY (1993) Five-tube most-probable-number method using the Fung-Yu tube for enumeration of *Listeria monocytogenes* in restructured meat products during refrigerated storage. Int J Food Microbiol 18:97–106

Abbreviations

L. monocytogenes	*Listeria monocytogenes*
L. innocua	*Listeria innocua*
L. seeligeri	*Listeria seeligeri*
L. ivanovii	*Listeria ivanovii*
MPN	most probable number
CFU	colony forming unit
MLEE	multilocus enzyme electrophoresis
PFGE	pulsed-field gel electrophoresis
MLST	multilocus sequence-based typing
spp.	species
EC	epidemic clone
RTE	ready-to-eat

7
Regulation of *Listeria monocytogenes* Virulence Genes

Maurine D. Miner, Gary C. Port, and Nancy E. Freitag

Seattle Biomedical Research Institute, University of Washington, 307 Westlake Ave N., Seattle, WA 98109-5219, USA

e-mail: nancy.freitag@sbri.org

Abstract: *Listeria monocytogenes* is a gram-positive bacterium that lives within soil and decaying plant material but is also capable of transitioning into a deadly pathogen following ingestion by mammals. The bacterium makes the transition from outside environment to host via the coordinate regulation of virulence gene products that enable bacterial replication within host cells. PrfA is a transcriptional activator that is required for the expression of nearly all identified *L. monocytogenes* virulence gene products. The expression of the *prfA* gene and the activity of the PrfA protein are carefully regulated by multiple mechanisms within *L. monocytogenes*. Two promoters function to provide the initial levels of PrfA that direct bacterial escape from host cell vacuoles, whereas a third promoter contributes to high level expression of *prfA* to promote spread of intracellular bacteria to adjacent host cells. The synthesis of PrfA protein is temperature-regulated such that *prfA* mRNA secondary structure prevents protein translation at low temperatures. An additional mechanism of control exists to regulate PrfA activity as PrfA appears to require the binding of a cofactor for full activation. Strains containing PrfA mutant proteins that are locked into a PrfA-activated state are fully virulent in animal models of infection, but are compromised for bacterial fitness outside of the host. It therefore appears that *L. monocytogenes* must maintain a balance between life in the host and life in the outside environment. Analyses of the *L. monocytogenes* genome sequence and microarray analysis indicate that additional gene products are subject to PrfA regulation either via direct interaction with PrfA or indirectly as the result of another PrfA-activated factor.

7.1.   Introduction

Regulated expression of virulence genes appears to be crucial for pathogens that encounter multiple environments during their lifecycles. This may be particularly true for environmental pathogens such as *Listeria monocytogenes* which must adapt to life in the outside world as well as to life within host cells. For some environmental bacterial pathogens, such as *Vibrio cholera* and *Legionella pneumophila* (Hammer and Swanson 1999; Krukonis and DiRita 2003), specific signals have been identified that serve as triggers for virulence gene expression within the host. Regulation of bacterial gene expression in response to changes in temperature, the presence or absence of metal ions, amino acid availability, and iron are some examples of signals used to identify the host environment from other environments for different pathogens (Garcia et al. 1996; Hammer and Swanson 1999; Konkel and Tilly 2000; Krukonis and DiRita 2003) and to direct the production of virulence gene products. How does the soil-loving *L. monocytogenes* mediate its transition from the dirt or drain pipe to the host? What changes in gene expression allow this bacterium to transform from a sedate life in silage into a potentially lethal assailant in humans?

Transcriptional regulation of gene expression represents an important mechanism for bacterial adaptation to new environments. Of the 201 putative transcriptional regulators identified via bioinformatics analyses of the *L. monocytogenes* genome sequence (Glaser et al. 2001), positive regulatory factor A (PrfA) stands out for its central and essential role in regulating the expression of virulence gene products. PrfA was first identified as a regulatory factor required for *hly* transcription (Leimeister-Wachter et al. 1990), and it has since been shown to regulate the expression of a growing number of bacterial gene products directly associated with virulence. The absolute requirement of PrfA for *L. monocytogenes* pathogenesis was demonstrated utilizing strains with deletions or loss-of-function mutations within the *prfA* gene: such strains failed to replicate in the cytosol of host cells or to spread to adjacent cells, and were severely attenuated for virulence in animal models of infection (Leimeister-Wachter et al. 1990; Mengaud et al. 1991; Freitag et al. 1993). As PrfA is a critical component for the regulated expression of virulence factors in *L. monocytogenes*, it is not surprising that the expression and production of this regulator is controlled at multiple levels, including transcriptional, posttranscriptional, and post translational mechanisms of regulation. This chapter will provide a brief overview of what is currently known regarding how *L. monocytogenes* controls the expression and activity of its central virulence regulator PrfA and discuss how this regulation impacts the expression of bacterial virulence factors. In addition to PrfA-dependent control mechanisms, recently identified regulatory elements that contribute to virulence gene expression in *L. monocytogenes* will also be described.

7.2. Regulation of PrfA Activity

7.2.1. The First Step in Regulating the Regulator: Transcriptional Control of prfA Expression

Transcriptional control of the *prfA* gene is mediated by three separate promoter elements (Figure 7.1A). Two promoters, *prfA*P1 and *prfA*P2, are located directly upstream of the *prfA* translation initiation codon and are separated by

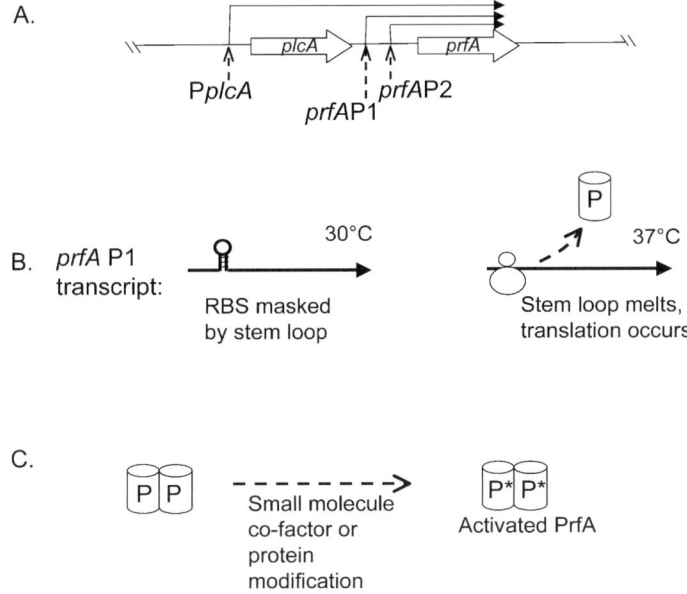

FIGURE 7.1. *Transcriptional, post-transcriptional, and post-translational modes of PrfA regulation. prfA transcripts are indicated with solid arrows.* **A** Transcriptional regulation of *prfA*. The three promoters that contribute to *prfA* expression are indicated (P*plcA*, *prfA*P1, and *prfA*P2). PrfA positively regulates its expression by directly activating expression at the P*plcA* promoter resulting in the generation of a *plcA-prfA* bicistronic transcript. PrfA negatively influences its own expression via elements within the *prfA*P2 promoter, but it is not known if this effect is direct or indirect. **B** Post-transcriptional control of *prfA* expression. The transcript originating from the *prfA*P1 promoter contains a region that forms a stem-loop structure which masks the ribosome-binding site (RBS) at temperatures below 30°C. At 37°C the stem-loop is destabilized, allowing ribosome binding and translation of the transcript. The cylinder labeled "P" represents PrfA protein. **C.** Post-translational regulation of PrfA activity. PrfA exists in a low-activity state until it presumably binds to an unknown cofactor or is somehow modified. Activated PrfA protein (cylinders with "P*") is required for optimal expression of *L. monocytogenes* virulence genes

approximately 80 nucleotides (Freitag et al. 1993). *prfA*P1 and *prfA*P2 each direct monocistronic transcripts of *prfA*. The third promoter that contributes to *prfA* expression, the *plcA* promoter, leads to the production of a bicistronic *plcA-prfA* transcript, as well as a monocistronic *plcA* transcript (Camilli et al. 1993; Freitag and Portnoy 1994). The *prfA*P1 and *prfA*P2 promoters are required to generate the initial levels of PrfA protein that can subsequently activate expression from the *plcA* promoter and increase overall levels of *prfA* expression. The *prfA*P1 and *prfA*P2 promoters are required for the efficient escape of *L. monocytogenes* from host cell vacuoles (Freitag and Portnoy 1994). The bicistronic *plcA-prfA* transcript provides for an additional increase of PrfA within the cytosol and is required for spread of intracellular bacteria to adjacent host cells (Camilli et al. 1993). Although the *prfA*P1 and *prfA*P2 promoters appear to be functionally redundant in vivo, they are differentially regulated in vitro (Freitag and Portnoy 1994). The *prfA*P2 promoter has been recently shown to be transcribed by RNA polymerase associated with the stress response sigma factor σ^B (Schwab et al. 2005). This regulation may permit the expression of *prfA* during periods of bacterial stress. In contrast to the *plcA* promoter which is up-regulated by PrfA, the expression of both *prfA*P1 and *prfA*P2 is negatively influenced by PrfA, although it is unclear as to whether PrfA has a direct or indirect role in the down-regulation of these two promoters (Freitag et al. 1993). Negative regulation of *prfA* expression by PrfA does not appear to be essential for virulence in a mouse model of infection (Greene and Freitag 2003) but it is possible that it serves a role outside of the host.

7.2.2. The Second Step in Regulation of the Regulator: Post-transcriptional Regulation of prfA

As mentioned above, changes in temperature have been shown to modulate the expression of virulence genes for several bacterial pathogens (Konkel and Tilly 2000). Temperature has been shown to significantly influence the expression of virulence factors in *L. monocytogenes*, and it appears that this is accomplished (at least in part) via transcripts directed by the *prfA*P1 promoter. Johansson et al. (Johansson et al. 2002) have demonstrated that the 5' untranslated region (UTR) of the *prfA*P1-directed mRNA forms a secondary structure at temperatures of 30°C or lower that masks the ribosome-binding region of *prfA* and inhibits *prfA* mRNA translation (Figure 7.1.B). This structure becomes unstable at higher temperatures thus enabling translation of *prfA* mRNA and for the production of PrfA at temperatures of 37°C or higher. In contrast, the *plcA* and *prfA*P2 promoters do not appear subject to this type of thermoregulation, and therefore are likely to contribute to the expression of some PrfA-dependent virulence genes at temperatures at or below 30° C (for example, within cultured insect cells and fly larvae at 25°C) (Cheng and Portnoy 2003; Mansfield et al. 2003). These data indicate that while temperature may be an important environmental signal mediating the expression of *L. monocytogenes* virulence factors, additional signals that promote expression must exist within host cells.

7.2.3. The Third Step in Regulation of the Regulator: Post-translational Control of PrfA

A third and perhaps most critical mechanism exists for control of PrfA activity: post-translational regulation. The initial evidence which suggested that PrfA protein activity was regulated on a post-translational level followed the experimental observation that PrfA-dependent gene expression was reduced in the presence of readily metabolized carbon sources, such as glucose and cellobiose, yet there was no significant change in the amount of PrfA protein present in the bacterial cell (Renzoni et al. 1997). Recognition of PrfA as a member of the Crp/Fnr family of transcriptional regulators, a family of approximately 400 members (Korner et al. 2003), provided further support for the supposition that a small molecule cofactor or some form of post-translational modification is required to convert PrfA to an active state (Vega et al. 1998) (Figure 7.1.C). Members of the Crp/Fnr family are activated either via interactions with small molecules (for example, Crp with cAMP) or through changes in the presence of prosthetic groups that interact with a signal (such as the binding of carbon monoxide by the heme moiety of CooA) (Korner et al. 2003).

Additional evidence supporting the hypothesis that PrfA requires post-translational modification for full activity followed the identification of a *L. monocytogenes* strain that contained a single mutation within *prfA* coding sequences that resulted in constitutive expression of PrfA-dependent virulence genes in broth culture (Ripio et al. 1997b). The substitution of a serine for a glycine at position 145 within PrfA was found to be similar to mutations identified within Crp (Crp* mutants) that enabled Crp to act as an activator in the absence of its cAMP cofactor (Vega et al. 1998). Crp* mutations alter protein conformation in ways that mimic the allosteric changes that occur following the binding of cAMP and that result in increased DNA-binding affinity for the wild-type protein (Harman 2001). The PrfA G145S protein was subsequently shown to have an approximately 18-fold increase in binding affinity for the *hly* promoter in comparison with wild-type PrfA, suggesting that a conformational change had occurred in PrfA that was likely similar in nature to those conferring the Crp* phenotype (Eiting et al. 2005). Since the initial identification of the PrfA G145S mutant, several other mutationally activated forms of PrfA (or PrfA* mutants) have been identified. PrfA E77K and PrfA G155S were identified by Shetron-Rama et al. (2003) following ethyl methane sulfonate mutagenesis of *L. monocytogenes* and the selection of bacterial isolates that exhibited increased levels of *actA* expression following growth on culture plates. Strains containing PrfA E77K or PrfA G155S were found to have levels of *actA* expression that were 90-fold and 270-fold greater (respectively) than the levels of expression observed for wild-type bacteria grown in broth culture and also produced increased levels of the PrfA-dependent gene products LLO, PlcA, PlcB,InlA, and InlB (Shetron-Rama et al. 2003). Similar to PrfA G145S strains, PrfA E77K and PrfA G155S strains were resistant to the repression of PrfA-dependent gene expression by readily metabolized sugars (Shetron-Rama et al. 2003).

Interestingly, despite being locked into an activated state, none of the PrfA* mutant strains characterized to date has exhibited any virulence defect in animal models of infection. Both PrfA G145S and PrfA E77K strains are fully virulent and strains containing the PrfA G155S mutation appear enhanced for virulence, such that 5- to 10-fold fewer bacteria are required for 50% mortality following intravenous infection of mice (Ripio et al. 1996; Shetron-Rama et al. 2003). These results suggest that once PrfA becomes activated during infection, no additional modulation of protein activity is required to promote virulence—once the switch occurs, it is full steam ahead for bacteria within the host. Why then do not more isolates of *L. monocytogenes* display a PrfA-activated phenotype? The answer may lie in the environmental lifestyle of *L. monocytogenes* and the fact that, to be successful, this bacterium must survive in the outside environment as well as within the host. Strains containing activated PrfA have been shown to be severely compromised for flagella-mediated swimming motility, a trait that would appear to be crucial for nutrient acquisition outside of host cells (Shetron-Rama et al. 2003; Wong and Freitag 2004).

Two additional mutations that confer PrfA*-like activity have been recently identified. PrfA I45S strains had increased hemolysin and lecithinase levels in vitro, and exhibited resistance to catabolite and low-temperature repression of PrfA-regulated genes (Vega et al. 2004). PrfA L140F, identified from the same EMS-based mutant selection that yielded PrfA E77K and PrfA G155S (Wong and Freitag 2004), as well as in an independent screen (Vega et al. 2004), increased levels of *actA* expression over 200-fold in comparison to wild type ((Wong and Freitag 2004) and M. Miner, unpublished results) and also increased production of LLO and PlcB. In contrast to the PrfA G145S, PrfA E77K, and PrfA G155S mutations, PrfA L140F appeared somewhat toxic to *L. monocytogenes* in that the usual allelic exchange methods used to produce an isogenic *prfA* L140F strain were repeatedly unsuccessful. Instead, reconstitution of the PrfA L140F mutation required the insertion of the pPL2 plasmid integration vector containing the *prfA* L140F allele into the chromosome of a *prfA* deletion strain and continuous antibiotic selection for the pPL2-encoded *cat* gene (Lauer et al. 2002; Wong and Freitag 2004). Interestingly, when the PrfA L140F mutation was integrated in the presence of the wild-type chromosomal copy of *prfA*, the mutant form of the protein was found to be completely dominant to the wild type in stimulating high levels of *actA* expression (Wong and Freitag 2004). Because the active form of PrfA is believed to be a dimer, the dominance of PrfA L140F suggests that either the mutant protein is capable of forming dimers with the wild-type protein and converting the wild type to an activated state, or alternatively that dimers of PrfA L140F may form more readily and out-compete the wild-type dimers for promoter-binding sites. The apparent toxicity of the PrfA L140F allele for *L. monocytogenes* is not yet understood.

7.2.3.1.   Structural Locations of PrfA* Mutations

On the amino acid level, PrfA shares relatively low (about 20%) identity with Crp (Korner et al. 2003; Eiting et al. 2005). The three-dimensional structure of PrfA was recently determined, facilitating structural and functional comparisons with Crp (Figure 7.2) (Eiting et al. 2005). PrfA shares extensive structural similarity with Crp, including an N-terminal beta-sheet domain, C-terminal helix-turn-helix (HTH) domain, and an extended alpha-helix connecting both domains (Eiting et al. 2005). The structure of the PrfA G145S mutant was also determined, and the authors found that the G145S substitution led to repositioning of the DNA-binding HTH domain of PrfA similar to the shift observed for the HTH

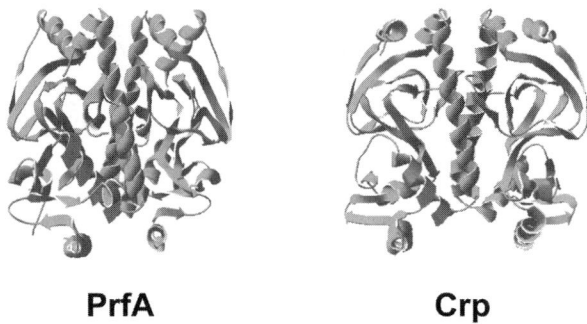

PrfA   **Crp**

FIGURE 7.2. *Comparison of the crystal structures of PrfA and Crp.* Both PrfA and Crp are shown as dimers; the long alpha helical domain of each monomer that mediates dimer interactions are visible in the center of the structures. Protein structures are adapted from Eiting et al. (2005).

domain of activated Crp bound to cAMP, a structural alteration that would seem to correlate with the increased DNA-binding affinity observed for the PrfA G145S mutant (Eiting et al. 2005) (Figure 7.2).

As evident from their amino acid positions and as seen in Figure 7.3, the PrfA G155S and PrfA L140F mutations are located in regions near the PrfA G145S mutation. It is anticipated that these mutations may induce a PrfA conformational change similar to that observed for PrfA G145S. Interestingly, PrfA E77K and

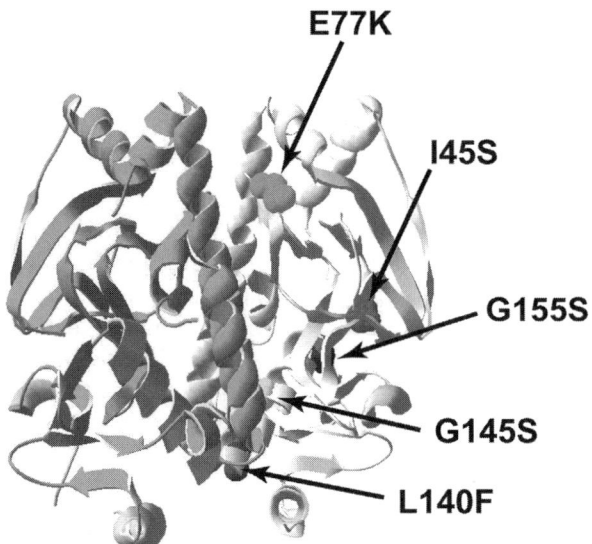

FIGURE 7.3. *Location of PrfA* mutations on protein crystal structure.* The PrfA homodimer is shown with one monomer in grey and one monomer in light blue. Each PrfA* mutation is indicated by an arrow. (A color version of this figure appears between pages 196 and 197.)

PrfA I45S in comparison map to locations within the protein that are distal to the HTH region, and may therefore alter different aspects of PrfA enzymatic function. Structural and functional comparisons of these mutant proteins are currently underway.

7.2.4. Additional Environmental Stimuli That May Influence PrfA Activity

As discussed above, the potential PrfA cofactor or post-translational modification of PrfA remains unknown. Based on the observation that readily metabolized carbon sources repress PrfA-dependent gene expression, Herro et al (2005) recently explored whether components of carbon catabolite repression might influence PrfA activity. Using a heterologous *Bacillus subtilis*-based expression system, the authors found that the accumulation of the serine phosphorylated form of the catabolite co-repressor HPr (P-Ser-HPr) strongly inhibited PrfA-dependent transcriptional activation. It was suggested that the presence of phosphoenolpyruvate (PEP):carbohydrate phosphotransferase system (PTS) transported sugars led to an increase in the amount of P-Ser-HPr present within *L. monocytogenes*, and P-Ser-HPr in turn inhibited PrfA from activating transcription at target gene promoters. It remains unclear whether inhibition of PrfA activity is mediated through direct interactions between P-Ser-HPr and PrfA or via PrfA interactions with another PTS component.

Two additional loci have been identified that contribute to the repression of PrfA-dependent gene expression via carbon sources. The *bvr* locus (Brehm et al. 1999) and the *csr* locus (Milenbachs et al. 2004) contribute to the repression of PrfA activity by β-glucosides such as cellobiose and arbutin; these sugars are abundant in silage, a common environmental source of *L. monocytogenes*. The *bvr* locus has been postulated to encode an environmental sensor of β-glucosides (Brehm et al. 1999), whereas the *csr* locus appears to encode a transport system for β-glucosides (Milenbachs et al. 2004). It has been postulated that *csr* and *bvr* influence PrfA-dependent gene expression via a pathway that is distinct but interconnected with the regulation of virulence gene expression by the general carbon catabolite repression pathway (Milenbachs et al. 2004).

The link between carbon source availability and PrfA-dependent virulence gene expression in *L. monocytogenes* is intriguing, but the precise mechanistic basis of this link remains elusive. It has been hypothesized that *L. monocytogenes* may determine its environmental location by sensing sugars: for example, plant sugars or β-glucosides to detect its presence in the outside environment and phosphorylated sugars such as glucose-6-phosphate or glucose-1-phosphate, present within the cytosol of infected mammalian cells (Ripio et al. 1997a; Chico-Calero et al. 2002). Unlike glucose or cellobiose, glucose-6-phosphate does not lead to the repression of PrfA-dependent gene expression (Chico-Calero et al. 2002). That *L. monocytogenes* is capable of distinguishing distinct environments within the host cell is clear; when the bacteria reach the cytosol of host cells, PrfA-dependent expression of *actA* increases more than 200-fold

(Moors et al. 1999; Shetron-Rama et al. 2002). Recently, reactive oxygen intermediates have also been reported to enhance PrfA-dependent gene expression in murine macrophage cell lines (Makino et al. 2005).

Much has thus been illuminated regarding how the expression and the activity of PrfA, the central regulatory factor of *L. monocytogenes*, is regulated and yet much still remains unknown. The environmental signals that trigger PrfA activation remain unidentified, but evidence suggests a link with the availability of specific carbon sources. The existence of multiple regulatory check-points to control PrfA activity would seem to indicate that regulation must be crucial for bacterial survival within the host and in the outside environment. Having established that modulation of PrfA activity occurs and is important, it is time to turn to the cast of gene products that make up the PrfA regulon.

7.3.   Touched by the Regulator: Gene Products Whose Expression is Modulated by PrfA

As mentioned, *L. monocytogenes* is a bacterium with the potential for a diverse and disparate life-style. From a saprophytic life in silage to invasion and infiltration of a mammalian host, PrfA serves as a key switch point for transitioning the bacterium from environment to host. Most attention thus far in the literature has focused on the identification of PrfA-regulated gene products that are expressed following bacterial entry into the host. Indeed, PrfA contributes to the expression of virulence factors required for all facets of *L. monocytogenes'* life within the host, including invasion of host cells, vacuolar escape, cytosolic replication, and bacterial spread to adjacent cells (Kreft and Vazquez-Boland 2001). A brief description of PrfA-dependent gene products that have been well characterized for their roles in *L. monocytogenes* pathogenesis is discussed below.

PrfA-dependent gene products that have been shown to contribute to bacterial invasion of mammalian cells include the internalins InlA and InlB that mediate attachment to host cell receptors and promote bacterial uptake by hijacking receptor-mediated endocytosis mechanisms (Cabanes et al. 2002). *inlC*, which encodes a small, secreted internalin-like protein, is also induced by PrfA although its contribution to *L. monocytogenes* invasion remains less well defined (Engelbrecht et al. 1996; Lingnau et al. 1996). The *actA* gene product, whose expression is entirely dependent upon PrfA, has also been shown to contribute to bacterial invasion when expressed at high levels (Mueller and Freitag 2005; Suarez et al. 2005).

Following uptake, three PrfA-dependent gene products have been shown to promote disruption of the host cell vacuolar membrane and bacterial entry into the cytosol:listeriolysin O (LLO), a pore-forming hemolysin encoded by *hly* (Portnoy et al. 1988; Mengaud et al. 1989) and two phospholipases encoded by the *plcA* and *plcB* genes (Leimeister-Wachter et al. 1991; Mengaud et al. 1991; Marquis et al. 1995). The *plcB* gene product is secreted as an inactive precursor that requires activation via proteolytic cleavage by another PrfA-regulated gene product, Mpl, a zinc-dependent metalloprotease (Poyart et al. 1993) (or alternatively by host cell cysteine proteases within the vacuole (Marquis et al. 1997)). Following entry into the cytosol, *L. monocytogenes* expresses the PrfA-dependent gene product encoded by *hpt*, a hexose-phosphate transporter, that facilitates the uptake of glucose-1-phosphate as well as other hexose phosphates (Chico-Calero et al. 2002). Host actin-based intracellular bacterial motility is also regulated by PrfA via control of the expression of the *actA* gene product (Kocks et al. 1992). A threshold level of ActA must be produced for motility to occur (Wong et al. 2004), and overexpression of the gene product appears detrimental, thus indicating that expression of *actA* must be carefully regulated for effective movement through the cytosol.

In addition to gene products that facilitate host cell invasion, cytosolic bacterial replication, and spread of *L. monocytogenes* to surrounding cells, PrfA has also been found to regulate the expression of gene products that contribute to bile resistance, including the bile salt hydrolase gene or *bsh* (Dussurget et al. 2002; Begley et al. 2005) and a bile exclusion system (*bilE*) (Sleator et al. 2005). This is particularly interesting in light of the fact that the murine gall bladder, where bile salts are stored before release into the duodenum, was recently found to be an unexpected niche for *L. monocytogenes* in live animals (Hardy et al. 2004). PrfA thus contributes to multiple facets of *L. monocytogenes* pathogenesis, and it is therefore no surprise that mutations that alter the activity of this key regulator have pleiotropic effects on virulence.

7.3.1. Characteristics of PrfA Target Promoters: The PrfA Box

PrfA functions as a transcriptional activator via recognition of a palindromic DNA-binding site located in the –40 region of target promoters (Kreft and Vazquez-Boland 2001) (Table 7.1). PrfA recognizes a 14 bp palindromic repeat known as the PrfA box with a consensus sequence of TTAACAnnTGTTAA. DNA footprinting experiments indicate that PrfA binding protects from digestion an approximate 26 bp region, beginning 10 bp upstream and ending 2 bp downstream of the 14 bp PrfA box (Dickneite et al. 1998). Electrophoretic mobility shift assays and experiments using a *B. subtilis* heterologous expression system have indicated that PrfA has a higher binding affinity to perfect palindromic sequences (such as the PrfA box of *hly*) than those with one or two mismatches (such as the PrfA box of *inlA*) (Sheehan et al. 1995; Dickneite et al. 1998). However, the substitution of a perfect PrfA DNA-binding site for an imperfect palindrome does not necessarily lead to an increase in PrfA activation of transcription from that promoter; for example, the replacement of *actA*'s imperfect PrfA box with the perfect palindrome of *hly* resulted in no increase in *actA* expression (Shetron-Rama et al. 2002). This observation in conjunction with the identification of mutations outside of *prfA* that influence PrfA-dependent gene expression (Shetron-Rama et al. 2003; Milenbachs et al. 2004) indicates that additional factors may be required for full activation of some PrfA-dependent

TABLE 7.1. Location and position of PrfA boxes for *L. monocytogenes* genes known to be directly regulated by PrfA

Chromosome location	Gene number	Gene name	Gene product	PrfA box ITAACAnnTGTTAA	Location [a]	Reference
LIPI-1[b]	lmo0201	*plcA*[c,d]	Phosphatidylinositol-specific phospholipase c	TTAACAAATGTTAA	−41	Mengaud et al. 1989
	lmo0200	*prfA*[c]	Listeriolysin positive regulatory protein	TTAACAAATGTTAA		
		prfA[e]		CTAACATTTGTTGT	−40 or −39	Freitag et al. 1993
	lmo0202	*hly*[d]	Listeriolysin O	TTAACATTTGTTAA	−42	Mengaud et al. 1989
	lmo0203	*mpl*	Zinc metalloproteinase precursor	TTAACAAATGTAAA	−41	Mengaud et al. 1989
	lmo0204	*actA*[c]	Actin assembly-inducing protein	TTAACAAATGTTAG	−41	Vazquez-Boland et al. 1992
	lmo0205	*plcB*[c]	Phospholipase C	TTAACAAATGTTAG		

(*Continued*)

TABLE 7.1. (*Continued*)

Chromosome location	Gene number	Gene name	Gene product	PrfA box TTAACAnnTGTTAA	Location [a]	Reference
Outside LIPI-1	lmo0433	inlA[c]	Internalin A	ATAACATAAGTTAA	−41	Lingnau et al. 1995
	lmo0434	inlB[c]	Internalin B	ATAACATAAGTTAA		Lingnau et al. 1995
	lmo1786	inlC	Internalin C	TTAACGCTTGTTAA	−39	Engelbrecht et al. 1996; Luo et al. 2004
					−41	Domann et al. 1997
	lmo0838	hpt	Hexose phosphate transport protein	ATAACAAGTGTTAA	n.d.[f]	
	lmo2067	bsh	Bile Salt Hydrolase	TTAAAAATIATTAA	−36	Dussurget et al. 2002

[a]Location of PrfA box from center of 14 bp palindrome to transcriptional start site of gene.
[b]LIPI-1, Listeria Pathogenicity Island 1.
[c]Bicistronic transcript; PrfA box shared with downstream genes (*plcA/prfA*, *actA/plcB*, *inlA/inlB*).
[d]Divergently transcribed; PrfA box shared by divergently transcribed genes (*hly/plcA*).
[e] PrfA box upstream of *prfAP2* which negatively regulates PrfA expression.
[f] n.d. = not determined.

promoters. Thus, while the presence of a PrfA box may help to identify a PrfA-regulated promoter, additional regulatory factors may still be required for optimal activation of gene expression.

7.3.2. Identification of Additional PrfA-dependent Promoters Through L. monocytogenes *Genome Analyses*

As PrfA activity is central to the expression of virulence genes in *L. monocytogenes*, a complete list of PrfA-regulated genes would undoubtedly provide additional information regarding how *L. monocytogenes* recognizes its location within host cells and within host tissues. Several strains of *L. monocytogenes* have now been sequenced (Glaser et al. 2001; Nelson et al. 2004) and *in silico* analysis of the EGD-e strain has identified a number of ORFs containing putative PrfA boxes using the criterion of potential binding sites being located less than 1 kb upstream of the translational start site. The divergently transcribed *hly/plcA* promoter contains the only PrfA box in the *L. monocytogenes* genome with no mismatches. There are 27 potential PrfA-regulated promoters with binding sites containing one mismatch, and 257 potential promoter targets with two mismatches (Glaser et al. 2001). The majority of these potential PrfA targets are putative ORFs with unknown function, and the presence of a 14 bp palindrome may not necessarily indicate a bona fide PrfA-regulated promoter. Additional analyses will be required based on the location of the PrfA boxes with respect to the transcriptional start sites of these ORFs to further determine if they are indeed likely to function in the regulation of gene expression by PrfA.

7.3.3. Transcriptome Analyses of PrfA-dependent Gene Expression

In addition to genome sequence-based approaches, whole-genome microarrays have been used to identify genes which are directly and indirectly under PrfA control. Transcript profiles of wild-type PrfA G145S and Δ*prfA* strains have been obtained following bacterial growth in broth culture (Milohanic et al. 2003). The expression of at least 73 genes was found to be influenced by PrfA, including 63 genes not previously associated with PrfA-dependent regulation, yet only 15 of the 73 appeared to have PrfA boxes within their promoter regions. Several genes whose expression was activated by PrfA but which lacked a PrfA box were observed to contain σ^B-dependent promoter elements (Milohanic et al. 2003). σ^B, the stress responsive alternative sigma factor for RNA polymerase, has been confirmed in a separate study to contribute to the expression of a subset of *L. monocytogenes* virulence genes, including the internalins *inlA* and *inlB*, and the *bsh*-encoded bile salt hydrolase as well as the *bilE* bile exclusion system (Kazmierczak et al. 2003).

Recently, in vivo transcript analyses have been undertaken to identify changes in *L. monocytogenes* gene expression during intracellular bacterial replication

(Chatterjee et al. 2006; Joseph et al. 2006). By comparing the transcription profile of broth-grown bacteria versus intracellular bacteria, 17–19% of the genes were found to be differentially expressed, including some previously unidentified members of the PrfA regulon (*orfX, orfZ*, and *prsA*) (Chatterjee et al. 2006). While these gene products were shown to contribute to intracellular growth of *L. monocytogenes* within tissue culture cells, the functions of the encoded gene products have yet to be defined.

7.4.   Regulatory Factors, Other Than PrfA, Thus Far Implicated in the Control of *L. monocytogenes* Virulence Gene Expression

PrfA is undeniably a major force governing diverse aspects of *L. monocytogenes* virulence gene expression. However, other bacterial factors that contribute to virulence regulation distinct from PrfA have also been identified. Examples of these virulence regulatory components will be briefly summarized below.

Genomic analysis has identified 16 two-component systems encoded in *L. monocytogenes* (Glaser et al. 2001). Two-component systems are a well-established means by which both gram-positive and gram-negative bacteria sense their external environment and respond by modulating gene expression (Stock et al. 2000) Two-component systems typically consist of a membrane-bound sensor kinase and a cytoplasmic response regulator. Upon stimulation by an external signal the sensor kinase is autophosphorylated on a conserved histidine residue. The phosphate group is then transferred to the response regulator which then becomes activated and binds to DNA to stimulate or repress the expression of target genes (Stock et al. 2000).

Kallipolitis et al. (Kallipolitis and Ingmer 2001) initially identified seven putative response regulators in *L. monocytogenes* using an approach based on degenerate PCR. Mutational analysis indicated that three of the seven regulators were important for virulence during an intragastric model of mouse infection (Kallipolitis and Ingmer 2001). One of the identified response regulators was *cesR*, which appears to encode a gene product that shares homology to the cephalosporin-sensitive response regulator present in *Enterococci* (Kallipolitis et al. 2003). The CesRK two-component system was further shown to play a role in ethanol tolerance (a cell-membrane stress agent) as well as β-lactam resistance, but its direct contribution to *L. monocytogenes* virulence is unknown. Another response regulator identified was *lisR*, a gene previously identified based on the acid tolerant phenotype conferred to strains containing a transposon insertion within the gene (Cotter et al. 1999). The LisRK two-component system contributes to bacterial growth in the presence of ethanol and to the ability of *L. monocytogenes* to tolerate antimicrobials such as nisin and cephalosporins (Cotter et al. 2002). LisK also has a role in pathogenesis as indicated by the reduced numbers of *lisK* mutants recovered from the spleens of infected mice in comparison to wild type following intraperitoneal injection (Cotter et al. 1999, 2002).

Recently, another apparent member of a two-component system within *L. monocytogenes* was shown to be important for virulence: *virR* (Mandin et al. 2005). VirR appears to regulate the expression of 12 genes by binding to a 16-bp DNA palindrome present in target promoter regions. Among the genes regulated by VirR include those of the *dlt* operon, whose gene products are necessary for the incorporation of D-alanine residues onto cell-wall-associated lipoteichoic acids. Strains lacking VirR were found to be severely attenuated in a mouse model of infection and exhibited reduced adhesion to tissue culture cells (Mandin et al. 2005).

Finally, a homologue to the AgrA regulator of *Staphylococcus aureus* has been identified in *L. monocytogenes* (Autret et al. 2003). AgrA plays an essential role in *S. aureus* virulence (Novick and Muir 1999), however loss of the *agrA* gene product in *L. monocytogenes* resulted in a moderate attenuation in *L. monocytogenes* virulence. Mutants lacking *agrA* exhibited no obvious phenotype in tissue culture assays, but were approximately 10-fold less virulent than wild-type strains following intravenous injection of mice (Autret et al. 2003). The nature of the regulatory role played by AgrA within *L. monocytogenes* remains to be defined.

7.5. Concluding Remarks

Much of what is currently known regarding the regulation of virulence in *L. monocytogenes* revolves around the central virulence gene transcriptional activator, PrfA. This key regulator serves as the major switch enabling *L. monocytogenes* to transition from the outside environment to inside the mammalian host. PrfA is known to regulate a variety of genes whose products contribute to host cell invasion, cytosolic bacterial replication, and dissemination of *L. monocytogenes* throughout cell monolayers and within host tissues (Kreft and Vazquez-Boland 2001). The list of genes whose expression is either directly dependent or indirectly influenced by PrfA continues to increase in number, and future experiments will undoubtedly sort out the potential contributions of these factors to *L. monocytogenes* pathogenesis. In addition, the list of other regulatory components that may feed into the PrfA regulon or function independently is also likely to continue to increase as we gain a better understanding of the integration of virulence gene expression with bacterial physiology during growth inside and outside of the host. Overall, *L. monocytogenes* continues to be a fascinating system for understanding how an environmental organism transitions into a pathogen by distinguishing the host from the outside environment.

Acknowledgments. The authors would like to acknowledge the excellent work done by many labs in the field of *L. monocytogenes* virulence gene regulation and apologize for any work not mentioned as a result of space considerations.

Work was supported by Public Health Service grants AI41816 and AI055651 (N.E.F), by a NIAID Bacterial Pathogenesis training grant fellowship AI55396 (M.D.M), and by a National Science Foundation Graduate Research Fellowship (G.C.P.).

References

Autret N, Raynaud C, Dubail I, Berche P, and Charbit A. 2003. Identification of the *agr* locus of *Listeria monocytogenes*: role in bacterial virulence. Infect Immun 71:4463–71.

Begley M, Sleator RD, Gahan CG, and Hill C. 2005. Contribution of three bile-associated loci, *bsh*, *pva*, and *btlB*, to gastrointestinal persistence and bile tolerance of *Listeria monocytogenes*. Infect Immun 73:894–904.

Brehm K, Ripio MT, Kreft J, and Vazquez-Boland JA. 1999. The *bvr* locus of *Listeria monocytogenes* mediates virulence gene repression by beta-glucosides. J Bacteriol 181:5024–32.

Cabanes D, Dehoux P, Dussurget O, Frangeul L, and Cossart P. 2002. Surface proteins and the pathogenic potential of *Listeria monocytogenes*. Trends Microbiol 10:238–45.

Camilli A, Tilney LG, and Portnoy DA. 1993. Dual roles of *plcA* in *Listeria monocytogenes* pathogenesis. Mol Microbiol 8:143–57.

Chatterjee SS, Hossain H, Otten S, Kuenne C, Kuchmina K, Machata S, Domann E, Chakraborty T, and Hain T. 2006. Intracellular Gene Expression Profile of Listeria monocytogenes. Infect Immun 74:1323–38.

Cheng LW, and Portnoy DA. 2003. Drosophila S2 cells: an alternative infection model for *Listeria monocytogenes*. Cell Microbiol 5:875–85.

Chico-Calero I, Suarez M, Gonzalez-Zorn B, Scortti M, Slaghuis J, Goebel W, European Listeria Genome Consortium T, and Vazquez-Boland JA. 2002. Hpt, a bacterial homolog of the microsomal glucose-6-phosphate translocase, mediates rapid intracellular proliferation in *Listeria*. Proc Natl Acad Sci U S A 99:431–6.

Cotter PD, Emerson N, Gahan CG, and Hill C. 1999. Identification and disruption of *lisRK*, a genetic locus encoding a two-component signal transduction system involved in stress tolerance and virulence in *Listeria monocytogenes*. J Bacteriol 181:6840–3.

Cotter PD, Guinane CM, and Hill C. 2002. The LisRK signal transduction system determines the sensitivity of *Listeria monocytogenes* to nisin and cephalosporins. Antimicrob Agents Chemother 46:2784–90.

Dickneite C, Böckmann R, Spory A, Goebel W, and Sokolovic Z. 1998. Differential interaction of the transcription factor PrfA and the PrfA-activating factor (Paf) of *Listeria monocytogenes* with target sequences. Mol Microbiol 27:915–28.

Domann E, Zechel S, Lingnau A, Hain T, Darji A, Nichterlein T, Wehland J, and Chakraborty T. 1997. Identification and characterization of a novel PrfA-regulated gene in *Listeria monocytogenes* whose product, IrpA, is highly homologous to internalin proteins, which contain leucine-rich repeats. Infect Immun 65:101–9.

Dussurget O, Cabanes D, Dehoux P, Lecuit M, Buchrieser C, Glaser P, and Cossart P. 2002. *Listeria monocytogenes* bile salt hydrolase is a PrfA-regulated virulence factor involved in the intestinal and hepatic phases of listeriosis. Mol Microbiol 45:1095–106.

Eiting M, Hageluken G, Schubert WD, and Heinz DW. 2005. The mutation G145S in PrfA, a key virulence regulator of *Listeria monocytogenes*, increases DNA-binding affinity by stabilizing the HTH motif. Mol Microbiol 56:433–46.

Engelbrecht F, Chun SK, Ochs C, Hess J, Lottspeich F, Goebel W, and Sokolovic Z. 1996. A new PrfA-regulated gene of *Listeria monocytogenes* encoding a small, secreted protein which belongs to the family of internalins. Mol Microbiol 21:823–37.

Freitag NE, Rong L, and Portnoy DA. 1993. Regulation of the *prfA* transcriptional activator of *Listeria monocytogenes*: multiple promoter elements contribute to intracellular growth and cell-to-cell spread. Infect Immun 61:2537–44.

Freitag NE, and Portnoy DA. 1994. Dual promoters of the *Listeria monocytogenes prfA* transcriptional activator appear essential in vitro but are redundant in vivo. Mol Microbiol 12:845–53.

Garcia Vescovi E, Soncini FC, and Groisman EA. 1996. Mg2+ as an extracellular signal: environmental regulation of *Salmonella* virulence. Cell 84:165–74.

Glaser P, Frangeul L, Buchrieser C, Rusniok C, Amend A, Baquero F, Berche P, Bloecker H, Brandt P, Chakraborty T, Charbit A, Chetouani F, Couve E, de Daruvar A, Dehoux P, Domann E, Dominguez-Bernal G, Duchaud E, Durant L, Dussurget O, Entian KD, Fsihi H, Portillo FG, Garrido P, Gautier L, Goebel W, Gomez-Lopez N, Hain T, Hauf J, Jackson D, Jones LM, Kaerst U, Kreft J, Kuhn M, Kunst F, Kurapkat G, Madueno E, Maitournam A, Vicente JM, Ng E, Nedjari H, Nordsiek G, Novella S, de Pablos B, Perez-Diaz JC, Purcell R, Remmel B, Rose M, Schlueter T, Simoes N, Tierrez A, Vazquez-Boland JA, Voss H, Wehland J, and Cossart P. 2001. Comparative genomics of *Listeria* species. Science 294:849–52.

Greene SL, and Freitag NE. 2003. Negative regulation of PrfA, the key activator of *Listeria monocytogenes* virulence gene expression, is dispensable for bacterial pathogenesis. Microbiology 149:111–20.

Hammer BK, and Swanson MS. 1999. Co-ordination of *Legionella pneumophila* virulence with entry into stationary phase by ppGpp. Mol Microbiol 33:721–31.

Hardy J, Francis KP, DeBoer M, Chu P, Gibbs K, and Contag CH. 2004. Extracellular replication of *Listeria monocytogenes* in the murine gall bladder. Science 303:851–3.

Harman JG. 2001. Allosteric regulation of the cAMP receptor protein. Biochim Biophys Acta 1547:1–17.

Herro R, Poncet S, Cossart P, Buchrieser C, Gouin E, Glaser P, and Deutscher J. 2005. How Seryl-Phosphorylated HPr inhibits PrfA, a transcription activator of *Listeria monocytogenes* virulence genes. J Mol Microbiol Biotechnol 9:224–34.

Johansson J, Mandin P, Renzoni A, Chiaruttini C, Springer M, and Cossart P. 2002. An RNA Thermosensor Controls Expression of Virulence Genes in *Listeria monocytogenes*. Cell 110:551.

Joseph B, Przybilla K, Stuhler C, Schauer K, Slaghuis J, Fuchs TM, and Goebel W. 2006. Identification of *Listeria monocytogenes* genes contributing to intracellular replication by expression profiling and mutant screening. J Bacteriol 188:556–68.

Kallipolitis BH, and Ingmer H. 2001. *Listeria monocytogenes* response regulators important for stress tolerance and pathogenesis. FEMS Microbiol Lett 204:111–15.

Kallipolitis BH, Ingmer H, Gahan CG, Hill C, and Sogaard-Andersen L. 2003. CesRK, a two-component signal transduction system in *Listeria monocytogenes*, responds to the presence of cell wall-acting antibiotics and affects beta-lactam resistance. Antimicrob Agents Chemother 47:3421–9.

Kazmierczak MJ, Mithoe SC, Boor KJ, and Wiedmann M. 2003. *Listeria monocytogenes sigma B* regulates stress response and virulence functions. J Bacteriol 185:5722–34.

Kocks C, Gouin E, Tabouret M, Berche P, Ohayon H, and Cossart P. 1992. L. monocytogenes-induced actin assembly requires the *actA* gene product, a surface protein. Cell 68:521–31.

Konkel ME, and Tilly K. 2000. Temperature-regulated expression of bacterial virulence genes. Microbes Infect 2:157–66.

Korner H, Sofia HJ, and Zumft WG. 2003. Phylogeny of the bacterial superfamily of Crp-Fnr transcription regulators: exploiting the metabolic spectrum by controlling alternative gene programs. FEMS Microbiol Rev 27:559–92.

Kreft J, and Vazquez-Boland JA. 2001. Regulation of virulence genes in *Listeria*. Int J Med Microbiol 291:145–57.

Krukonis ES, and DiRita VJ. 2003. From motility to virulence: Sensing and responding to environmental signals in *Vibrio cholerae*. Curr Opin Microbiol 6:186–90.

Lauer P, Chow MY, Loessner MJ, Portnoy DA, and Calendar R. 2002. Construction, characterization, and use of two *Listeria monocytogenes* site-specific phage integration vectors. J Bacteriol 184:4177–86.

Leimeister-Wachter M, Haffner C, Domann E, Goebel W, and Chakraborty T. 1990. Identification of a gene that positively regulates expression of listeriolysin, the major virulence factor of *Listeria monocytogenes*. Proc Natl Acad Sci U S A 87:8336–40.

Leimeister-Wachter M, Domann E, and Chakraborty T. 1991. Detection of a gene encoding a phospatidylinositol-specific phospholipase C that is co-ordinately expressed with listeriolysin in *Listeria monocytogenes*. Mol Microbiol 5:361–6.

Lingnau A, Domann E, Hudel M, Bock M, Nichterlein T, Wehland J, and Chakraborty T. 1995. Expression of the *Listeria monocytogenes* EGD *inlA* and *inlB* genes, whose products mediate bacterial entry into tissue culture cell lines, by PrfA-dependent and -independent mechanisms. Infect Immun 63:3896–903.

Lingnau A, Chakraborty T, Niebuhr K, Domann E, and Wehland J. 1996. Identification and purification of novel internalin-related proteins in *Listeria monocytogenes* and *Listeria ivanovii*. Infect Immun 64:1002–6.

Luo Q, Rauch M, Marr AK, Muller-Altrock S, and Goebel W. 2004. In vitro transcription of the *Listeria monocytogenes* virulence genes *inlC* and *mpl* reveals overlapping PrfA-dependent and -independent promoters that are differentially activated by GTP. Mol Microbiol 52:39–52.

Makino M, Kawai M, Kawamura I, Fujita M, Gejo F, and Mitsuyama M. 2005. Involvement of reactive oxygen intermediate in the enhanced expression of virulence-associated genes of *Listeria monocytogenes* inside activated macrophages. Microbiol Immunol 49:805–11.

Mandin P, Fsihi H, Dussurget O, Vergassola M, Milohanic E, Toledo-Arana A, Lasa I, Johansson J, and Cossart P. 2005. VirR, a response regulator critical for *Listeria monocytogenes* virulence. Mol Microbiol 57:1367–80.

Mansfield BE, Dionne MS, Schneider DS, and Freitag NE. 2003. Exploration of host-pathogen interactions using *Listeria monocytogenes* and *Drosophila melanogaster*. Cell Microbiol 5:901–11.

Marquis H, Doshi V, and Portnoy DA. 1995. The broad-range phospholipase C and a metalloprotease mediate listeriolysin O-independent escape of *Listeria monocytogenes* from a primary vacuole in human epithelial cells. Infect Immun 63:4531–4.

Marquis H, Goldfine H, and Portnoy DA. 1997. Proteolytic pathways of activation and degradation of a bacterial phospholipase C during intracellular infection by *Listeria monocytogenes*. J Cell Biol 137:1381–92.

Mengaud J, Vicente MF, and Cossart P. 1989. Transcriptional mapping and nucleotide sequence of the *Listeria monocytogenes* *hly*A region reveal structural features that may be involved in regulation. Infect Immun 57:3695–701.

Mengaud J, Braun-Breton C, and Cossart P. 1991. Identification of phosphatidylinositol-specific phospholipase C activity in *Listeria monocytogenes*: a novel type of virulence factor? Mol Microbiol 5:367–72.

Mengaud J, Dramsi S, Gouin E, Vazquez-Boland JA, Milon G, and Cossart P. 1991. Pleiotropic control of *Listeria monocytogenes* virulence factors by a gene that is autoregulated. Molec Microbiol 5:2273–83.

Milenbachs Lukowiak A, Mueller KJ, Freitag NE, and Youngman P. 2004. Deregulation of *Listeria monocytogenes* virulence gene expression by two distinct and semi-independent pathways. Microbiology 150:321–33.

Milohanic E, Glaser P, Coppee JY, Frangeul L, Vega Y, Vazquez-Boland JA, Kunst F, Cossart P, and Buchrieser C. 2003. Transcriptome analysis of *Listeria monocytogenes* identifies three groups of genes differently regulated by PrfA. Mol Microbiol 47:1613–25.

Moors MA, Levitt B, Youngman P, and Portnoy DA. 1999. Expression of listeriolysin O and ActA by intracellular and extracellular *Listeria monocytogenes*. Infect Immun 67:131–9.

Mueller KJ, and Freitag NE. 2005. Pleiotropic enhancement of bacterial pathogenesis resulting from the constitutive activation of the *Listeria monocytogenes* regulatory factor PrfA. Infect Immun 73:1917–26.

Nelson KE, Fouts DE, Mongodin EF, Ravel J, DeBoy RT, Kolonay JF, Rasko DA, Angiuoli SV, Gill SR, Paulsen IT, Peterson J, White O, Nelson WC, Nierman W, Beanan MJ, Brinkac LM, Daugherty SC, Dodson RJ, Durkin AS, Madupu R, Haft DH, Selengut J, Van Aken S, Khouri H, Fedorova N, Forberger H, Tran B, Kathariou S, Wonderling LD, Uhlich GA, Bayles DO, Luchansky JB, and Fraser CM. 2004. Whole genome comparisons of serotype 4b and 1/2a strains of the food-borne pathogen *Listeria monocytogenes* reveal new insights into the core genome components of this species. Nucleic Acids Res 32:2386–95.

Novick RP, and Muir TW. 1999. Virulence gene regulation by peptides in staphylococci and other Gram-positive bacteria. Curr Opin Microbiol 2:40–5.

Portnoy DA, Jacks PS, and Hinrichs DJ. 1988. Role of hemolysin for the intracellular growth of *Listeria monocytogenes*. J Exp Med 167:1459–71.

Poyart C, Abachin E, Razafimanantsoa I, and Berche P. 1993. The zinc metalloprotease of *Listeria monocytogenes* is required for maturation of phosphatidylcholine phospholipase C: direct evidence obtained by gene complementation. Infect Immun 61:1576–80.

Renzoni A, Klarsfeld A, Dramsi S, and Cossart P. 1997. Evidence that PrfA, the pleitropic activator of virulence genes in *Listeria monocytogenes* can be present but inactive. Infect Immun 65:1515–18.

Ripio MT, Dominguez-Bernal G, Suarez M, Brehm K, Berche P, and Vazquez-Boland JA. 1996. Transcriptional activation of virulence genes in wild-type strains of *Listeria monocytogenes* in response to a change in the extracellular medium composition. Res Microbiol 147:371–84.

Ripio M-T, Brehm K, Lara M, Suárez M, and Vázquez-Boland J-A. 1997a. Glucose-1-phosphate utilization by *Listeria monocytogenes* is PrfA-dependent and coordinately expressed with virulence factors. J Bacteriol 179:7174–80.

Ripio M-T, Dominguez-Bernal G, Lara M, Suarez M, and Vazquez-Boland J-A. 1997b. A Gly145Ser substitution in the transcriptional activator PrfA causes constitutive overexpression of virulence factors in *Listeria monocytogenes*. J Bacteriol 179:1533–40.

Schwab U, Bowen B, Nadon C, Wiedmann M, and Boor KJ. 2005. *The Listeria monocytogenes prfA*P2 promoter is regulated by sigma B in a growth phase dependent manner. FEMS Microbiol Lett 245:329–36.

Sheehan B, Klarsfeld A, Msadek T, and Cossart P. 1995. Differential activation of virulence gene expression by PrfA, the *Listeria monocytogenes* virulence regulator. J Bacteriol 177:6469–76.

Shetron-Rama LM, Marquis H, Bouwer HGA, and Freitag NE. 2002. Intracellular induction of *Listeria monocytogenes actA* expression. Infect Immun 70:1087–96.

Shetron-Rama LM, Mueller K, Bravo JM, Bouwer HG, Way SS, and Freitag NE. 2003. Isolation of *Listeria monocytogenes* mutants with high-level in vitro expression of host cytosol-induced gene products. Mol Microbiol 48:1537–51.

Sleator RD, Wemekamp-Kamphuis HH, Gahan CG, Abee T, and Hill C. 2005. A PrfA-regulated bile exclusion system (BilE) is a novel virulence factor in *Listeria monocytogenes*. Mol Microbiol 55:1183–95.

Stock AM, Robinson VL, and Goudreau PN. 2000. Two-component signal transduction. Annu Rev Biochem 69:183–215.

Suarez M, Gonzalez-Zorn B, Vega Y, Chico-Calero I, and Vazquez-Boland JA. 2001. A role for ActA in epithelial cell invasion by *Listeria monocytogenes*. Cell Microbiol 3:853–64.

Vazquez-Boland J, Kocks C, Dramsi S, Ohayon H, Geoffroy C, Mengaud J, and Cossart P. 1992. Nucleotide sequence of the lecithinase operon of *Listeria monocytogenes* and possible role of lecithinase in cell-to-cell spread. Infect. Immun. 60:219–30.

Vega Y, Dickneite C, Ripio M-T, Böckmann R, Gonzalez-Zorn B, Novella S, Gominguez-Bernal G, Goebel W, and Vazquez-Boland W. 1998. Functional similarities between the *Listeria monocytogenes* virulence regulator PrfA and cyclic AMP receptor protein: the PrfA* (Gly145Ser) mutation increases binding affinity for target DNA. J Bacteriol 180:6655–60.

Vega Y, Rauch M, Banfield MJ, Ermolaeva S, Scortti M, Goebel W, and Vazquez-Boland JA. 2004. New *Listeria monocytogenes prfA** mutants, transcriptional properties of PrfA* proteins and structure-function of the virulence regulator PrfA. Mol Microbiol 52:1553–65.

Wong KK, Bouwer HG, and Freitag NE. 2004. Evidence implicating the 5' untranslated region of *Listeria monocytogenes actA* in the regulation of bacterial actin-based motility. Cell Microbiol 6:155–66.

Wong KK, and Freitag NE. 2004. A novel mutation within the central *Listeria monocytogenes* regulator PrfA that results in constitutive expression of virulence gene products. J Bacteriol 186:6265–76.

8
Invasion of Host Cells
by *Listeria monocytogenes*

Javier Pizarro-Cerdá and Pascale Cossart

Institut Pasteur, Unité des Interactions Bactéries-Cellules, Paris, F-75015 France;
INSERM, U604, Paris, F-75015 France; INRA, USC2020, Paris, F-75015 France
e-mail: pizarroj@pasteur.fr, pcossart@pasteur.fr

Abstract: Critical for the pathophysiology of the disease caused by *Listeria monocytogenes* is the bacterial ability to promote its entry into nonphagocytic cells. Different members of the internalin family of listerial surface proteins have been shown to be required for cellular invasion, subverting cellular functions and structures such as adherens junctions in the case of InlA, or the hepatocyte growth factor receptor signaling in the case of InlB. Additional noninternalin molecules, including the LPXTG-anchored cell wall protein VIP, the glycine/tryptophane-anchored amidases Ami and Auto, and the secreted pore-forming cholesterol-dependent cytolysin listeriolysin O are also necessary for efficient cell entry, revealing the complexity of the bacterial molecular program required for gaining access to the host intracellular environment.

8.1. Introduction

Central for listeriosis physiopathology is the ability of *L. monocytogenes* to cross three tissue barriers—the intestinal barrier, the hemato-encephalic barrier and the feto-placental barrier—by subversion of specific cellular effectors and functions that will allow bacterial internalization within target host cells (Figure 8.1). Several *L. monocytogenes* proteins required for invasion have been identified, and the cellular signaling pathways that are subverted by these bacterial virulence factors have been partially unraveled, highlighting often that *L. monocytogenes* takes advantage of existing molecular machineries in the cell (the adherens junctions or the clathrin-dependent endocytosis, for example) that are sabotaged from their original function to promote the active and pathogen-directed internalization of listeriae within cells. Many of the original *L. monocytogenes* virulence determinants involved in cellular invasion were characterized by classical transposon mutagenesis approaches, but the publication of the *L. monocytogenes* and *L. innocua* genomes (Glaser et al., 2001) has permitted to identify new

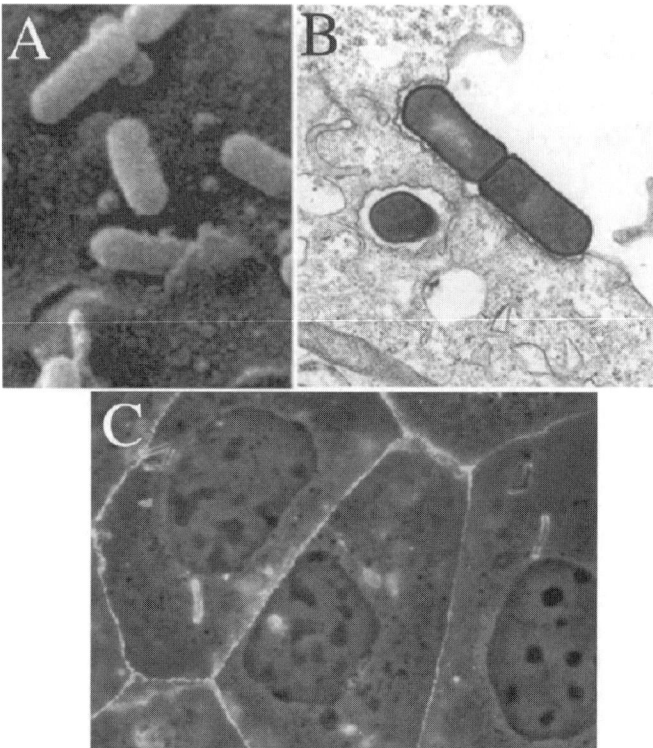

FIGURE 8.1. Electron microscopy and immunofluoresce micrographs of the *L. monocytogenes* entry into target cells. **A** Scanning and **B** transmission electron micrographs of *L. monocytogenes* during the invasion of human epithelial Caco-2 cells, illustrating that the host cell membrane is subtly rearranged around internalized bacteria. Reprinted from *Cell*, Vol. 84, J. Mengaud et al., E-cadherin is the receptor for internalin, a surface protein required for entry of *L. monocytogenes* into epithelial cells, pp. 923–932, copyright 1996, with permission from Elsevier; and *EMBO J*, Vol. 17, P. Cossart and M. Lecuit, Interactions of *Listeria monocytogenes* with mammalian cells during entry and actin-based movement: bacterial factors, cellular ligands and signaling, pp. 3797–3806, copyright 1998, with permission from Nature Publishing Group. **C** Immunfluorescence image of *L. monocytogenes* invading Caco-2 cells, depicting the recruitment of β-catenin and α-catenin to the bacterial entry site. These two catenins are also required for the formation of adherens junctions. Reprinted from *Trends Cell Biol.*, Vol. 13, P. Cossart et al., Invasion of mammalian cells by *Listeria monocytogenes*: functional mimicry to subvert cellular functions, pp. 23–31, copyright 2003, with permission from Elsevier. (A color version of this figure appears between pages 196 and 197.)

potential invasion proteins *in silico* by comparing sequences present on the virulent *L. monocytogenes* genome but absent in the *L. innocua* one. In this review, we will address the state of the art of the main molecular mechanisms employed by *L. monocytogenes* to promote entry into mammalian host cells.

8.2.   Internalins: Main Listeria monocytogenes Effectors for Cell Invasion

The *L. monocytogenes* internalin family of proteins comprises 24 members, characterized by the presence of amino-terminal regions with tandemly arranged leucine-rich repeats (LRRs) (Pizarro-Cerda and Cossart, 2006). The LRR regions have been found in prokaryotic and eukaryotic proteins, providing a versatile framework for the formation of ligand-binding sites (Kobe and Kajava, 2001). Nineteen members of the internalin family including InlA, the internalin prototype, contain a carboxy-terminal LPXTG motif that is recognized by the sortase A and allows their covalent binding to the cell wall peptidoglycan (Bierne et al., 2002). One member, InlB, is loosely attached to the bacterial cell wall by electrostatic interactions between glycine/tryptophane-rich (GW) modules on its carboxy-terminal domain and lipoteichoic acids on the bacterial cell wall (Jonquieres et al., 1999). Four members of the internalin family (including InlC) do not present anchor motifs and are released as soluble proteins. The inter-action of several surface-associated internalins with eukaryotic plasma membrane receptors has been well characterized and represents the main signaling pathways that allow *L. monocytogenes* internalization within target cells.

8.2.1.   InlA-Mediated Pathway

8.2.1.1.   InlA Structure and Function

InlA is an 800-amino acid protein that displays a classical signal peptide sequence at its amino-terminal domain, followed by a short conserved cap and an LRR region comprising 15-and-a-half repeats of 22 amino acids (with the exception of repeat 6 which contains 21 amino acids). Each LRR starts with a β-strand of five residues, followed by a seven-residue loop, a five-residue helix and a second five-residue loop; the β-strands combine to form together a 16 stranded parallel β-sheet (Schubert et al., 2002). An interrepeat (IR) region separates the LRR from a second repeat region called the B-repeat region. The carboxy-terminal domain exhibits an LPTTG motif that is recognized by the sortase A for covalent anchoring of the protein to the bacterial cell wall (Figure 8.2.) (Bierne et al., 2002).

InlA was initially identified in a transposon mutagenesis screen study for noninvasive *L. monocytogenes* mutants, in which it was demonstrated that this protein is required to induce bacterial entry into epithelial human Caco-2 cells (Gaillard et al., 1991). Deletion mutational analysis showed that the LRR and IR

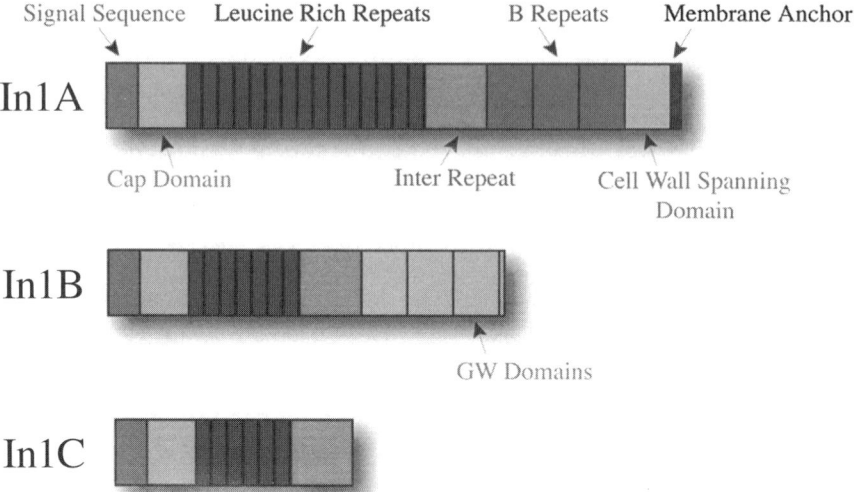

FIGURE 8.2. Schematic representation of different types of internalins involved in invasion of target cells by *L. monocytogenes*. InlA is an LPXTG-anchored membrane protein that promotes bacterial entry into polarized epithelial cells by interacting with the adhesion molecule E-cadherin through its leucine-rich repeats. InlB is loosely anchored to the bacterial cell wall through its GW domains and can be detached from the *L. monocytogenes* surface; soluble InlB interacts with host cell glycosaminoglycans (GAGs) and the globular receptor for the complement gC1 molecule (gC1q-R) through its GW domains, while its leucine-rich repeats induce invasion of a large panel of target cell by stimulation of the hepatocyte growth factor receptor Met. InlC is a soluble internalin that potentiates the InlA-mediated entry pathway in the presence of InlB by interacting with a still unknown ligand.

regions are sufficient to induce the entry of inert latex beads or of the noninvasive *L. innocua* species into target cells (Lecuit et al., 1997). Recent in vivo studies have highlighted the critical role played by InlA during the traversal of the human intestinal and materno-fetal barriers (Lecuit et al., 2001, 2004) (see below). The importance of InlA has also been estimated at the population level: although certain *L. monocytogenes* strains carry a truncated form of InlA that can be released from the bacterial cell wall (Jonquieres et al., 1998), an epidemiological study carried out in France demonstrated that 96% of clinical *L. monocytogenes* isolates present a full-length protein; moreover, 100% of isolates from placental infections present the full-length form of InlA, confirming the critical role of InlA in the listeriosis pathogenesis, as well as its potential value as an indicator of virulence in food-safety assessment programs (Jacquet et al., 2004).

8.2.1.2. E-cadherin: the InlA Receptor

By affinity chromatography on an InlA-column, E-cadherin was identified as the cellular receptor for InlA (Mengaud et al., 1996). E-cadherin belongs to the cadherin superfamily of calcium-dependent cell adhesion molecules, which includes other classical cadherins such as the N-, P-, or H-cadherins, but also includes more than 80 protocadherins (the largest subgroup in the cadherin super-family) involved in the morphogenesis and formation of neuronal circuits, and in modulation of synaptic transmission (Junghans et al., 2005). Classical cadherins in neighboring cells establish homophilic interactions between their extracel-lular domains, while their cytoplasmic tails interact with several molecules, including proteins of the Armadillo repeat family such as β-catenin and p120 catenin, which mediate association of cell adhesion complexes to the cytoskeleton, and regulate cadherin stability/retention at the plasma membrane (Nelson and Nusse, 2004). E-cadherin, in particular, is involved in the formation of adherens junctions in polarized epithelial cells of different tissues such as the intestine or the feto-placental barrier and is regarded as the main organizer of the epithelial phenotype; indeed, E-cadherin dysfunction or down-regulation is closely linked to cancer (Gumbiner, 2005). Distribution of E-cadherin in polarized epithelial layers is normally restricted to basolateral membranes where E-cadherin is not accessible to the lumen; however, within an epithelial barrier, sites of senescent cell extrusion exist in which E-cadherin is transiently exposed to the luminal surface, and these sites have been shown to be used by *L. monocytogenes* to access epithelial junctions, promoting cellular invasion (Pentecost et al., 2006).

The initial interaction between bacterial InlA and human E-cadherin is mediated by the LLRs present on InlA and the first extracellular domain of E-cadherin. Interestingly, this interaction is species-specific: a proline at position 16 on E-cadherin (such as in humans and guinea pigs) is necessary for InlA binding, and mutation of this proline to glutamic acid (such as in the mouse or rat) not only inhibits adhesion of *L. monocytogenes* to E-cadherin-expressing cells but also inhibits invasion (Lecuit et al., 1999). The crystal structure of the InlA LRRs in complex with the first extracellular domain of human E-cadherin illustrates the exquisite fine interaction between these two molecules: an hydrophobic pocket between the LRR 5 and the LRR 7 (due to the absence of one amino acid on LLR 6) accommodates precisely the proline at position 16 (Figure 8.3.) (Schubert et al., 2002). The generation of transgenic mice expressing the human E-cadherin at the intestinal level revealed that only the interaction of *L. monocytogenes* with the transgenic human InlA-binding E-cadherin (and not with the endogenous mouse E-cadherin) allowed bacterial translocation across the intestinal barrier, highlighting the crucial role of this interaction for the initial steps of the disease (Lecuit et al., 2001). E-cadherin is also present in syncytiotrophoblasts and villious cytotrophoblasts of the placenta, and it has been recently shown that the InlA/E-cadherin interaction is as well required for traversal of the human materno-fetal barrier (Lecuit et al., 2004). E-cadherin is also present at epithelial cells in contact with the encephalorachidean liquid, and it is suspected that InlA plays a role during bacterial translocation through the blood–brain barrier.

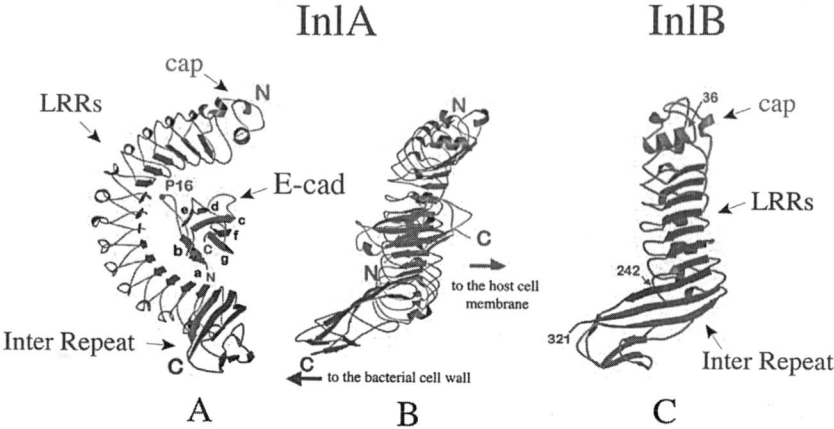

FIGURE 8.3. Structure of the N-terminal domains of InlA and InlB. **A** Leucine-rich repeats (LRRs) from InlA are composed of 16 β-sheets that accommodate the first extracellular domain of E-cadherin; an hydrophobic pocket created between LRR 5 and 7 (due to the absence of one amino acid in LRR 6) accommodates E-cadherin proline 16, which is critical for InlA–E-cadherin interaction. **B** 90°C rotation of the figure depicted in **A**. **C** View toward the concave surface of the InlB LRRs. Reprinted from *Cell*, Vol. 111, W.D. Schubert et al., Structure of internalin, a major invasion protein of *Listeria monocytogenes*, in complex with its human receptor E-cadherin, pp. 825–836, copyright 2001, and *J Mol Biol*, Vol. 312, W.D. Schubert et al., Internalins of the human pathogen *L. monocytogenes* combine three distinct folds into a contiguous internalin domain, pp. 783–794, copyright 2002, with permission from Elsevier. (A color version of this figure appears between pages 196 and 197.)

8.2.1.3.   Adherens Junctions and the Molecular Machinery Involved
           in the InlA-Entry Pathway

As stated above, the cytoplasmic tail of E-cadherin is able to interact with several proteins of the catenin family that link the adherens junction complex to the actin cytoskeleton. In particular, β-catenin is able to bind to the last 35 amino acids of the E-cadherin cytoplasmic domain and also interacts with α-catenin, which in turn is able to directly bind actin (Jamora and Fuchs, 2002). Taking into account this model of interaction in which α-catenin provides a stable link between E-cadherin and the cytoskeleton, a chimera was constructed which contained the E-cadherin ecto-domain fused to the actin-binding site of α-catenin: this fusion molecule allowed similar levels of bacterial infection as wild-type cells (as opposed to E-cadherin molecules that lack the β-catenin-binding site and which are nonpermisive for infection), suggesting that *L. monocytogenes* exploits the same molecular scaffold used for adherens junction formation to induce entry into target cells (Figure 8.4.) (Lecuit et al., 2000). Recently, the model depicting the static binding of E-cadherin to the actin cytoskeleton via α-catenin has been challenged; in fact, it has been shown that the interaction of α-catenin with either E-cadherin or β-catenin is exclusive (Drees et al., 2005; Yamada et al., 2005). These authors have suggested that the direct connection between adherens junctions and the actin cytoskeleton could be mediated by

In1A entry pathway

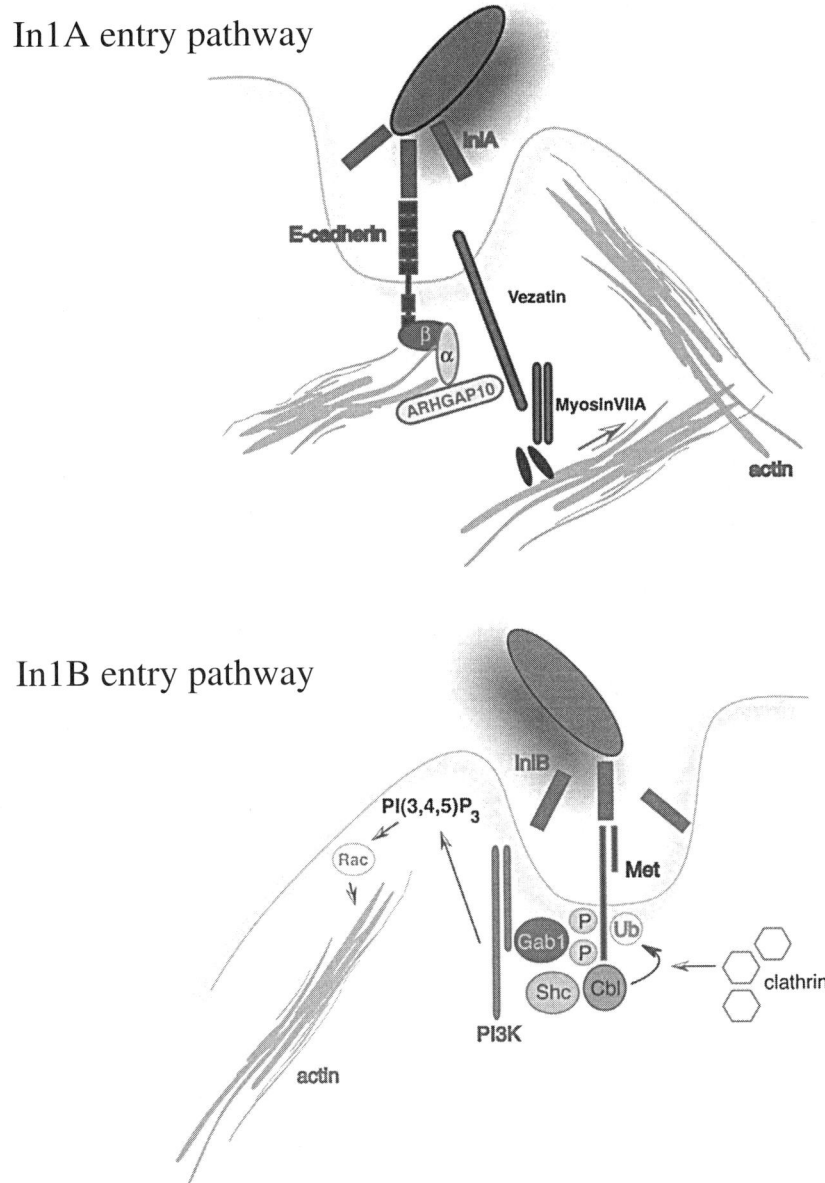

In1B entry pathway

FIGURE 8.4. Signaling pathways triggered by InlA and InlB. *InlA entry pathway:* InlA interacts with E-cadherin, recruiting to the bacterial entry site the adherens junction proteins β- and α-catenin, promoting rearrangements in the actin cytoskeleton required for invasion. The protein ARHGAP10 is a GTPase activating protein for RhoA and Cdc42 that interacts with α-catenin and is required for entry as well as adherens junction formation. The unconventional myosin VIIa is also required for entry, providing probably the

(continued)

other junction proteins such as nectin or afadin; alternatively, several weak transient and cumulative interactions between actin-binding proteins such as vinculin, afadin, or spectrin with the junction components could be the basis of the cytoskeleton connection to the adherens junctions. In the case of the *L. monocytogenes* entry, it has been determined that other proteins besides catenins also participate in the entry process which could also regulate the adherens junction–cytoskeleton interaction: through a two-hybrid screen using α-catenin as a bait, the molecule ARHGAP10—which is a GTPase-activating protein (GAP) for the small Rho GTPases RhoA and Cdc42—has been found to be required for efficient bacterial invasion as well as for adherens junction formation (Sousa et al., 2005). The unconventional myosin VIIA and its ligand, the membrane-associated protein vezatin, are also required for entry, probably generating the contractile force necessary to promote bacterial internalization (Sousa et al., 2004). Polymerization of actin downstream of the InlA/E-cadherin interaction is mediated by the small Rho GTPase Rac, by the Src substrate cortactin and by the actin nucleating Arp2/3 complex (a cortactin substrate itself), but how these molecules are activated to favor cytoskeletal rearrangements during invasion is not known yet.

8.2.1.4.   Effect of Rafts on the InlA–E-cadherin Interaction

It is important to note that a functional interaction between InlA and E-cadherin leading to bacterial entry cannot take place if the host membrane organization in lipid domains is altered. Indeed, formation of lipid membrane micro-domains or rafts provides a mechanism for the segregation of molecular effectors into functional subunits for efficient signaling and sorting processes (Simons and Toomre, 2000). The drug methyl-β-cyclodextrin (MβCD) that sequesters the major membrane lipid cholesterol and disrupts lipid rafts has been used to study the function of lipid micro-domains in the *L. monocytogenes* infection process; it has been demonstrated that MβCD does not alter the amounts of E-cadherin present at the plasma membrane of cholesterol-depleted cells, but affects the

FIGURE 8.4. *(Continued)* contractile force that drives bacterial internalization. The protein Vezatin is also localized at the bacterial entry site, but its function in the *L. monocytogenes* entry process is still unknown. *InlB entry pathway:* InlB stimulates the hepatocyte growth factor Met, inducing its phosphorylation and the recruitment to the bacterial entry site of the molecular adaptors Cbl, Shc, and Gab1, which in turn recruit the PI3K type I: this enzyme generates PI(3,4,5)P$_3$, which is a potent second messenger upstream of Rac required for the induction of actin rearrangements associated with bacterial invasion. Cbl is also an ubiquitin ligase that targets the endocytic machinery to the bacterial entry site, favoring Met and *L. monocytogenes* internalization into target cells. Reprinted from Subversion of cellular functions by *Listeria monocytogenes*, J. Pizarro-Cerdá and P. Cossart, *J Pathol Vol* 208, pp. 215–223, 2006, copyright Pathological Society of Great Britain and Ireland. Reproduced with permission. Permission granted by John Wiley & Sons Ltd on behalf of the Pathological Society.

efficient recruitment of E-cadherin complexes to the site of bacterial attachment (Seveau et al., 2004). Lipid raft markers such as glycosylphosphatidylinositol-conjugated proteins are also recruited at the bacterial entry site in an InlA-dependent manner. These results suggest that the function of E-cadherin, not only in the case of *L. monocytogenes* invasion but probably also in the context of native adherens junctions, requires the organization of the plasma membrane in intact lipid micro-domains that will favor the mobility and clustering of E-cadherin into functional units favoring downstream signaling.

8.2.2. InlB-Mediated Pathway

8.2.2.1. InlB Structure and Function

InlB is a 630-amino acids protein from the internalin family encoded in the same operon as InlA (Gaillard et al., 1991). InlB presents an amino-terminal signal peptide followed by the short conserved internalin cap and seven LRRs; the IR region that follows the LRR region has a minimal immunoglobulin (Ig)-like fold that, together with the LRRs and with the cap (which presents itself a truncated EF-hand fold structure), constitutes aunique domain with a continuous hydrophobic core and an extended β-sheet (Figure 8.3.) (Schubert et al., 2001). This amino-terminal domain is followed by a central B repeat, while the carboxy-terminal domain is constituted by three C repeats of approximately 80 amino acids, each of them starting with the residues glycine/tryptophane (GW) and hence known as the GW region (Braun et al., 1997). The GW domains mediate loose attachment of InlB with the lipoteichoic acids of the *L. monocytogenes* cell wall (Jonquieres et al., 1999).

As in the case of InlA, InlB has been shown to be required for invasion of human epithelial cells; however, InlB allows entry within a larger panel of target cells than InlA (Braun et al., 1998; Dramsi et al., 1995). The LRR region of InlB has been shown to be sufficient for inducing entry into cells of latex beads covalently bound to this domain (Braun et al., 1999). Exogenously added InlB can also favor entry of noninvasive *L. innocua* or *Streptococcus carnosus* (Braun et al., 1998). Interestingly, InlB can be detached from the bacterial cell wall, and it has been shown that the GW domains of its carboxy terminal can also interact with cellular targets (Jonquieres et al., 2001), indicating that both the amino- and carboxy-terminal domains of InlB cooperate to promote efficient invasion of nonphagocytic cells (see below).

8.2.2.2. Met, GAGs, and gC1qR: InlB Receptors

The main signaling receptor for InlB is the hepatocyte growth factor (HGF) receptor, also known as Met (Shen et al., 2000). Met belongs to the family of receptor tyrosine kinases (RTK), one of the largest and most important families of transmembrane signaling receptors (Hubbard and Till, 2000). Upon binding to their cognate ligands, RTKs can transduce extracellular signals into the cytoplasm by autophosphorylation and by phosphorylation of downstream molecules, activating multiple pathways involved in cell proliferation, migration, and/or differentiation. Met, in particular, plays an important role during development, controlling epithelial cell migration and growth during embryogenesis; Met also controls growth, invasion and metastasis in cancer cells, and activating Met mutations predispose to human cancer (Birchmeier et al., 2003).

The crystal structure of the InlB LRRs reveals that this region adopts an elongated curved conformation (Figure 8.3.) (Marino et al., 1999), and it has been recently shown that four aromatic residues within the concave surface of the LRR domain are crucial for binding to Met (Machner et al., 2003). The InlB–Met interaction is necessary for *L. monocytogenes* entry into target cells (Shen et al., 2000). As is the case for the interaction between the HGF and the Met, interaction between InlB and Met is potentiated by glycosaminoglycans (GAGs) of the extracellular matrix (Jonquieres et al., 2001). It is relevant to note that InlB is able to bind GAGs directly through its GW domains (Banerjee et al., 2004; Jonquieres et al., 2001). The GW repeats of InlB actually present similarity to eukaryotic Src-homologytype 3 (SH3)-like domains; however, the potential proline-binding sites typical from functional SH3 domains are absent from the GW repeats (Marino et al., 2002). The GW domains of InlB also interact with the receptor for the globular head of the complement C1q molecule (known as gC1q-R) (Braun et al., 2000; Marino et al., 2002); however, the functional relevance of this binding during interaction of *L. monocytogenes* with target cells has not been established yet.

8.2.2.3.   Signaling Downstream of InlB–Met Interaction

Met is a heterodimer formed by an extracellular α- and a β-subunit that comprises extracellular, transmembrane and cytoplasmic domains. The cytoplasmic tail of the β-subunit contains several potential sites for tyrosine phosphorylation: tyrosines 1234 and 1235 are referred as the activation loop since they increase Met kinase activity (Birchmeier and Gherardi, 1998). The juxtamembrane tyrosine 1003 serves as a binding site for the ubiquitin ligase Cbl (Peschard et al., 2001), while tyrosines 1349 and 1356 are referred as the multidocking site since they are involved in the recruitment of several molecules including the adaptor proteins Shc and Gab1, which in turn can be phosphorylated in several tyrosine residues and bind other signaling proteins (Ponzetto et al., 1994). It has been reported that upon stimulation by InlB, phosphorylation of Cbl, Shc, and Gab1 occurs on target cells (Ireton et al., 1999; Shen et al., 2000). Of note, recruitment/activation of Gab1 by InlB can take place by two redundant pathways that require either phosphorylation of Met tyrosines 1349/1356 and Gab1-binding to the Met multidocking site, or formation of the phosphoinositide phosphatidylinositol-3,4-5-triphosphate (PI[3,4,5]P$_3$) at the plasma membrane and Gab1-recruitment via its pleckstrin homology domain, able to directly bind PI(3,4,5)P$_3$ (Basar et al., 2005). Gab1, Shc, and Cbl are involved in the recruitment of the phosphatidylinositol 3-kinase (PI3K) that precisely promotes PI[3,4,5]P$_3$ production at the site of *L. monocytogenes* entry

(Ireton et al., 1996, 1999) and favors cytoskeletal rearrangements required for bacterial engulfment (Figure 8.4.).

Regulation of actin modification downstream of PI3K during InlB stimulation is complex and varies in different cell lines. In the green monkey hepatocyte cell line Vero, it has been demonstrated that the small GTPase Rac1 is involved in the WAVE2-dependent activation of the Arp2/3 complex (Bierne et al., 2001, 2005), which nucleates and polymerizes actin filaments in branched networks (Pollard and Beltzner, 2002; Stradal and Scita, 2006). Proteins of the Ena/VASP family are required for elongation of actin filaments (Bierne et al., 2005). Formation and disruption of the phagocytic cup during *L. monocytogenes* entry require a fine-tuning of the activity of the actin depolymerizing factor cofilin, its activity in turn being regulated by the LIM kinase (Bierne et al., 2001). In HeLa cells, activation of the Arp2/3 complex by InlB requires Rac1 upstream of WAVE2 and WAVE1, and also Cdc42 upstream of N-WASP (Bierne et al. 2005).

8.2.2.4. Clathrin-Mediated Endocytosis of Met and Invasion

As mentioned above, the ubiquitin ligase Cbl is recruited to Met upon cellular stimulation by InlB (Ireton et al., 1999). It has been recently determined that during this stimulation, Cbl ubiquitinates Met and triggers activation of the endocytosis machinery, favoring the clathrin-dependent internalization of Met (Li et al., 2005) and of *L. monocytogenes* to produce infection (Li et al., 2005; Veiga and Cossart, 2005). This result is highly surprising since it was thought that the endocytic machinery supported internalization of vesicles presenting a size inferior only to 100 nm (Gao et al., 2005). However, it seems that the potential use of the endocytic machinery for phagocytosis is a broad phenomenon also observed with other bacterial pathogens (E. Veiga, 2007, personal communication).

8.2.2.5. Effect of Rafts on the Downstream Signaling of Met upon InlB Stimulation

As in the case of the InlA-dependent pathway, the contribution of lipid microdomains in the context of the InlB-dependent entry of *L. monocytogenes* has been analyzed. As opposed to the InlA–E-cadherin interaction, which requires the presence of intact rafts to take place, interaction between InlB and Met can occur in cells which have been depleted of cholesterol (Seveau et al., 2004). However, despite recruitment of Met to the bacterial entry site, downstream signaling leading to actin polymerization was disrupted in MβCD-treated cells. Activation of PI3K is not affected in these cells, but activation of Rac1 is compromised, suggesting that it is the recruitment of $PI(3,4,5)P_3$-binding protein(s) involved in Rac1 activation that requires integrity of plasma membrane microdomains (S. Séveau and J. Swanson, 2006, personal communication).

8.2.3. Role of Other Internalins in L. monocytogenes Entry

As stated above, besides InlA and InlB, 22 more internalins have been identified in the *L. monocytogenes* genome. For example, InlC (also known as IrpA) is an abundantly secreted internalin that does not display LPXTG or GW cell wall-anchoring motifs, but which is characterized by the typical internalin cap, the LRR, and the IR regions (Domann et al., 1997). Four other internalins (InlC2, InlD, InlE, and InlF) which present LPXTG motifs were identified by screening DNA libraries of a *L. monocytogenes* EGD strain using the *inlA* gene as a probe; however, it has been shown that these proteins do not play a direct role in cellular invasion (Dramsi et al., 1997). Interestingly, on a different *L. monocytogenes* EGD isolate, a different gene cluster was identified, coding for the proteins that have been named InlG, InlH, and InlE, which are required for host tissue colonization in the mouse model (Raffelsbauer et al., 1998). A recent study (Bergmann et al., 2002) suggests that InlA needs the support of other internalins for efficient entry into nonphagocytic cells: the InlA-mediated entry is increased in the presence of InlB and InlC, and in the absence of InlB, it requires the presence of InlG, InlH, and InlE.

8.3. Other Molecules Involved in Invasion

8.3.1. Vip: An LPXTG-Anchored Protein Involved in Invasion

Comparison of the genomes of the pathogenic *L. monocytogenes* and the nonphatogenic *L. innocua* has led to the identification of several new *L. monocytogenes* virulence factors that are absent from the *L. innocua* genome. One of these new proteins is Vip, an LPXTG-anchored cell wall protein required for the invasion of Caco-2 and L2071 cell lines (Cabanes et al., 2005). The cellular receptor of Vip is Gp96, an endoplasmic reticulum-resident chaperone that can also be present at the plasma membrane (Cabanes et al., 2005). Gp96 has been implicated in modulation of the immune response by affecting the cellular trafficking of several molecules, including Toll-like receptors (Tsan and Gao, 2004). Thus, it has been suggested that Vip could use Gp96 not only as a receptor for invasion but it could also sequester Gp96 to subvert immunological response during the course of infection (Cabanes et al., 2005).

8.3.2. Ami and Auto: GW Proteins Involved in Adhesion and Invasion

Several *L. monocytogenes* autolysins have been shown to be involved in virulence, including the amidase Ami and Auto. Ami is a 917-amino acids protein with an N-terminal domain that presents similarities to the amidase domain of the *Staphylococcus aureus* Atl autolysin, while its C-terminal domain is anchored to the bacterial cell wall by eight GW domains (Milohanic et al., 2001). Ami exhibits lytic activity on *L. monocytogenes* cell walls (McLaughlan and Foster, 1998), but also mediates bacterial adhesion to target cells within an Δ*inlAB* background (Milohanic et al., 2001). *ami* mutants are attenuated in a mouse model of infection, indicating that Ami plays an important role in virulence (Milohanic et al., 2001). Auto is another *L. monocytogenes* GW-anchored autolysin that is absent from the *L. innocua* genome (Cabanes et al., 2004). Inactivation of *aut* decreases *L. monocytogenes* invasiveness in nonprofessional phagocytes; however, expression of Auto in *L. innocua* does not confer an invasive phenotype, indicating that Auto is necessary but not sufficient for inducing entry (Cabanes et al., 2004). Interestingly, over-expression of Auto on wild-type *L. monocytogenes* also impairs invasion, suggesting that its autolytic activity could be involved in the fine regulation of the bacterial surface architecture required for invasion (Cabanes et al., 2004).

8.3.3. LLO and Calcium Signaling

Listeriolysin O (LLO) is a member of the pore-forming cholesterol-dependent cytolysin (CDC) family (Tweten et al., 2001). Once *L. monocytogenes* has been internalized in target cells, LLO plays a critical role in the permeabilization of the bacterial-containing vacuoles favoring bacterial relocation to the host cell cytosol (Dramsi and Cossart, 2002). Interestingly, it has been shown that LLO secreted by extracellular *L. monocytogenes* forms Ca^{2+}-permeable pores leading to intracellular Ca^{2+} oscillations (Repp et al., 2002), and these oscillations have been associated with cellular responses including interleukin-6 persistent production in Caco-2 epithelial cells (Tsuchiya et al., 2005). In addition, influx of extracellular Ca^{2+} (but not Ca^{2+} released from intracellular stores) potentiates entry of *L. monocytogenes* into Hep-2 cells (Dramsi and Cossart, 2003). The cellular receptor for LLO is cholesterol, and as expected, LLO can be associated to lipid rafts in the plasma membrane of host cells (Coconnier et al., 2000).

8.3.4. ActA

ActA is the surface protein responsible for the actin-based motility system that enables *L. monocytogenes* to move from one infected cell to an uninfected neighboring cell, favoring bacterial tissue spreading without being exposed to the extracellular environment (Gouin et al., 2005; Kocks et al., 1992). Besides its intracellular role in cell-to-cell spreading, ActA has been implicated in *L. monocytogenes* cellular invasion: the N-terminal region of ActA presents several clusters of positively charged amino acids that could be implicated in heparan sulfate binding, and *L. monocytogenes* mutant defective in ActA is significantly impaired in cellular attachment and entry due to altered heparan sulfate recognition (Alvarez-Dominguez et al., 1997). Recently, these results have been complemented by another report indicating that expression of ActA in the noninvasive *L. innocua* is sufficient to induce bacterial entry into epithelial cell lines (Suarez et al., 2001).

8.4. Conclusions

Translocation to the intracellular space is a critical step during the *L. monocytogenes* infectious process. Several bacterial surface molecules have been implicated in the invasion of target cells by *L. monocytogenes*, among which internalins InlA and InlB remain the best-characterized examples. Comparative genomics has permitted the identification of new virulence factors in the *L. monocytogenes* genome that are necessary but not sufficient for inducing entry, highlighting the fact that cellular invasion requires the interplay of many molecular actors for the generation of an optimal bacterial phenotype suited for interaction with host cell surfaces and subversion of cellular functions. Understanding how *L. monocytogenes* coordinates the expression of these different virulence factors during infection remains a topic of intense current interest.

Acknowledgments. We would like to thank members of the Cossart Lab for helpful discussions. We would like to apologize to all those authors whose work we may not have adequately presented or even omitted due to space limitations. PC is an international research scholar of the Howard Hughes Medical Institute.

References

Alvarez-Dominguez C, Vazquez-Boland J A, Carrasco-Marin E, Lopez-Mato P, and Leyva-Cobian F (1997). Host cell heparan sulfate proteoglycans mediate attachment and entry of *Listeria monocytogenes*, and the listerial surface protein ActA is involved in heparan sulfate receptor recognition. Infect Immun *65*, 78–88.

Banerjee M, Copp J, Vuga D, Marino M, Chapman T, van der Geer P, and Ghosh P (2004). GW domains of the *Listeria monocytogenes* invasion protein InlB are required for potentiation of Met activation. Mol Microbiol *52*, 257–271.

Basar T, Shen Y, and Ireton K (2005). Redundant roles for Met docking site tyrosines and the Gab1 pleckstrin homology domain in InlB-mediated entry of *Listeria monocytogenes*. Infect Immun *73*, 2061–2074.

Bergmann B, Raffelsbauer D, Kuhn M, Goetz M, Hom S, and Goebel W (2002). InlA-but not InlB-mediated internalization of *Listeria monocytogenes* by non-phagocytic mammalian cells needs the support of other internalins. Mol Microbiol *43*, 557–570.

Bierne H, Gouin E, Roux P, Caroni P, Yin H L, and Cossart P (2001). A role for cofilin and LIM kinase in Listeria-induced phagocytosis. J Cell Biol *155*, 101–112.

Bierne H, Mazmanian S K, Trost M, Pucciarelli M G, Liu G, Dehoux P, Jansch L, Garcia-del Portillo F, Schneewind O, and Cossart P (2002). Inactivation of the srtA gene in *Listeria monocytogenes* inhibits anchoring of surface proteins and affects virulence. Mol Microbiol *43*, 869–881.

Bierne H, Miki H, Innocenti M, Scita G, Gertler F B, Takenawa T, and Cossart P (2005). WASP-related proteins, Abi1 and Ena/VASP are required for Listeria invasion induced by the Met receptor. J Cell Sci *118*, 1537–1547.

Birchmeier C, Birchmeier W, Gherardi E, and Vande Woude G F (2003). Met, metastasis, motility and more. Nat Rev Mol Cell Biol *4*, 915–925.

Birchmeier C, and Gherardi E (1998). Developmental roles of HGF/SF and its receptor, the c-Met tyrosine kinase. Trends Cell Biol *8*, 404–410.

Braun L, Dramsi S, Dehoux P, Bierne H, Lindahl G, and Cossart P (1997). InlB: an invasion protein of *Listeria monocytogenes* with a novel type of surface association. Mol Microbiol *25*, 285–294.

Braun L, Ghebrehiwet B, and Cossart P (2000). gC1q-R/p32, a C1q-binding protein, is a receptor for the InlB invasion protein of *Listeria monocytogenes*. Embo J *19*, 1458–1466.

Braun L, Nato F, Payrastre B, Mazie J C, and Cossart P (1999). The 213-amino-acid leucine-rich repeat region of the *Listeria monocytogenes* InlB protein is sufficient for entry into mammalian cells, stimulation of PI 3-kinase and membrane ruffling. Mol Microbiol *34*, 10–23.

Braun L, Ohayon H, and Cossart P (1998). The InlB protein of *Listeria monocytogenes* is sufficient to promote entry into mammalian cells. Mol Microbiol *27*, 1077–1087.

Cabanes D, Dussurget O, Dehoux P, and Cossart P (2004). Auto, a surface associated autolysin of *Listeria monocytogenes* required for entry into eukaryotic cells and virulence. Mol Microbiol *51*, 1601–1614.

Cabanes D, Sousa S, Cebria A, Lecuit M, Garcia-del Portillo F, and Cossart P (2005). Gp96 is a receptor for a novel *Listeria monocytogenes* virulence factor, Vip, a surface protein. Embo J *24*, 2827–2838.

Coconnier M H, Lorrot M, Barbat A, Laboisse C, and Servin A L (2000). Listeriolysin O-induced stimulation of mucin exocytosis in polarized intestinal mucin-secreting cells: evidence for toxin recognition of membrane-associated lipids and subsequent toxin internalization through caveolae. Cell Microbiol *2*, 487–504.

Cossart P, and Lecuit M (1998). Interactions of *Listeria monocytogenes* with mammalian cells during entry and actin-based movement: bacterial factors, cellular ligands and signaling. EMBO J 17(14), 3797–3806.

Cossart P, Pizarro-Cerda J, and Lecuit M (2003). Invasion of mammalian cells by *Listeria monocytogenes*: functional mimicry to subvert cellular functions. Trends Cell Biol *13*, 23–31.

Domann E, Zechel S, Lingnau A, Hain T, Darji A, Nichterlein T, Wehland J, and Chakraborty T (1997). Identification and characterization of a novel PrfA-regulated gene in *Listeria monocytogenes* whose product, IrpA, is highly homologous to internalin proteins, which contain leucine-rich repeats. Infect Immun *65*, 101–109.

Dramsi S, Biswas I, Maguin E, Braun L, Mastroeni P, and Cossart P (1995). Entry of *Listeria monocytogenes* into hepatocytes requires expression of inIB, a surface protein of the internalin multigene family. Mol Microbiol *16*, 251–261.

Dramsi S, and Cossart P (2002). Listeriolysin O: a genuine cytolysin optimized for an intracellular parasite. J Cell Biol *156*, 943–946.

Dramsi S, and Cossart P (2003). Listeriolysin O-mediated calcium influx potentiates entry of *Listeria monocytogenes* into the human Hep-2 epithelial cell line. Infect Immun *71*, 3614–3618.

Dramsi S, Dehoux P, Lebrun M, Goossens P L, and Cossart P (1997). Identification of four new members of the internalin multigene family of *Listeria monocytogenes* EGD. Infect Immun *65*, 1615–1625.

Drees F, Pokutta S, Yamada S, Nelson W J, and Weis W I (2005). Alpha-catenin is a molecular switch that binds E-cadherin-beta-catenin and regulates actin-filament assembly. Cell *123*, 903–915.

Gaillard J L, Berche P, Frehel C, Gouin E, and Cossart P (1991). Entry of *L. monocytogenes* into cells is mediated by internalin, a repeat protein reminiscent of surface antigens from gram-positive cocci. Cell *65*, 1127–1141.

Gao H, Shi W, and Freund L B (2005). Mechanics of receptor-mediated endocytosis. Proc Natl Acad Sci U S A *102*, 9469–9474.

Glaser P, Frangeul L, Buchrieser C, Rusniok C, Amend A, Baquero F, Berche P, Bloecker H, Brandt P, Chakraborty T, Charbit A, Chetouani F, Couve, E, de DaruvarA, Dehoux P, Domann E, Dominguez-Bernal G, Duchaud E, Durant L, Dussurget O, Entian K-D, Fsihi H, Garcia-del Portillo F, Garrido P, Gautier L, Goebel W, Gomez-Lopez N, Hain T, Hauf J, Jackson D, Jones L-M, Kaerst U, Kreft J, Kuhn M, Kunst F, Kurapkat G, Madueno E, Maitournam A, Mata Viciente J, Ng E, Nedjari H, Nordsiek G, Novella S, de Pablos B, Perez-Diaz J-C, Purcell R, Remmel B, Rose M, Schlueter T, Simoes N, Tierrez A, Vazquez-Boland J-A, Voss H, Wehland J and Cossart P (2001). Comparative genomics of Listeria species. Science *294*, 849–852.

Gouin E, Welch M D, and Cossart P (2005). Actin-based motility of intracellular pathogens. Curr Opin Microbiol *8*, 35–45.

Gumbiner B M (2005). Regulation of cadherin-mediated adhesion in morphogenesis. Nat Rev Mol Cell Biol *6*, 622–634.

Hubbard S R, and Till J H (2000). Protein tyrosine kinase structure and function. Annu Rev Biochem *69*, 373–398.

Ireton K, Payrastre B, Chap H, Ogawa W, Sakaue H, Kasuga M, and Cossart P. (1996). A role for phosphoinositide 3-kinase in bacterial invasion. Science *274*, 780–782.

Ireton K, Payrastre B, and Cossart P (1999). The *Listeria monocytogenes* protein InlB is an agonist of mammalian phosphoinositide 3-kinase. J Biol Chem *274*, 17025–17032.

Jacquet C, Doumith M, Gordon J I, Martin P M, Cossart P, and Lecuit M (2004). A molecular marker for evaluating the pathogenic potential of foodborne *Listeria monocytogenes*. J Infect Dis *189*, 2094–2100.

Jamora C, and Fuchs E (2002). Intercellular adhesion, signalling and the cytoskeleton. Nat Cell Biol *4*, 101–108.

Jonquieres R, Bierne H, Fiedler F, Gounon P, and Cossart P (1999). Interaction between the protein InlB of *Listeria monocytogenes* and lipoteichoic acid: a novel mechanism of protein association at the surface of gram-positive bacteria. Mol Microbiol *34*, 902–914.

Jonquieres R, Bierne H, Mengaud J, and Cossart P (1998). The inlA gene of *Listeria monocytogenes* LO28 harbors a nonsense mutation resulting in release of internalin. Infect Immun *66*, 3420–3422.

Jonquieres R, Pizarro-Cerda J, and Cossart P (2001). Synergy between the N- and C-terminal domains of InlB for efficient invasion of non-phagocytic cells by *Listeria monocytogenes*. Mol Microbiol *42*, 955–965.

Junghans D, Haas I G, and Kemler R (2005). Mammalian cadherins and protocadherins: about cell death, synapses and processing. Curr Opin Cell Biol *17*, 446–452.

Kobe B, and Kajava A V (2001). The leucine-rich repeat as a protein recognition motif. Curr Opin Struct Biol *11*, 725–732.

Kocks C, Gouin E, Tabouret M, Berche P, Ohayon H, and Cossart P (1992). *L. monocytogenes*-induced actin assembly requires the actA gene product, a surface protein. Cell *68*, 521–531.

Lecuit M, Dramsi S, Gottardi C, Fedor-Chaiken M, Gumbiner B, and Cossart P (1999). A single amino acid in E-cadherin responsible for host specificity towards the human pathogen *Listeria monocytogenes*. Embo J *18*, 3956–3963.

Lecuit M, Hurme R, Pizarro-Cerda J, Ohayon H, Geiger B, and Cossart P (2000). A role for alpha-and beta-catenins in bacterial uptake. Proc Natl Acad Sci U S A *97*, 10008–10013.

Lecuit M, Nelson D M, Smith S D, Khun H, Huerre M, Vacher-Lavenu M C, Gordon J I, and Cossart P (2004). Targeting and crossing of the human maternofetal barrier by *Listeria monocytogenes*: role of internalin interaction with trophoblast E-cadherin. Proc Natl Acad Sci U S A *101*, 6152–6157.

Lecuit M, Ohayon H, Braun L, Mengaud J, and Cossart P (1997). Internalin of *Listeria monocytogenes* with an intact leucine-rich repeat region is sufficient to promote internalization. Infect Immun *65*, 5309–5319.

Lecuit M, Vandormael-Pournin S, Lefort J, Huerre M, Gounon P, Dupuy C, Babinet C, and Cossart P (2001). A transgenic model for listeriosis: role of internalin in crossing the intestinal barrier. Science *292*, 1722–1725.

Li N, Xiang G S, Dokainish H, Ireton K, and Elferink L A (2005). The Listeria protein internalin B mimics hepatocyte growth factor-induced receptor trafficking. Traffic *6*, 459–473.

Machner M P, Frese S, Schubert W D, Orian-Rousseau V, Gherardi E, Wehland J, Niemann H H, and Heinz D W (2003). Aromatic amino acids at the surface of InlB are essential for host cell invasion by *Listeria monocytogenes*. Mol Microbiol *48*, 1525–1536.

Marino M, Banerjee M, Jonquieres R, Cossart P, and Ghosh P (2002). GW domains of the *Listeria monocytogenes* invasion protein InlB are SH3-like and mediate binding to host ligands. Embo J *21*, 5623–5634.

Marino M, Braun L, Cossart P, and Ghosh P (1999). Structure of the lnlB leucine-rich repeats, a domain that triggers host cell invasion by the bacterial pathogen *L. monocytogenes*. Mol Cell *4*, 1063–1072.

McLaughlan A M, and Foster S J (1998). Molecular characterization of an autolytic amidase of *Listeria monocytogenes* EGD. Microbiology *144 (Pt 5)*, 1359–1367.

Mengaud J, Ohayon H, Gounon P, Mege R M, and Cossart P (1996). E-cadherin is the receptor for internalin, a surface protein required for entry of *L. monocytogenes* into epithelial cells. Cell *84*, 923–932.

Milohanic E, Jonquieres R, Cossart P, Berche P, and Gaillard J L (2001). The autolysin Ami contributes to the adhesion of *Listeria monocytogenes* to eukaryotic cells via its cell wall anchor. Mol Microbiol *39*, 1212–1224.

Nelson W J, and Nusse R (2004). Convergence of Wnt, beta-catenin, and cadherin pathways. Science *303*, 1483–1487.

Pentecost M, Otto G, Theriot J A, and Amieva M R (2006). *Listeria monocytogenes* invades the epithelial junctions at sites of cell extrusion. PLoS Pathog *2*, e3.

Peschard P, Fournier T M, Lamorte L, Naujokas M A, Band H, Langdon W Y, and Park M (2001). Mutation of the c-Cbl TKB domain binding site on the Met receptor tyrosine kinase converts it into a transforming protein. Mol Cell *8*, 995–1004.

Pizarro-Cerda J, and Cossart P (2006). Subversion of cellular functions by *Listeria monocytogenes*. J Pathol *208*, 215–223.

Pollard T D, and Beltzner C C (2002). Structure and function of the Arp2/3 complex. Curr Opin Struct Biol *12*, 768–774.

Ponzetto C, Bardelli A, Zhen Z, Maina F, dalla Zonca P, Giordano S, Graziani A, Panayotou G, and Comoglio P M (1994). A multifunctional docking site mediates signaling and transformation by the hepatocyte growth factor/scatter factor receptor family. Cell *77*, 261–271.

Raffelsbauer D, Bubert A, Engelbrecht F, Scheinpflug J, Simm A, Hess J, Kaufmann S H, and Goebel W (1998). The gene cluster inlC2DE of *Listeria monocytogenes* contains additional new internalin genes and is important for virulence in mice. Mol Gen Genet *260*, 144–158.

Repp H, Pamukci Z, Koschinski A, Domann E, Darji A, Birringer J, Brockmeier D, Chakraborty T, and Dreyer F (2002). Listeriolysin of *Listeria monocytogenes* forms Ca^{2+}-permeable pores leading to intracellular Ca^{2+} oscillations. Cell Microbiol *4*, 483–491.

Schubert W D, Gobel G, Diepholz M, Darji A, Kloer D, Hain T, Chakraborty T, Wehland J, Domann E, and Heinz D W (2001). Internalins from the human pathogen *Listeria monocytogenes* combine three distinct folds into a contiguous internalin domain. J Mol Biol *312*, 783–794.

Schubert W D, Urbanke C, Ziehm T, Beier V, Machner M P, Domann E, Wehland J, Chakraborty T, and Heinz D W (2002). Structure of internalin, a major invasion protein of *Listeria monocytogenes*, in complex with its human receptor E-cadherin. Cell *111*, 825–836.

Seveau S, Bierne H, Giroux S, Prevost M C, and Cossart P. (2004). Role of lipid rafts in E-cadherin— and HGF-R/Met—mediated entry of *Listeria monocytogenes* into host cells. J Cell Biol *166*, 743–753.

Shen Y, Naujokas M, Park M, and Ireton K (2000). InIB-dependent internalization of Listeria is mediated by the Met receptor tyrosine kinase. Cell *103*, 501–510.

Simons K, and Toomre D (2000). Lipid rafts and signal transduction. Nat Rev Mol Cell Biol *1*, 31–39.

Sousa S, Cabanes D, Archambaud C, Colland F, Lemichez E, Popoff M, Boisson-Dupuis S, Gouin E, Lecuit M, Legrain P, and Cossart P (2005). ARHGAP10 is necessary for alpha-catenin recruitment at adherens junctions and for Listeria invasion. Nat Cell Biol *7*, 954–960.

Sousa S, Cabanes D, El-Amraoui A, Petit C, Lecuit M, and Cossart P (2004). Unconventional myosin VIIa and vezatin, two proteins crucial for Listeria entry into epithelial cells. J Cell Sci *117*, 2121–2130.

Stradal T E, and Scita G (2006). Protein complexes regulating Arp2/3-mediated actin assembly. Curr Opin Cell Biol *18*, 4–10.

Suarez M, Gonzalez-Zorn B, Vega Y, Chico-Calero I, and Vazquez-Boland JA (2001). A role for ActA in epithelial cell invasion by *Listeria monocytogenes*. Cell Microbiol *3*, 853–864.

Tsan M F, and Gao B (2004). Heat shock protein and innate immunity. Cell Mol Immunol *1*, 274–279.

Tsuchiya K, Kawamura I, Takahashi A, Nomura T, Kohda C, and Mitsuyama M (2005). Listeriolysin O-induced membrane permeation mediates persistent interleukin-6 production in Caco-2 cells during *Listeria monocytogenes* infection in vitro. Infect Immun *73*, 3869–3877.

Tweten R K, Parker M W, and Johnson A E (2001). The cholesterol-dependent cytolysins. Curr Top Microbiol Immunol *257*, 15–33.

Veiga E, and Cossart P (2005). Listeria hijacks the clathrin-dependent endocytic machinery to invade mammalian cells. Nat Cell Biol 7, 894–900.

Yamada S, Pokutta S, Drees F, Weis W I, and Nelson WJ (2005). Deconstructing the cadherin-catenin-actin complex. Cell 123, 889–901.

9
Escape of *Listeria monocytogenes* from a Vacuole

Howard Goldfine[1] and Hélène Marquis[2]

[1]*Department of Microbiology, University of Pennsylvania School of Medicine, Philadelphia, PA 19104-6076, USA*
[2]*Department of Microbiology and Immunology, Cornell University, Ithaca, NY 14853-6401, USA*

Abstract: The ability of *Listeria monocytogenes* to escape from vacuoles of infected cells and subsequently to replicate in the cytosol and spread from cell to cell is one of the distinctive features of this facultative intracellular pathogen. The process of escape is mediated by several proteins that are encoded by genes in the PrfA regulon cluster. These include listeriolysin O, a pore-forming, cholesterol-dependent cytolysin, a phosphatidylinositol-specific phospholipase C (PI-PLC), and a broad range phospholipase whose proteolytic activation is mediated by a metalloprotease. These proteins are described and their specific roles in escape from the vacuole are discussed.

9.1.   Introduction

The ability of *Listeria monocytogenes* to escape from an endocytic vacuole after engulfment by either professional phagocytic cells or parenchymal cells invaded as a result of the action of specific listerial surface proteins (Chap. 8) is of paramount importance for its survival in its hosts and one of the features that distinguishes it from almost all other facultative intracellular pathogens (Gaillard et al. 1987; Mounier et al. 1990; Tilney and Portnoy 1989). The PrfA regulon cluster consisting of the genes *prfA*, *plcA*, *hly*, *mpl*, *actA*, and *plcB* is organized to provide sequentially all of the proteins that assist in this process (Chap. 7). The properties of the products of these genes and their role in escape from a vacuole will be discussed in this chapter.

9.2.   Escape from the Primary Vacuole

9.2.1.   Macrophages

The fate of *L. monocytogenes* in a macrophage depends upon both the cell lineage and its history. Primary peritoneal macrophages are capable of killing approximately 90% of entering *L. monocytogenes* (Portnoy et al. 1989), bone marrow–derived macrophages kill approximately half the bacteria, whereas cell lines such as J774 and RAW 264.7 murine macrophage–derived cells are weakly cidal (Camilli et al. 1993; De Chastellier 1994). Treatment of peritoneal macrophages with γ interferon (IFN-γ) prior to infection resulted in complete suppression of bacterial growth (Portnoy et al. 1989), consistent with the ability of activated macrophages to kill *L. monocytogenes* (Cossart and Portnoy 2000). The generation of reactive oxygen and nitrogen intermediates within the vacuole contributes to the retention of *L. monocytogenes* within phagosomes in activated macrophages (Myers et al. 2003).

9.2.1.1.   Requirements

9.2.1.1.1.   Listeriolysin O

Two of the gene products of the PrfA regulon, listeriolysin O (LLO), the product of *hly*, and a phosphatidylinositol-specific phospholipase C (PI-PLC), the product of *plcA*, play key roles in escape from the primary vacuole of a macrophage. LLO is a member of a large family of cholesterol-dependent pore-forming toxins or cytolysins (CDC) which includes streptolysin O (SLO) secreted by *Streptococcus pyogenes*, perfringolysin O (PFO) secreted by *Clostridium perfringens*, anthrolysin O (ALO) secreted by *Bacillus anthracis,* and pneumolysin, which is produced by *Streptococcus pneumoniae*. These multidomain proteins insert into cholesterol-containing eukaryotic cell membranes forming large pores of 25–30 nm in diameter by oligomerization of approximately 50 monomers (Alouf 1999; Giddings et al. 2004). LLO deletion mutants are avirulent and fail to escape from the primary vacuole of a macrophage (Gaillard et al. 1987; Kuhn et al. 1988; Portnoy et al. 1988). Expression of *hly* in *Bacillus subtilis* and *Salmonella* conferred the ability to escape from the phagosome of J774 cells to this species; however, the expression of LLO did not convert *B. subtilis* into a pathogen (Bielecki et al. 1990; Gentschev et al. 1995). Although the three-dimensional structure of LLO is not known, that of PFO has been established (Rossjohn et al. 1997) (Figure 9.1.). Molecular simulation based on the structure of PFO suggests that the structure of LLO is similar, consisting of four domains. Domain four anchors the oligomers to the membrane (Ramachandran et al. 2002) whereas domain three regulates polymerization (Ramachandran et al. 2004) and pore formation (Shatursky et al. 1999).

LLO, 60 kDa, is uniquely structured to function in the acidic environment of the vacuole by virtue of its low pH optimum (pH 5.5). Leucine 461 was found to be critical for the acidic pH optimum of LLO. When leucine 461 was changed

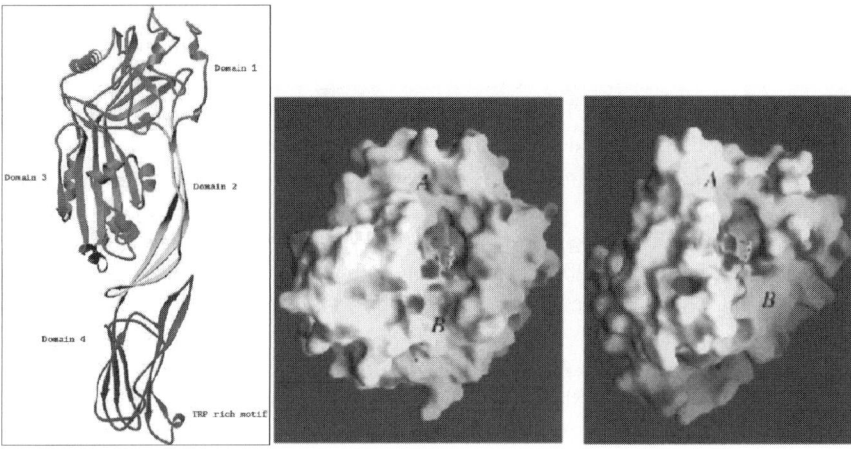

FIGURE 9.1. (Left) Structure of the perfringolysin O molecule. Taken from Rossjohn et al. (1997), with permission from Elsevier. The comparative structures of *B. cereus* PI-PLC (center) and *L. monocytogenes* PI-PLC (right) with the active site pocket containing *myo*-inositol in the center. The hydrophobic ridge is designated by A and the extended GPI-binding site by B. Taken from (Moser et al. 1997), with permission of the publisher. Reprinted from Journal of Molecular Biology, Vol 273:1, J. Moser et al, Crystal Structure of the Phosphatidylinositol, pages 269–282, Copyright 1997, with permission from Elsevier.

to threonine, LLO became about 10-fold more active at neutral pH (Glomski et al. 2003). This increase in activity results from an increase in protein stability at neutral pH (Schuerch et al. 2005). Substitution of either PFO or ALO for LLO in *L. monocytogenes* permits escape of bacteria from the primary phagosome of a macrophage (Jones and Portnoy 1994; Wei et al. 2005a). Both of these proteins are much more active than LLO at neutral pH, consequently their expression leads to membrane permeabilization upon growth of *L. monocytogenes* in the cytosol (Jones and Portnoy 1994; Wei et al. 2005a). Like PFO or ALO, which have threonine at the equivalent position, LLO L461T was able to mediate escape of bacteria from the primary vacuole, but it became cytotoxic to the host cell by permeabilizing the cell membrane (Glomski et al. 2002; Glomski et al. 2003).

LLO synthesis is upregulated in the phagosome (Bubert et al. 1999; Freitag and Jacobs 1999; Gray et al. 2006; Klarsfeld et al. 1994; Moors et al. 1999). Recently evidence has emerged indicating that the nucleotide sequence of the hly transcript is involved in translational repression of LLO in the host cytosol. This regulation is dictated by sequences within the coding region of hly mRNA (Schnupf et al. 2006). Thus, the potential cytotoxicity of LLO in the cytosol is modulated both by its lower activity at neutral pH and by repression of its synthesis.

The CDCs are characterized as oxygen-labile or thiol-activated. A characteristic motif in the carboxyl terminal of domain 4 is a conserved undecapeptide, ECTGLAWEWWR. This sequence contains the only cysteine residue present in most of the toxins in this family (Alouf 1999). Although reduction of this cysteine by sulfhydryl reagents is necessary for the activity of CDCs, replacement of Cys-484 by Ala in LLO showed that a thiol group is not essential for hemolytic activity (Michel et al. 1990).

LLO and other members of the CDC family interact with receptors on mammalian cell surfaces. Intermedilysin, a CDC toxin secreted by *Streptococcus intermedius*, interacts with human CD59 on erythrocytes (Giddings et al. 2004). ALO, a recently discovered member of the CDC family secreted by *Bacillus anthracis* (Shannon et al. 2003), interacts with the macrophage Toll-like receptor 4 (TLR4), known as a specific receptor for lipopolysaccharide of gram-negative bacteria. The interaction of ALO with TLR-4 resulted in typical signaling through p38 MAPK. In addition to ALO, activation of macrophages through TLR-4 was also observed with LLO, SLO, PFO, and pneumolysin (Park et al. 2004). These recent findings are consistent with the known ability of noncytolytic cholesterol complexes of LLO to induce cytokine expression (Nishibori et al. 1996), lipid second messengers (Sibelius et al. 1996), and an IL-1 response (Yoshikawa et al. 1993), and suggest that LLO is multifunctional by virtue of its ability to form pores and its ability to interact with surface receptors on eukaryotic cells. Binding of LLO to membrane cholesterol has been attributed to domain 4, whereas cytokine induction was attributed to domains 1–3 (Jacobs et al. 1999; Kohda et al. 2002).

9.2.1.1.2. *Phosphatidylinositol-Specific Phospholipase C*

Phosphatidylinositol-specific phospholipase C of *L. monocytogenes* (LmPI-PLC), encoded by *plcA* (Camilli et al. 1991; Leimeister-Wächter et al. 1991; Mengaud et al. 1991a), is highly specific for phosphatidylinositol (PI) with relatively weak activity on glycosyl-PI-(GPI)-anchored proteins (Gandhi et al. 1993; Goldfine and Knob 1992). Mutants with deletions of *plcA* escape from the primary vacuole of macrophages less efficiently than the wild type. At 90 min after infection between 30 and 65%, fewer of these mutants have escaped from the phagosome compared to wild type (Bannam and Goldfine, 1999; Camilli et al. 1993; Smith et al. 1995; Wadsworth and Goldfine 1999).

The weak activity of LmPI-PLC differentiates it from the classical bacterial PI-PLCs from *Bacillus* species and *Staphylococcus aureus*, which have strong activity on GPI-anchored proteins (Low 1989; Wei et al. 2005b). Although LmPI-PLC shares only 24% identity with PI-PLC from *B. cereus* (BcPI-PLC), the overall three-dimensional structures are highly homologous (Moser et al. 1997). Both consist of a single $(\beta\alpha)_8$-barrel domain. The active site pocket is highly conserved with only two differences in amino acids involved in inositol binding, but complete conservation of the residues thought to be involved in catalysis. An important structural difference is the absence in LmPI-PLC of the Vb β-strand of BcPI-PLC, which supports the edge of a shallow groove extending from the active site pocket and is predicted to promote interactions with the glycan of GPI-anchored proteins (Figure 9.1.) (Moser et al. 1997). Removal of the Vb β-strand of BcPI-PLC resulted in somewhat decreased activity on PI and essentially complete loss of activity on the GPI-anchored protein Thy-1 on mouse splenocytes (Wei et al. 2005b). Like other bacterial PI-PLCs, LmPI-PLC does not cleave polyphosphoinositides such as PI 4,5-bisphosphate (PI-4,5-P$_2$), an important substrate of eukaryotic phospholipases involved in intracellular signaling (Goldfine and Knob 1992). Although both types of PI-PLC produce the second messenger diacylglycerol (DAG), the action of LmPI-PLC on PI produces inositol-1-P and not inositol 1,4,5-trisphosphate (IP$_3$), another second messenger that releases Ca^{2+} from the endoplasmic reticulum. Another significant difference between LmPI-PLC and those from eukaryotic cells is the absence of a divalent cation requirement. Instead, LmPI-PLC shows a strong dependence on high ionic strength salts, e.g., 100–200 mM NaCl, KCl, NH$_4$Cl, or (NH$_4$)$_2$SO$_4$, which is needed for disaggregation of multimers (Goldfine and Knob 1992).

9.2.1.2.   Permeabilization of the Phagosome

Upon internalization of *L. monocytogenes*, the J774 phagosome is rapidly acidified to pH 4.8–6.5. During this time the synthesis of LLO is greatly increased (Chap. 7). Soon after internalization the phagosomal membrane is permeabilized to the dye HPTS (8-hydroxypyrene-1,3,6-trisuolfonic acid), and the pH of the vacuole rapidly increases. Agents that inhibit acidification, such as bafilomycin, inhibit perforation and escape (Conte et al. 1996; Beauregard et al. 1997; Glomski et al. 2002). Permeabilization of the vacuole is absolutely dependent on the expression of LLO (Beauregard et al. 1997). These findings are consistent with a model in which LLO forms pores in the phagosome and permits a two-way exchange of small molecules. It appears, however, that host factors play a role in this process. A mutant lacking PI-PLC produced about 65% fewer permeabilized vacuoles than the wild type between 30 and 60 min after infection of J774 cells (Poussin and Goldfine 2005). This finding could be consistent with a model in which PI-PLC, by hydrolyzing a minor component of the vacuolar membrane, assists in the degradation of the membrane (Villar et al. 2000). However, inhibition of host PKC β isoforms by treatment of J774 cells with the inhibitors RO-31-8425 and Gö-6983, produced 29 and 62% fewer permeabilized vacuoles, respectively, than in untreated cells during the same time frame. Inhibition of host calcium signaling by treatment with thapsigargin or SK&F 96365 produced even greater inhibition of vacuolar permeabilization by wild type *L. monocytogenes* (Poussin and Goldfine 2005). As will be discussed below, activation of PKC β isoforms is dependent on expression of both LLO and PI-PLC and on the elevation of host intracellular calcium. Inhibition of both calcium signaling and PKC β activation also inhibits escape from the primary vacuole of a macrophage.

Perforation of the Lm phagosome in macrophage-like cells which permits the escape of small molecules like HPTS (524 MW) (Beauregard et al. 1997; Poussin and Goldfine 2005) and Lucifer Yellow (522 MW) into the cytosol is followed 5–9 min later by larger pores that permit the escape of molecules like fluorescent dextrans (average MW 10,000). This exchange of small molecules, protons, and calcium ions is postulated to inhibit vacuole fusion with lysosomes (Shaughnessy et al. 2006).

9.2.1.3.   Mechanism of Escape

It is well known that LLO is indispensable for escape from the primary
phagosome of a macrophage. From this fact a predominant hypothesis has
emerged which states that the perforation of the phagosomal membrane by LLO
leads to escape. It is possible that the physical act of perforation is the mechanical
means of disruption of the membrane. Further considerations argue against this
hypothesis. The phagosomal membrane must be larger than the bacterium it
contains; therefore, its surface area should be approximately $1–4\,\mu m^2$. A pore
formed by LLO is approximately 25 nm in diameter or about $500\,nm^2$. To cover
the surface of the phagosome completely would require approximately 2000
pores. Yet, once the phagosome is permeabilized by LLO, there is a rapid
increase in pH (Beauregard et al. 1997) which presumably results in greatly
reduced LLO activity. Unless there is a concerted action of LLO resulting in
the simultaneous formation of multiple pores, these calculations strongly argue
against mechanical disruption of the phagosome by LLO.

   LLO along with PI-PLC leads to the rapid activation of host functions in
macrophages including opening of calcium channels and release of Ca^{2+} from
stores, activation of host polyphosphoinositide-specific PLC, phospholipase D
(PLD), and PKC isoforms. Inhibition of calcium elevation or activation of PLD
or PKC β leads to strong, but not complete inhibition of escape from the
phagosome (Wadsworth and Goldfine 1999; Goldfine et al. 2000; Wadsworth
and Goldfine 2002; Poussin and Goldfine 2005). PKC β I and II are found on
early endosomes which fuse to form large vesicles, within minutes of infection
of J774 macrophage-like cells with wild type *L. monocytogenes*. LLO and
PI-PLC expression are needed for mobilization of PKC β II, and LLO expression
is required for mobilization of PKC β I (Wadsworth and Goldfine 2002).
The activation of host polyphosphoinositide-specific PLC and phospholipase
D (PLD) also requires expression of LLO, but not PI-PLC. These findings
suggest a model in which LLO activates host functions that are needed for
disruption of the phagosome. At this time, there is no available information
on how PKC β influences phagosomal perforation and disruption. Hannun
and colleagues have recently shown that classical PKC isoforms α and β II
appear on a juxtanuclear subset of recycling endosomes, called the pericen-
trion, after treatment of a variety of cell types with PMA, a known activator of
classical PKC isoforms (Becker and Hannun 2003; Becker and Hannun 2004).
Compared to the mobilization of PKC β II upon infection of J774 cells with
L. monocytogenes, which takes place within the first minute of infection, juxtanuclear translocation after treatment with PMA requires 30–60 min treatment.
PKC translocation may regulate the clustering of recycling endosomes in the
perinuclear region (Idkowiak-Baldys et al. 2006). These findings suggest that
L. monocytogenes subverts another normal process in cells which results in a
specific outcome, i.e., delay of maturation of the phagosome and subsequent
escape.

   It appears that *L. monocytogenes* controls vesicular trafficking in the host
by controlling the activity of Rab5, a small GTPase involved in the regulation

of phagosome–endosome fusion and phagosomal maturation (Desjardins 1995; Alvarez-Dominguez et al. 1996). The ability of LLO-negative bacteria to survive requires inhibition of phagosome maturation by a mechanism involving Rab5a (Alvarez-Dominguez et al. 1997; Alvarez-Dominguez and Stahl 1999). In these two studies, the authors used a LLO-minus strain. Treatment of J774 cells with IFN-γ increases the association of lysosomal markers such as cathepsin-D, lysosome-associated membrane protein-1 (LAMP1), and Limp-II with phagosomes and this also appears to be controlled by Rab5a, which is increased on phagosomes from cells treated with IFN-γ (Prada-Delgado et al. 2001). Evidence has been presented implicating inhibition of Rab5a exchange activity by *L. monocytogenes* as the means for controlling phagosome–lysosome trafficking (Prada-Delgado et al. 2005). On the other hand, Henry et al. observed that *L. monocytogenes* escapes from Rab5a-negative, LAMP1-negative, Rab7-positive, and PI-3-P-positive vacuoles in a manner that is LLO-dependent. When *hly*, the gene coding for LLO, was expressed under control of an inducible promoter and its expression in the vacuole was delayed until the bacteria were in LAMP1-positive vacuoles, escape was less efficient than when *hly* was expressed normally (Henry et al. 2006).

The exclusion of Rab5a from *L. monocytogenes* vacuoles may be related to the absence of the GTP-exchange mechanism needed for bringing Rab5a to the vacuolar membrane (Prada-Delgado et al. 2005). Overexpression of Rab5Q79L, which is locked in the GTP-bound state, resulted in association of Rab5a with the *L. monocytogenes* vacuole, but did not affect *L. monocytogenes* escape from vacuoles (Henry et al. 2006). Although Rab5a is important for vesicular trafficking, its role in macrophage killing of *L. monocytogenes* is at present unclear.

9.2.2. Nonphagocytic Cells

9.2.2.1. Requirements

9.2.2.1.1. LLO and PI-PLC

The two bacterial factors contributing to escape from primary vacuoles of macrophages, LLO and PI-PLC, also contribute to escape from primary vacuoles of non-phagocytic cells (Marquis et al. 1995; Dancz et al. 2002), with the exception that LLO is not essential in human epithelial and fibroblast cells, as well as in *Potoroo tridactylis* kidney (Ptk2) cells (Portnoy et al. 1988; Marquis et al. 1995; Dancz et al. 2002; Mueller and Freitag 2005). This phenomenon was also observed in human dendritic cells (Paschen et al. 2000). In Henle 407 cells, LLO-negative bacteria show a twofold reduction in escape from primary vacuoles, whereas a double LLO-PI-PLC mutant shows a fourfold defect. In absence of LLO and PI-PLC, the broad-range phospholipase C (PC-PLC) and a metalloprotease (Mpl) of *L. monocytogenes* mediate escape from vacuoles (Marquis et al. 1995; Grundling et al. 2003).

9.2.2.1.2. PC-PLC and Mpl

L. monocytogenes secretes a PLC that has the ability to hydrolyze a large variety of phospholipids including phosphatidylcholine (PC), phosphatidylethanolamine, phosphatidylserine, phosphatidylinositol, and sphingomyelin (Sm) (Geoffroy et al. 1991; Goldfine et al. 1993). The activity of this PLC is zinc-dependent and optimum at pH 5.5–8. The enzyme was initially named PC-PLC because of its similarity to other bacterial PLCs for which PC is the preferred substrate. However, the *L. monocytogenes* enzyme is often referred to as BR-PLC because of its activity on a broad range of substrates. In addition, it is occasionally named PlcB in reference to the gene coding for it, *plcB*.

The three-dimensional structure of the *L. monocytogenes* PC-PLC (LmPC-PLC) is not known, but that of the *Bacillus cereus* (BcPC-PLC) and of the *Clostridium perfringens* (α-toxin) orthologs have been established. The compact catalytic domain of these PLCs is composed uniquely of α-helices, and the nine zinc-coordinating amino acid residues are conserved among them (Hough et al. 1989; Naylor et al. 1998; Zückert et al. 1998). Exceptionally, BcPC-PLC and LmPC-PLC have an extra N-terminal propeptide-regulating enzyme activity (Johansen et al. 1988; Vazquez-Boland et al. 1992), whereas the *C. perfringens* α-toxin contains an additional C-terminal domain implicated in calcium-dependent membrane binding (Guillouard et al. 1997). This extra C-terminal domain is essential for α-toxin sphingomyelinase and hemolytic activities, conferring toxicity. BcPC-PLC and LmPC-PLC are not considered toxins.

The LmPC-PLC is made as a preproenzyme of 289 amino acid residues. It contains a canonical Sec-dependent signal sequence of 27 residues and a propeptide of 24 residues (Figure 9.2.) (Vazquez-Boland et al. 1992). The N-terminus of the active enzyme was sequenced and the first three residues (WSA) are identical to that of BcPC-PLC (Niebuhr et al. 1993). The secreted proenzyme has no enzymatic activity (Marquis et al. 1997). Interestingly, an LmPC-PLC mutant with a complete deletion of the propeptide is secreted as an

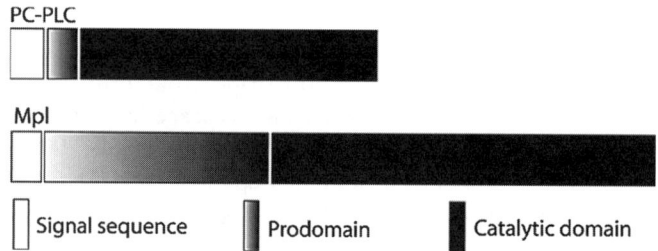

FIGURE 9.2. Schematic representation of LmPC-PLC and Mpl domain organization. Both proteins are comprised of a signal sequence, a prodomain, and a catalytic domain. LmPC-PLC is 289 aa long, with a signal sequence of 27 aa, a prodomain of 24 aa, and a catalytic domain of 238 aa. Mpl is 510 aa long, and the size of the respective domains are predicted to be 24 aa for the signal sequence, 180 aa for the prodomain, and 306 aa for the catalytic domain.

active enzyme suggesting that the propeptide does not contribute to folding of the catalytic domain (Yeung et al. 2005). Presumably, the function of the propeptide is to prevent activity by interfering with substrate binding in the active site.

The Mpl, a secreted zinc-dependent metalloprotease of *L. monocytogenes*, is involved in the proteolytic activation of LmPC-PLC (Poyart et al. 1993; Yeung et al. 2005). Mpl is made as a preproenzyme of 510 aa residues. It is predicted to contain a Sec-dependent signal sequence of 24 residues, a prodomain of 180 residues, and a catalytic domain of 306 residues (Figure 9.2.) (Mengaud et al. 1991b). The large prodomain is typical of metalloproteases, and is thought to function as a protease inhibitor and as an intramolecular chaperone facilitating folding of the catalytic domain (Braun and Tommassen 1998). The three-dimensional structure of thermolysin, an ortholog of Mpl and the prototype for this class of bacterial metalloproteases, indicates that the catalytic domain is comprised of two spherical subdomains, which together form a deep cleft containing the active site (Holmes and Matthews 1982). In addition to the active-site zinc ion, thermolysin contains four calcium-binding sites, which confer high thermal stability. Similarly to thermolysin, Mpl exhibits high thermostability. It is active at pH 5–9, but its activity is optimum at pH 7 (Coffey et al. 2000).

The genes coding for Mpl and PC-PLC, *mpl* and *plcB*, localize to the PrfA regulon (Portnoy et al. 1992). The *mpl* gene is immediately upstream of the *actA* and *plcB* genes, and promoters from the *mpl* and *actA* genes contribute to transcription of the promoterless *plcB* gene (Vazquez-Boland et al. 1992). However, intracellular expression of *plcB* is primarily under the control of the *actA* promoter (Shetron-Rama et al. 2002) (see Chap. 7).

Listeria ivanovii, a ruminant pathogen, expresses an additional sphingomyelinase encoded by *smcL,* which is flanked by genes encoding members of the internalin family. The sphingomyelinase appears to improve by twofold to –fourfold bacterial escape from vacuoles and intracellular growth in bovine epithelial cells. It is possible that this additional sphingomyelinase activity is important in ruminants because of the high level of sphingomyelin in their cell membranes (González-Zorn et al. 1999).

9.2.2.2. Mechanism

The LLO-independent escape from vacuoles of human epithelial cells and Ptk2 cells requires high-level expression of PC-PLC (Grundling et al. 2003; Mueller and Freitag 2005). Expression from the *actA* promoter, which is responsible for the intracellular transcription of *plcB*, is very low in broth culture (Grundling et al. 2003; Moors et al. 1999) and in primary vacuoles of macrophages (Freitag and Jacobs 1999), but increases by ≈200-fold later during infection. Based on these observations, it is difficult to conceive that PC-PLC would be able to mediate LLO-independent escape from primary vacuoles. Perhaps the vacuolar makeup and/or environment of human epithelial cells are different than that of macrophages. To address this question, Cheng et al. (Cheng et al. 2005) used RNA interference to identify host knockdowns that bypass the need for LLO in vacuolar escape of *Drosophila* S2 cells. Knockdowns in components controlling trafficking to and from multivesicular bodies/late endosomes were identified as being permissible for escape of LLO-negative bacteria from primary vacuoles of *Drosophila* S2 cells. Concomitantly, it was shown that the efficacy of bacteria to escape primary vacuoles in mouse macrophages correlates with the inhibition of vacuolar maturation to LAMP1-positive compartments, and that in the absence of LLO, bacteria-containing vacuoles mature slightly faster than conventional phagosomes (Henry et al. 2006). Perhaps, factors other than LLO influence the kinetics of vesicular trafficking in human epithelial cells making time for increased expression of PC-PLC prior to reaching a stage that is no longer permissible for vacuolar escape of *L. monocytogenes*.

9.3. Escape from the Secondary Vacuole

9.3.1.1. Kinetics

As part of its intracellular growth cycle, *L. monocytogenes* spreads directly from cell to cell using an actin-based mechanism of motility (Tilney and Portnoy 1989; Mounier et al. 1990). Moving bacteria induce the formation of membrane protrusions that are taken up by neighboring cells generating double membrane vacuoles, also called secondary vacuoles. Perpetuation of the intracellular growth cycle requires that bacteria escape from secondary vacuoles.

In tissue culture cells, bacteria begin to spread to neighboring cells 3–4 h after initial infection. The kinetics of bacterial cell-to-cell spread was not studied on a population basis because the process cannot be synchronized. As an alternative, the kinetics of cell-to-cell spread was studied by following the spread of individual bacteria within polarized monolayers of MDCK cells by video microscopy (Robbins et al. 1999). This process can be divided into four stages lasting between 35–40 min. Initially, bacteria in membrane protrusions make their way by pushing against the membrane of neighboring cells in a fitful movement. This period persists for ≈15 min and is followed by a ≈20-min period of immobility, during which the morphology of the protrusion is maintained. This change in bacterial movement presumably results from closure of the donor membrane and depletion of host-cell ATP within the protrusion. As the recipient membrane closes, the protrusion collapses into a spherical vesicle, which is permeabilized within 5 min of this transition.

9.3.1.2. Requirements

Secondary vacuoles differ from the primary vacuole in that the cytosolic face of the inner membrane is closest to the bacterial cell (Figure 9.3.). Thus the two membranes of the secondary vacuole have opposite protein and phospholipid asymmetry. Bacterial factors contributing to escape from secondary vacuoles include LLO, PC-PLC, Mpl, and PI-PLC, and expression of these factors is amplified during intracellular infection (Freitag and Jacobs 1999; Moors et al. 1999).

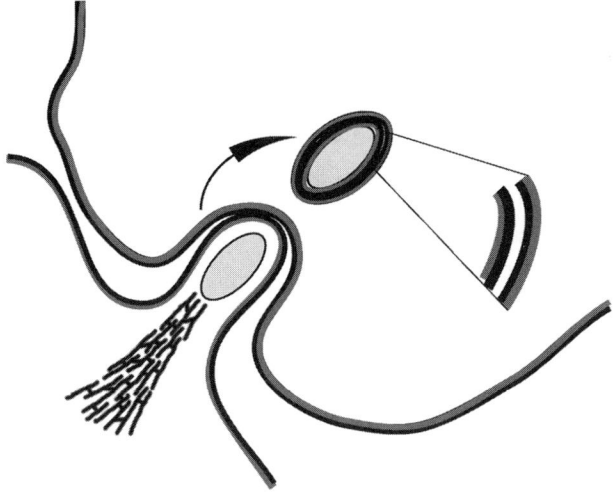

FIGURE 9.3. Schematic representation of membrane organization during *L. monocytogenes* cell-to-cell spread. A moving bacterium (light grey) faces the inner leaflet of the host cytoplasmic membrane during a cell-to-cell spread event and after formation of the double membrane vacuole. The double membrane vacuole is organized in a manner that the outer leaflets of the donor and receiver cell membranes (black) are juxtaposed and sandwiched between the inner leaflets of the donor and the receiver cell membranes (medium grey).

9.3.1.2.1. LLO

A role for LLO in *L. monocytogenes* cell-to-cell spread has been difficult to assess as LLO-negative mutants fail to escape primary vacuoles formed upon initial invasion of most cell types (Portnoy et al. 1988; Tilney and Portnoy 1989). Two very different approaches were used to answer this question, and the results clearly showed that LLO contributes to the ability of *L. monocytogenes* to escape from secondary vacuoles. Using a biochemical approach, Gedde et al. (Gedde et al. 2000) infected murine macrophage–like J774 cells with LLO-minus bacteria coated with purified LLO molecules, enabling bacteria to lyse the primary vacuole and to access the cytosol of infected cells. Intra-cytosolic LLO-negative bacteria multiplied and spread to neighboring cells similarly to wild-type bacteria. However, in the absence of LLO synthesis these bacteria failed to escape from secondary vacuoles indicating that LLO is required for escape from secondary vacuoles of mouse macrophages. Using a genetic approach, Dancz et al. (Dancz et al. 2002) constructed a *lac* repressor-/operator-based system to regulate the expression of *hly*, the gene coding for LLO. Bacterial expression of *hly* was shut down after escape from primary vacuoles to evaluate the requirement for LLO in escape from secondary vacuoles. The system was tested in mouse fibroblasts using a plaque-forming assay, in which the formation and size of plaques reflect the bacterial ability to spread to neighboring cells and escape from secondary vacuoles. Results from this approach also clearly indicated that LLO is required for escape from secondary vacuoles of mouse fibroblasts. However, in human epithelial cells, LLO is dispensable not only for bacterial escape from primary vacuoles, but also for escape from secondary vacuoles (Grundling et al. 2003; Marquis et al. 1995). In the absence of LLO, PC-PLC is required for *L. monocytogenes* escape from secondary vacuoles. Together, these results indicate that the requirement for LLO in bacterial escape from vacuoles differs between human and nonhuman cell lines, whether it is a primary or a secondary vacuole.

9.3.1.2.2.   PC-PLC and Mpl

The PC-PLC enhances the efficacy of *L. monocytogenes* to escape secondary vacuoles. In the absence of PC-PLC, bacteria form small plaques in mouse fibroblasts and are more frequently observed to be trapped in secondary vacuoles of mouse macrophages (Smith et al. 1995; Vazquez-Boland et al. 1992). Since expression of *plcB* is under the control of the *actA* promoter (Freitag and Jacobs 1999; Moors et al. 1999 Shetron-Rama et al. 2002), PC-PLC is made and secreted by bacteria multiplying in the cytosol of infected cells, although proteolytic activation of PC-PLC is restricted to vacuoles (Marquis et al. 1997). Interestingly, intracytosolic bacteria accumulate at their membrane–cell wall interface a pool of proenzyme, which is rapidly released as a bolus of active enzyme upon acidification of the vacuole (Marquis and Hager 2000; Snyder and Marquis 2003). This phenomenon is believed to enhance the kinetics of *L. monocytogenes* escape from vacuoles.

Mpl mediates the proteolytic activation of bacteria-associated PC-PLC (Poyart et al. 1993; Yeung et al. 2005), although in tissue culture cells activation of secreted PC-PLC can be mediated by a vacuolar cysteine protease (Marquis et al. 1997). Mpl is also essential for efficient translocation of PC-PLC across the bacterial cell wall. To determine whether Mpl-mediated PC-PLC activation is a prerequisite to rapid PC-PLC translocation across the cell wall, a PC-PLC cleavage site mutant was created. This mutant, PC-PLC S51DS53N, contains two amino acid substitutions at the propeptide cleavage site. This mutant was rapidly translocated across the bacterial cell wall upon a decrease in pH, in an Mpl-dependent manner (Yeung et al. 2005). Therefore, Mpl controls PC-PLC translocation across the bacterial cell wall in a manner that is independent of propeptide cleavage, but dependent on a decrease in vacuolar pH. Furthermore, to determine whether the control of Mpl over PC-PLC translocation is dependent on the PC-PLC propeptide, a PC-PLC mutant with a complete deletion of the propeptide was generated. The propeptide-less PC-PLC mutant was efficiently translocated across the bacterial cell wall independently of Mpl and of pH indicating that Mpl regulates only the proform of PC-PLC (Yeung et al. 2005). Collectively, these results indicate that Mpl contributes to the regulation of PC-PLC activity by two independent but related pH-sensitive mechanisms: proteolytic activation and cell-wall translocation.

9.3.1.2.3. PI-PLC

A role for PI-PLC in escape from secondary vacuoles remained elusive until the behavior of a double phospholipase mutant was studied. A single PI-PLC mutant forms plaques whose diameters are ≈90% the size of wild-type plaques (Camilli et al. 1993) and a single PC-PLC mutant forms plaques that are ≈62% as large as wild-type plaques (Smith et al. 1995). However, a double PI-PLC and PC-PLC mutant forms plaques whose diameters are ≈32% that of wild-type plaques (Smith et al. 1995). This result is consistent with overlapping functions for the two PLCs in escape from secondary vacuoles. It was also shown that the levels of DAG, a product of phospholipid hydrolysis, are lower in cells infected with the double mutant as compared to those infected with wild-type strain (Smith et al. 1995). Similarly, levels of ceramide, a product of sphingomyelin hydrolysis, are lower in cells infected with the double mutant as compared to those infected with wild-type strain. Generation of these lipid hydrolysis products in combination may influence the formation and lysis of secondary vacuoles as they predispose to membrane fusion (Montes et al. 2004).

9.3.1.3. Mechanism

The specific mechanism by which *L. monocytogenes* escapes from secondary vacuoles is unknown. Evidence has been presented showing that PI-PLC and PC-PLC can individually act on the inner membrane of the double membrane vacuole. In absence of both phospholipases, dissolution of the inner membrane is strongly reduced. LLO is required to perforate the outer membrane (Alberti-Segui et al. 2007). However, host cell factors may contribute to dissolution of the outer membrane as the situation would be analogous to that encountered in primary vacuoles (Section 9.2.1.2).

Both LLO and PC-PLC require a decrease in vacuolar pH to function. However, even if vacuolar maturation leads to acquisition of the proton pump ATPase and acidification of the outer vacuole, it is not clear how the inner vacuole becomes acidified for PC-PLC to function. Perhaps, proton leakage across membranes (Haines 2001) from the acidified outer vacuole into the inner vacuole causes a sufficient decrease in pH to enable PC-PLC activation. PC-PLC activation only requires a drop of 0.3 pH unit although it is more efficient at pH 6.0–6.5 (Marquis and Hager 2000). Dissolution of the inner membrane by the phospholipases would enable LLO to act on the membrane of the acidified outer vacuole.

9.4. Coda

Although much has been learned about the mechanism of escape of *L. monocytoegenes* from a vacuole in the past 20 years of the molecular era of research, much remains to be done. There are two major themes, which appear to be in conflict, but are not. The first is the classical breakdown of the vacuole mediated by pore formation through the action of LLO and enzymic degradation of the membrane barriers by the two phospholipases. The other is the induction of host cell signaling pathways as a consequence of the actions of these proteins. Further insights into the mechanisms of escape will undoubtedly come through careful and innovative research. These studies will surely teach us more about this wily pathogen and open up new understandings of eukaryotic cell biology.

Acknowledgments. We wish to thank Carmen Alverez-Dominguez, Darren Higgins, and Joel Swanson for their careful reading of this manuscript and their very helpful comments.

References

Albert-Segui C, Goeden KR, Higgins DE (2007) Differential function of *listeria monocytogenes* listeriolysin O and phospholipases C in vacuolar dissolution following cell-to-cell spread. Cell Microbiol 9:179–195

Alouf JE (1999) Introduction to the family of the structurally related cholesterol-binding cytolysins ('sulfhydryl-activated' toxins). In: Alouf JE, Freer JH (eds) Comprehensive Sourcebook of Bacterial Toxins. Academic Press, London, pp. 443–456

Alvarez-Dominguez C, Barbieri AM, Berón W, Wandinger-Ness A, Stahl PD (1996) Phagocytosed live *Listeria monocytogenes* influences rab5- regulated *in vitro* phagosome-endosome fusion. J Biol Chem 271:13834–13843

Alvarez-Dominguez C, Roberts R, Stahl PD (1997) Internalized *Listeria monocytogenes* modulates intracellular trafficking and delays maturation of the phagosome. J Cell Sci 110:731–743

Alvarez-Dominguez C, Stahl PD (1999) Increased expression of Rab5a correlates directly with accelerated maturation of *Listeria monocytogenes* phagosomes. J Biol Chem 274: 11459–11462

Bannam T, Goldfine H (1999) Mutagenesis of active-site histidines of *Listeria monocytogenes* phosphatidylinositol-specific phospholipase C: Effects on enzyme activity and biological function. Infect Immun 67:182–186

Beauregard KE, Lee KD, Collier RJ, Swanson JA (1997) pH-dependent perforation of macrophage phagosomes by listeriolysin O from *Listeria monocytogenes*. J Exp Med 186:1159–1163

Becker KP, Hannun YA (2003) CPKC-dependent sequestration of membrane-recycling components in a subset of recycling endosomes. J Biol Chem 278:52747–52754

Becker KP, Hannun YA (2004) Isoenzyme-specific translocation of protein kinase C (PKC) beta II and not PKC beta I to a juxtanuclear subset of recycling endosomes— Involvement of phospholipase D. J Biol Chem 279:28251–28256

Bielecki J, Youngman P, Connelly P, Portnoy DA (1990) *Bacillus subtilis* expressing a haemolysin gene from *Listeria monocytogenes* can grow in mammalian cells. Nature 345:175–176

Braun P, Tommassen J (1998) Function of bacterial propeptides. Trends Microbiol 6:6–8

Bubert A, Sokolovic Z, Chun SK, Papatheodorou L, Simm A, Goebel W (1999) Differential expression of *Listeria monocytogenes* virulence genes in mammalian host cells. Mol Gen Genet 261:323–336

Camilli A, Goldfine H, Portnoy DA (1991) *Listeria monocytogenes* mutants lacking phosphatidylinositol-specific phospholipase C are avirulent. J Exp Med 173:751–754

Camilli A, Tilney LG, Portnoy DA (1993) Dual roles of *plcA* in *Listeria monocytogenes* pathogenesis. Mol Microbiol 8:143–157

Cheng LW, Viala JPM, Stuurman N, Wiedemann U, Vale RD, Portnoy DA (2005) Use of RNA interference in Drosophila S2 cells to identify host pathways controlling compartmentalization of an intracelluilar pathogen. Proc Natl Acad Sci USA 102: 13646–13651

Coffey A, Van den Burg B, Veltman R, Abee T (2000) Characteristics of the biologically active 35-kDa metalloprotease virulence factor from *Listeria monocytogenes*. J Appl Microbiol 88:132–141

Conte MP, Petrone G, Longhi C, Valenti P, Morelli R, Superti F, Seganti L (1996) The effects of inhibitors of vacuolar acidification on the release of *Listeria monocytogenes* from phagosomes of Caco- 2 cells. J Med Microbiol 44:418–424

Cossart P, Portnoy DA (2000) The cell biology of invasion and intracellular growth by *Listeria monocytogenes*. In: Fischetti VA, Novick RP, Ferretti JJ, Portnoy DA, Rood JA (eds) Gram-positive pathogens. ASM Press, Washington, DC, pp. 507–515

Dancz CE, Haraga A, Portnoy DA, Higgins DE (2002) Inducible control of virulence gene expression in *Listeria monocytogenes:* Temporal requirement of listeriolysin O during intracellular infection. J Bacteriol 184:5935–5945

De Chastellier C, Berche P (1994) Fate of *Listeria monocytogenes* in murine macrophages: Evidence for simultaneous killing and survival of intracellular bacteria. Infect Immun 62:543–553

Desjardins M (1995) Biogenesis of phagolysosomes: the 'kiss and run' hypothesis. Trends Cell Biol 5:183–186

Freitag NE, Jacobs KE (1999) Examination of *Listeria monocytogenes* intracellular gene expression by using the green fluorescent protein of *Aequorea victoria*. Infect Immun 67:1844–1852

Gaillard JL, Berche P, Mounier J, Richard S, Sansonetti P (1987) In vitro model of penetration and intracellular growth of *Listeria monocytogenes* in the human enterocyte-like cell line Caco-2. Infect Immun 55:2822–2829

Gandhi AJ, Perussia B, Goldfine H (1993) *Listeria monocytogenes* phosphatidylinositol (PI)-specific phospholipase C has low activity on glycosyl-PI anchored proteins. J Bacteriol 175:8014–8017

Gedde MM, Higgins DE, Tilney LG, Portnoy DA (2000) Role of listeriolysin O in cell-to-cell spread of *Listeria monocytogenes*. Infect Immun 68:999–1003

Gentschev I, Sokolovic Z, Mollenkopf HJ, Hess J, Kaufmann SHE, Kuhn M, Krohne GF, Goebel M (1995) Salmonella strain secreting active listeriolysin changes its intracellular-localization. Infect Immun 63:4202–4205

Geoffroy C, Raveneau J, Beretti J-L, Lecroisey A, Vazquez-Boland J-A, Alouf JE, Berche P (1991) Purification and characterization of an extracellular 29-kilodalton phospholipase C. from *Listeria monocytogenes*. Infect Immun 59:2382–2388

Giddings KS, Zhao J, Sims PJ, Tweten RK (2004) Human CD59 is a receptor for the cholesterol-dependent cytolysin intermedilysin. Nat Struct Mol Biol 11: 1173–1178

Glomski IJ, Decatur AL, Portnoy DA (2003) *Listeria monocytogenes* mutants that fail to compartmentalize listerolysin O activity are cytotoxic, avirulent, and unable to evade host extracellular defenses. Infect Immun 71:6754–6765

Glomski IJ, Gedde MM, Tsang AW, Swanson JA, Portnoy DA (2002) The *Listeria monocytogenes* hemolysin has an acidic pH optimum to compartmentalize activity and prevent damage to infected host cells. J Cell Biol 156:1029–1038

Goldfine H, Johnston NC, Knob C (1993) The non-specific phospholipase C of *Listeria monocytogenes*: Activity on phospholipids in Triton X-100 mixed micelles and in biological membranes. J Bacteriol 175:4298–4306

Goldfine H, Knob C (1992) Purification and characterization of *Listeria monocytogenes* phosphatidylinositol-specific phospholipase C. Infect Immun 60:4059–4067

Goldfine H, Wadsworth SJ, Johnston NC (2000) Activation of host phospholipases C and D in macrophages after infection with *Listeria monocytogenes*. Infect Immun 68:5735–5741

González-Zorn B, Domínguez-Bernal G, Suárez M, Ripio MT, Vega Y, Novella S, Vázquez-Boland JA (1999) The *smcL* gene of *Listeria ivanovii* encodes a sphingomyelinase C that mediates bacterial escape from the phagocytic vacuole. Mol Microbiol 33:510–523

Gray MJ, Freitag NE, Boor KJ (2006) How the bacterial pathogen *Listeria monocytogenes* mediates the switch from environmental Dr. Jekyll to pathogenic Mr. Hyde. Infect Immun 74:2505–2512

Grundling A, Gonzalez MD, Higgins DE (2003) Requirement of the *Listeria monocytogenes* broad-range phospholipase PC-PLC during infection of human epithelial cells. J Bacteriol 185:6295–6307

Guillouard I, Alzari PM, Saliou B, Cole ST (1997) The carboxy-terminal C_2-like domain of the α-toxin from *Clostridium perfringens* mediates calcium-dependent membrane recognition. Mol Microbiol 26:867–876

Haines TH (2001) Do sterols reduce proton and sodium leaks through lipid bilayers? Prog Lipid Res 40:299–324

Henry R, Shaughnessy L, Loessner MJ, Alberti-Segui C, Higgins DE, Swanson JA (2006) Cytolysin-dependent delay of vacuole maturation in macrophages infected with *Listeria monocytogenes*. Cellular Microbiol 8:107–119

Holmes MA, Matthews BW (1982) Structure of Thermolysin Refined at 1.6-A Resolution. J Mol Biol 160:623–639

Hough E, Hansen LK, Birknes B, Jynge K, Hansen S, Hordvik A, Little C, Dodson E, Derewenda Z (1989) High-resolution (1.5 A) crystal structure of phospholipase C from *Bacillus cereus*. Nature 338:357–360

Idkowiak-Baldys J, Becker KP, Kitatani K, Hannun YA (2006) Dynamic sequestration of the recycling compartment by classical protein kinase C. J Biol Chem 281:22321–22331

Jacobs T, Darji A, Weiss S, Chakraborty T (1999) Listeriolysin, the thiol-activated haemolysin of *Listeria monocytogenes*. In: Alouf JE, Freer JH (eds) Comprehensive Sourcebook of Bacterial Protein Toxins. Academic Press, London, pp. 511–521

Johansen T, Holm T, Guddal PH, Sletten K, Haugli FB, Little C (1988) Cloning and sequencing of the gene encoding the phosphatidylcholine-preferring phospholipase C of *Bacillus cereus*. Gene 65:293–304

Jones S, Portnoy DA (1994) Characterization of *Listeria monocytogenes* pathogenesis in a strain expressing perfringolysin O in place of listeriolysin O. Infect Immun 62:5608–5613

Klarsfeld AD, Goossens PL, Cossart P (1994) Five *Listeria monocytogenes* genes preferentially expressed in infected mammalian cells: *plcA, purH, purD, pyrE* and an arginine ABC transporter gene, *arpJ*. Mol Microbiol 13:585–597

Kohda C, Kawamura I, Baba H, Nomura T, Ito Y, Kimoto T, Watanabe I, Mitsuyama M (2002) Dissociated linkage of cytokine-inducing activity and cytotoxicity to different domains of listeriolysin O from *Listeria monocytogenes*. Infect Immun 70:1334–1341

Kuhn M, Kathariou S, Goebel W (1988) Hemolysin supports survival but not entry of the intracellular bacterium *Listeria monocytogenes*. Infect Immun 56:79–82

Leimeister-Wächter M, Domann E, Chakraborty T (1991) Detection of a gene encoding a phosphatidylinositol specific phospholipase C that is co-ordinately expressed with listeriolysin in *Listeria monocytogenes*. Mol Microbiol 5:361–366

Low MG (1989) The glycosyl-phosphatidylinositol anchor of membrane proteins. Biochim Biophys Acta 988:427–454

Marquis H, Doshi V, Portnoy DA (1995) Broad range phospholipase C and metalloprotease mediate Listeriolysin O-independent escape of *Listeria monocytogenes* from a primary vacuole in human epithelial cells. Infect Immun 63:4531–4534

Marquis H, Goldfine H, Portnoy DA (1997) Proteolytic pathways of activation and degradation of a bacterial phospholipase C during intracellular infection by *Listeria monocytogenes*. J Cell Biol 137:1381–1392

Marquis H, Hager EJ (2000) pH-regulated activation and release of a bacteria-associated phospholipase C during intracellular infection by *Listeria monocytogenes*. Mol Microbiol 35:289–298

Mengaud J, Braun-Breton C, Cossart P (1991a) Identification of phosphatidylinositol-specific phospholipase C activity in *Listeria monocytogenes*: a novel type of virulence factor. Mol Microbiol 5:367–372

Mengaud J, Geoffroy C, Cossart P (1991b) Identification of a new operon involved in *Listeria monocytogenes* virulence: Its first gene encodes a protein homologous to bacterial metalloproteases. Infect Immun 59:1043–1049

Michel E, Reich KA, Favier R, Berche P, Cossart P (1990) Attenuated mutants of the intracellular bacterium *Listeria monocytogenes* obtained by single amino acid substitutions in listeriolysin O. Mol Microbiol 4:2167–2178

Montes LR, Goni FM, Johnston NC, Goldfine H, Alonso A (2004) Membrane fusion induced by the catalytic activity of a phospholipase C/sphingomyelinase from *Listeria monocytogenes*. Biochem 43:3688–3695

Moors MA, Levitt B, Youngman P, Portnoy DA (1999) Expression of listeriolysin O and ActA by intracellular and extracellular *Listeria monocytogenes*. Infect Immun 67:131–139

Moser J, Gerstel B, Meyer JEW, Chakraborty T, Wehland J, Heinz DW (1997) Crystal structure of the phosphatidylinositol-specific phospholipase C from the human pathogen *Listeria monocytogenes*. J Mol Biol 273:269–282

Mounier J, Ryter A, Coquis-Rondon M, Sansonetti PJ (1990) Intracellular and cell-to-cell spread of *Listeria monocytogenes* involves interaction with F-actin in the enterocyte-like cell line Caco-2. Infect Immun 58:1048–1058

Mueller KJ, Freitag NE (2005) Pleiotropic enhancement of bacterial pathogenesis resulting from the constitutive activation of the *Listeria monocytogenes* regulatory factor PrfA. Infect Immun 73:1917–1926

Myers JT, Tsang AW, Swanson JA (2003) Localized reactive oxygen and nitrogen intermediates inhibit escape of *Listeria monocytogenes* from vacuoles in activated macrophages. J Immunol 171:5447–5453

Naylor CE, Eaton JT, Howells A, Justin N, Moss DS, Titball RW, Basak AK (1998) Structure of the key toxin in gas gangrene. Nat Struct Biol 5:738–746

Niebuhr K, Chakraborty T, Kollner P, Wehland J (1993) Production of monoclonal antibodies to the phosphatidylcholine-specific phospholipase C of *Listeria monocytogenes*, a virulence factor for this species. Med Microbiol Lett 2:9–16

Nishibori T, Xiong H, Kawamura I, Arakawa M, Mitsuyama M (1996) Induction of cytokine gene expression by listeriolysin O and roles of macrophages and NK cells. Infect Immun 64:3188–3195

Park JM, Ng VH, Maeda S, Rest RF, Karin M (2004) Anthrolysin O and other Gram-positive cytolysins are Toll-like receptor 4 agonists. J Exp Med 200:1647–1655

Paschen A, Dittmar KEJ, Grenningloh R, Rohde M, Schadendorf D, Domann E, Chakraborty T, Weiss S (2000) Human dendritic cells infected by *Listeria monocytogenes*: induction of maturation, requirements for phagolysosomal escape and antigen presentation capacity. Eur J Immunol 30:3447–3456

Portnoy DA, Chakraborty T, Goebel W, Cossart P (1992) Molecular determinants of *Listeria monocytogenes* pathogenesis. Infect Immun 60:1263–1267

Portnoy DA, Jacks PS, Hinrichs DJ (1988) Role of hemolysin for the intracellular growth of *Listeria monocytogenes*. J Exp Med 167:1459–1471

Portnoy DA, Schreiber RD, Connelly P, Tilney LG (1989) Gamma interferon limits access of *Listeria monocytogenes* to the macrophage cytoplasm. J Exp Med 170:2141–2146

Poussin MA, Goldfine H (2005) Involvement of *Listeria monocytogenes* phosphatidyl-inositol-specific phospholipase C and host protein kinase C in permeabilization of the macrophage phagosome. Infect Immun 73:4410–4413

Poyart C, Abachin E, Razafimanantsoa I, Berche P (1993) The zinc metalloprotease of *Listeria monocytogenes* is required for maturation of phosphatidylcholine phospholipase C: Direct evidence obtained by gene complementation. Infect Immun 61:1576–1580

Prada-Delgado A, Carrasco-Marin E, Bokoch GM, Alvarez-Dominguez C (2001) Interferon-gamma listericidal action is mediated by novel Rab5a functions at the phagosomal environment. J Biol Chem 276:19059–19065

Prada-Delgado A, Carrasco-Marin E, Pena-Macarro C, Cerro-Vadillo E, Fresno-Escudero M, Leyva-Cobian F, Alvarez-Dominguez C (2005) Inhibition of Rab5a exchange activity is a key step for *Listeria monocytogenes* survival. Traffic 6:252–265

Ramachandran R, Heuck AP, Tweten RK, Johnson AE (2002) Structural insights into the membrane-anchoring mechanism of a cholesterol-dependent cytolysin. Nat Struct Biol 9:823–827

Ramachandran R, Tweten RK, Johnson AE (2004) Membrane-dependent conformational changes initiate cholesterol-dependent cytolysin oligomerization and intersubunit beta-strand alignment. Nat Struct Mol Biol 11:697–705

Robbins JR, Barth AI, Marquis H, De Hostos EL, Nelson WJ, Theriot JA (1999) *Listeria monocytogenes* exploits normal host cell processes to spread from cell to cell. J Cell Biol 146:1333–1349

Rossjohn J, Feil SC, McKinstry WJ, Tweten RK, Parker MW (1997) Structure of a cholesterol-binding, thiol-activated cytolysin and a model of its membrane form. Cell 89:685–692

Schnupf P, Hofmann J, Norseen J, Glomski IJ, Schwartzstein H, Decatur AL (2006) Regulated translation of listeriolysin O controls virulence of *Listeria monocytogenes*. Mol Microbiol 61:999–1012

Schuerch DW, Wilson-Kubalek EM, Tweten RK (2005) Molecular basis of listeriolysin O pH dependence. Proc Natl Acad Sci USA 102:12537–12542

Shannon JG, Ross CL, Koehler TM, Rest RF (2003) Characterization of anthrolysin O, the *Bacillus anthracis* cholesterol-dependent cytolysin. Infect Immun 71: 3183–3189

Shatursky O, Heuck AP, Shepard LA, Rossjohn J, Parker MW, Johnson AE, Tweten RK (1999) The mechanism of membrane insertion for a cholesterol-dependent cytolysin: A novel paradigm for pore-forming toxins. Cell 99:293–299

Shaughnessy LM, Hoppe AD, Christensen KA, Swanson JA (2006) Membrane perforations inhibit lysosome fusion by altering pH and calcium in *Listeria monocytogenes* vacuoles. Cell Microbiol 8:781–792

Shetron-Rama LM, Marquis H, Bouwer HG, Freitag NE (2002) Intracellular induction of *Listeria monocytogenes actA* expression. Infect Immun 70:1087–1096

Sibelius U, Rose F, Chakraborty T, Darji A, Wehland J, Weiss S, Seeger W, Grimminger F (1996) Listeriolysin is a potent inducer of the phosphatidylinositol response and lipid mediator generation in human endothelial cells. Infect Immun 64:674–676

Smith GA, Marquis H, Jones S, Johnston NC, Portnoy DA, Goldfine H (1995) The two distinct phospholipases C of *Listeria monocytogenes* have overlapping roles in escape from a vacuole and cell-to-cell spread. Infect Immun 63:4231–4237

Snyder A, Marquis H (2003) Restricted translocation across the cell wall regulates secretion of the broad-range phospholipase C of *Listeria monocytogenes*. J Bacteriol 185:5953–5958

Tilney LG, Portnoy DA (1989) Actin filaments and the growth, movement, and spread of the intracellular bacterial parasite, *Listeria monocytogenes*. J Cell Biol 109:1597–1608

Vazquez-Boland J-A, Kocks C, Dramsi S, Ohayon H, Geoffroy C, Mengaud J, Cossart P (1992) Nucleotide sequence of the lecithinase operon of *Listeria monocytogenes* and possible role of lecithinase in cell-to-cell spread. Infect Immun 60:219–230

Villar AV, Alonso A, Goñi FM (2000) Leaky vesicle fusion induced by phosphatidylinositol-specific phospholipase C: Observation of mixing of vesicular inner monolayers. Biochem 39:14012–14018

Wadsworth SJ, Goldfine H (1999) *Listeria monocytogenes* phospholipase C-dependent calcium signaling modulates bacterial entry into J774 macrophage-like cells. Infect Immun 67:1770–1778

Wadsworth SJ, Goldfine H (2002) Mobilization of protein kinase C in macrophages induced by *Listeria monocytogenes* affects its internalization and escape from the phagosome. Infect Immun 70:4650–4660

Wei Z, Schnupf P, Poussin MA, Zenewicz LA, Shen H, Goldfine H (2005a) Characterization of *Listeria monocytogenes* expressing anthrolysin O and phosphatidylinositol-specific phospholipase C from *Bacillus anthracis*. Infect Immun 73:6639–6646

Wei Z, Zenewicz LA, Goldfine H (2005b) *Listeria monocytogenes* phosphatidylinositol-specific phospholipase C has evolved for virulence by greatly reduced activity of GPI anchors. Proc Natl Acad Sci USA 102:12927–12931

Yeung PSM, Zagorski N, Marquis H (2005) The metalloprotease of *Listeria monocytogenes* controls cell wall translocation of the broad-range phospholipase c. J Bacteriol 187:2601–2608

Yoshikawa H, Kawamura I, Fujita M, Tsukada H, Arakawa M, Mitsuyama M (1993) Membrane damage and interleukin-1 production in murine macrophages exposed to listeriolysin O. Infect Immun 61:1334–1339

Zückert WR, Marquis H, Goldfine H (1998) Modulation of enzymatic activity and biological function of *Listeria monocytogenes* broad-range phospholipase C by amino acid substitutions and by replacement with the *Bacillus cereus* ortholog. Infect Immun 66:4823–4831

Color Plate Section

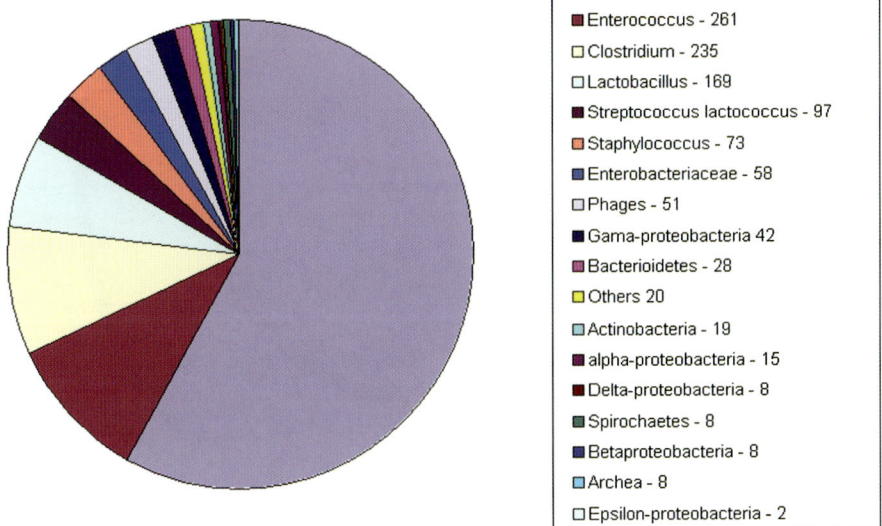

☐ Bacillus - 1523
■ Enterococcus - 261
☐ Clostridium - 235
☐ Lactobacillus - 169
■ Streptococcus lactococcus - 97
■ Staphylococcus - 73
■ Enterobacteriaceae - 58
☐ Phages - 51
■ Gama-proteobacteria 42
■ Bacterioidetes - 28
☐ Others 20
☐ Actinobacteria - 19
■ alpha-proteobacteria - 15
■ Delta-proteobacteria - 8
■ Spirochaetes - 8
■ Betaproteobacteria - 8
☐ Archea - 8
☐ Epsilon-proteobacteria - 2

FIGURE 3.1. Distribution of best BLASTp hits of *L. monocytogenes* EGDe proteins among bacterial genera or higher-order taxonomic groups.

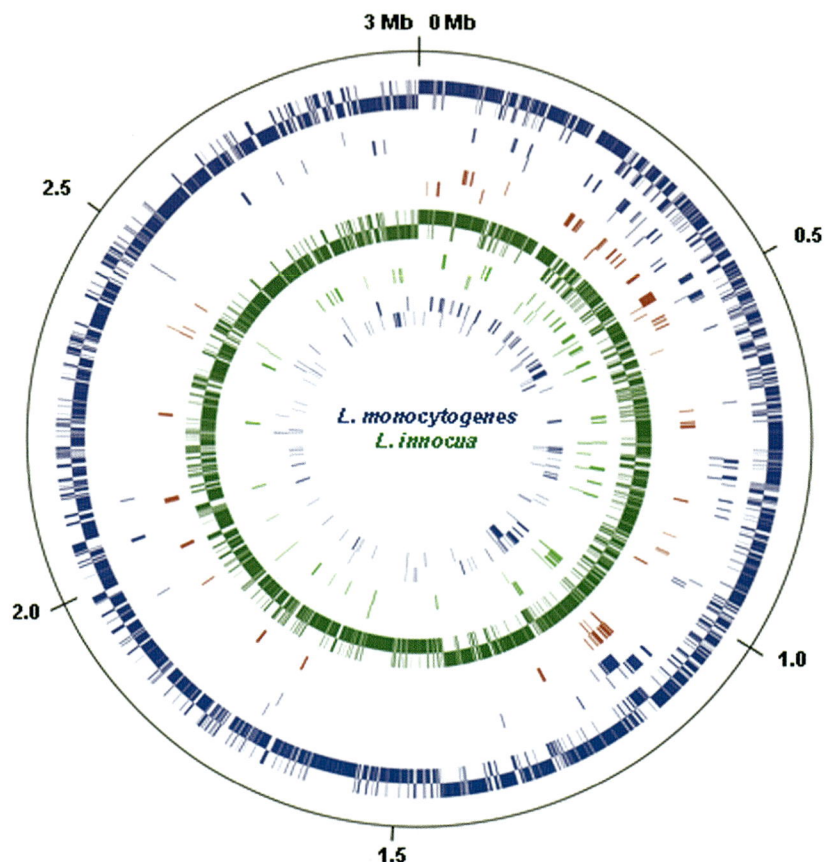

FIGURE 3.2. Circular genome maps of *L. monocytogenes* EGDe and *L. innocua* CLIP11262 showing the position and orientation of genes on the + and − strands, respectively. From the outside: circle 1, *L. monocytogenes* EGDe genes; circle 2, *L. monocytogenes* EGDe genes missing in strain F2365; circle 3, *L. monocytogenes* F2365 missing in EGDe; circle 4, *L. innocua* genes; circle 4, *L. innocua* genes missing in *L. monocytogenes* EGDe; circle 5, *L. monocytogenes* EGDe genes missing in *L. innocua*. The scale in Mb is indicated on the outside with the origin of replication being at position 0.

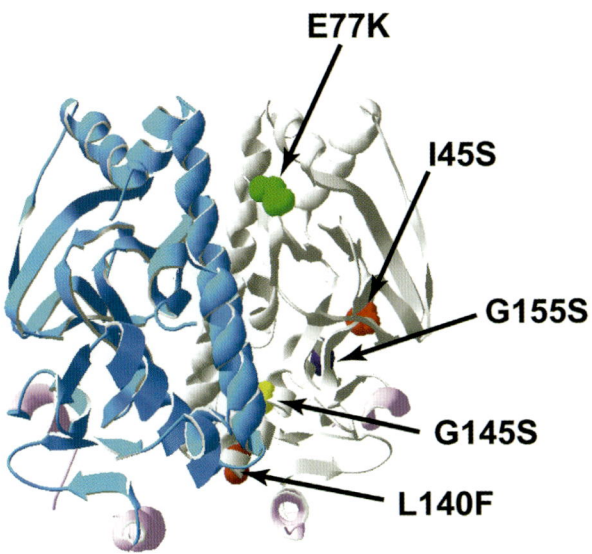

FIGURE 7.3. *Location of PrfA* mutations on protein crystal structure.* The PrfA homodimer is shown with one monomer in grey and one monomer in light blue. Each PrfA* mutation is indicated by an arrow.

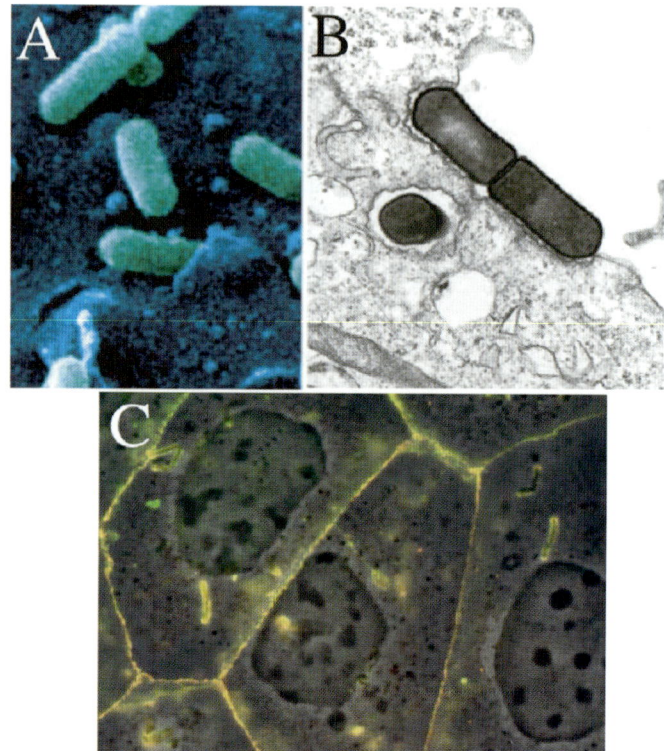

FIGURE 8.1. Electron microscopy and immunofluoresce micrographs of the *L. monocytogenes* entry into target cells. **A** Scanning and **B** transmission electron micrographs of *L. monocytogenes* during the invasion of human epithelial Caco-2 cells, illustrating that the host cell membrane is subtly rearranged around internalized bacteria. Reprinted from *Cell*, Vol. 84, J. Mengaud et al., E-cadherin is the receptor for internalin, a surface protein required for entry of *L. monocytogenes* into epithelial cells, pp. 923–932, copyright 1996, with permission from Elsevier; and *EMBO J*, Vol. 17, P. Cossart and M. Lecuit, Interactions of *Listeria monocytogenes* with mammalian cells during entry and actin-based movement: bacterial factors, cellular ligands and signaling, pp. 3797–3806, copyright 1998, with permission from Nature Publishing Group. **C** Immunfluorescence image of *L. monocytogenes* invading Caco-2 cells, depicting the recruitment of β-catenin (green) and α-catenin (red) to the bacterial entry site (overlay: yellow). These two catenins are also required for the formation of adherens junctions. Reprinted from *Trends Cell Biol.*, Vol. 13, P. Cossart et al., Invasion of mammalian cells by *Listeria monocytogenes*: functional mimicry to subvert cellular functions, pp. 23–31, copyright 2003, with permission from Elsevier.

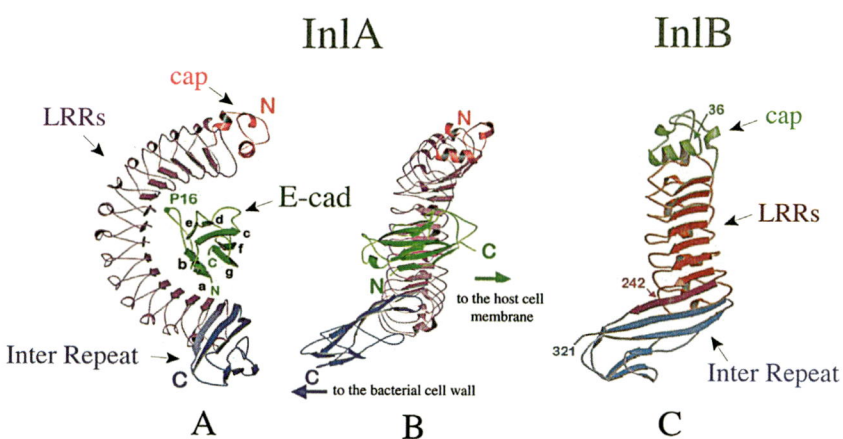

FIGURE 8.3. Structure of the N-terminal domains of InlA and InlB. **A** Leucine-rich repeats (LRRs) from InlA are composed of 16 β-sheets that accommodate the first extracellular domain of E-cadherin; an hydrophobic pocket created between LRR 5 and 7 (due to the absence of one amino acid in LRR 6) accommodates E-cadherin proline 16, which is critical for InlA–E-cadherin interaction. **B** 90°C rotation of the figure depicted in **A**. **C** View toward the concave surface of the InlB LRRs. Reprinted from *Cell*, Vol. 111, W.D. Schubert et al., Structure of internalin, a major invasion protein of *Listeria monocytogenes*, in complex with its human receptor E-cadherin, pp. 825–836, copyright 2001, and *J Mol Biol*, Vol. 312, W.D. Schubert et al., Internalins of the human pathogen *L. monocytogenes* combine three distinct folds into a contiguous internalin domain, pp. 783–794, copyright 2002, with permission from Elsevier.

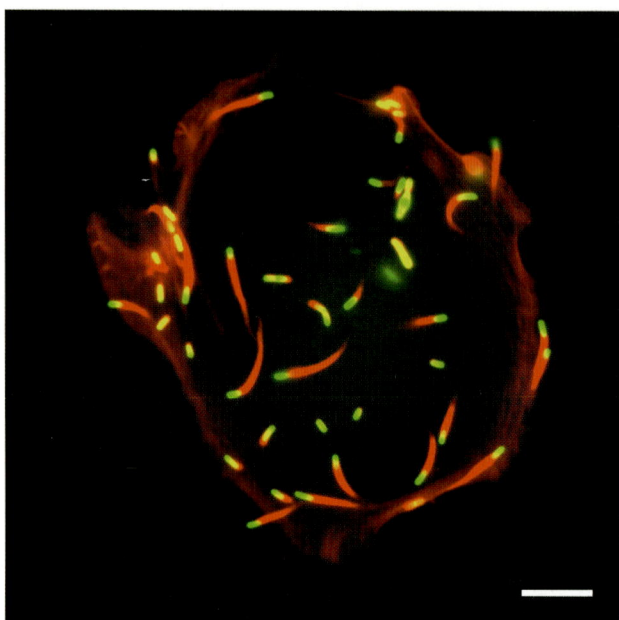

FIGURE 10.2. Image of actin comet tails formed by moving *L. monocytogenes* in infected tissue culture cells. Bacteria (green) were visualized by immunofluorescence using an anti-*L. monocytogenes* primary antibody and a FITC conjugated secondary antibody. Actin (red) was visualized using rhodamine-phalloidin. Image courtesy of Justin Skoble, Daniel Portnoy, and Matthew Welch. Scale bar 10 μm.

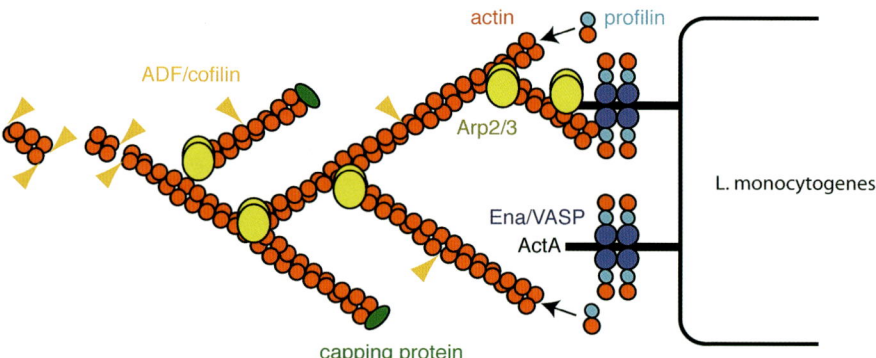

FIGURE 10.4. Cartoon diagram depicting the location and function of proteins known to participate in actin nucleation (Arp2/3 complex, yellow), elongation (VASP, dark blue; profilin, light blue), organization (Arp2/3 complex; capping protein, green), and depolymerization (ADF/cofilin, orange) during *L. monocytogenes* motility.

10
Actin-Based Motility and Cell-to-Cell Spread of *Listeria monocytogenes*

Matthew D. Welch

Department of Molecular & Cell Biology, University of California,
Berkeley, 301 Life Sciences Addition, Berkeley, CA 94720-3200, USA
e-mail: welch@berkeley.edu

Abstract: *Listeria monocytogenes* has evolved the ability to exploit its host's actin cytoskeleton to power movement within and between cells without exiting from the cell, enabling it to evade the immune response. This remarkable adaptation requires the expression of a single bacterial surface protein, called ActA, that performs two key functions. It activates the host Arp2/3 complex, which promotes the nucleation of actin filaments at the bacterial surface and the organization of filaments into branched networks. Moreover, it recruits host Ena/VASP proteins and profilin, which stimulate actin filament elongation. Together these cellular factors promote the assembly of actin comet tails that recruit additional host cytoskeletal proteins that control filament bundling, terminate polymerization, and promote depolymerization. The assembly of the comet tail is essential for coupling actin polymerization to the force that drives bacterial propulsion. The process of bacterial motility can be reconstituted in vitro, facilitating a relatively complete understanding of the biochemical and biophysical mechanisms of actin polymerization and force generation. This chapter presents a review of the experiments that have led to our current understanding of the molecular mechanisms of *L. monocytogenes*' motility and spread.

10.1.   Introduction

10.1.1.   *The Actin Cytoskeleton and Intercellular Spread of* Listeria monocytogenes

The ability to spread from cell to cell plays a critical role in infection by many intracellular pathogens. *L. monocytogenes* has evolved the ability to move directly from cell to cell without passing through the intercellular space, enabling it to evade the host immune response. The process of *L. monocytogenes*' spread was first observed by imaging infections in guinea pig corneal and intestinal epithelia using electron microscopy (Racz et al. 1970, 1972). These early studies documented that, after internalization into a vacuole and then escape into the cytoplasm, bacteria move into membrane protrusions that extend into invaginations of the membrane of a neighboring cell. Protrusions from an infected cell are then internalized by neighboring cells, completing the process of cell-to-cell transfer. Although these observations were remarkable, the process of *L. monocytogenes* cell-to-cell spread received little attention for many years.

The mechanism of spread was examined in greater detail nearly two decades later by Tilney and Portnoy (1989), who carefully imaged the process of infection in a macrophage-like cell line using electron microscopy. Their observations revealed the same basic phenomenon involving movement of *Listeria* into protrusions containing bacteria at their tips, protrusion internalization by neighboring cells to form a double membrane vacuole, and eventual escape from this vacuole (Figure 10.1.). Tilney and Portnoy also made the seminal observation that bacterial movement and protrusion formation involves an ability to associate with the host cell actin cytoskeleton (Figure 10.2.). This observation was also confirmed by other contemporary studies (Mounier et al. 1990). Interestingly, bacteria induce the formation of progressively more elaborate actin-containing structures as infection progresses. At short times (2 h) after infection, the majority of cytoplasmic bacteria are surrounded by an electron dense cloud of host actin filaments. At later times (4 h) postinfection, many bacteria in the cytoplasm are associated with long comet tails of actin filaments, with the bacterium at the

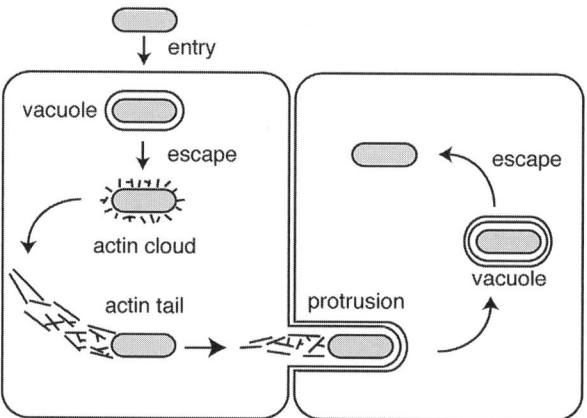

FIGURE 10.1. Cartoon diagram of the process of *L. monocytogenes* actin-based motility and cell-to-cell spread. Internalization of bacteria into a vacuole is followed by escape into the cytoplasm. At short times after escape, bacteria are surrounded by clouds of actin filaments. At later times, bacteria initiate movement, and clouds are converted into comet tails that trail the moving bacteria. Movement propels bacteria into the plasma membrane, causing the formation of a protrusion that can be engulfed by a neighboring cell, where the cycle repeats itself.

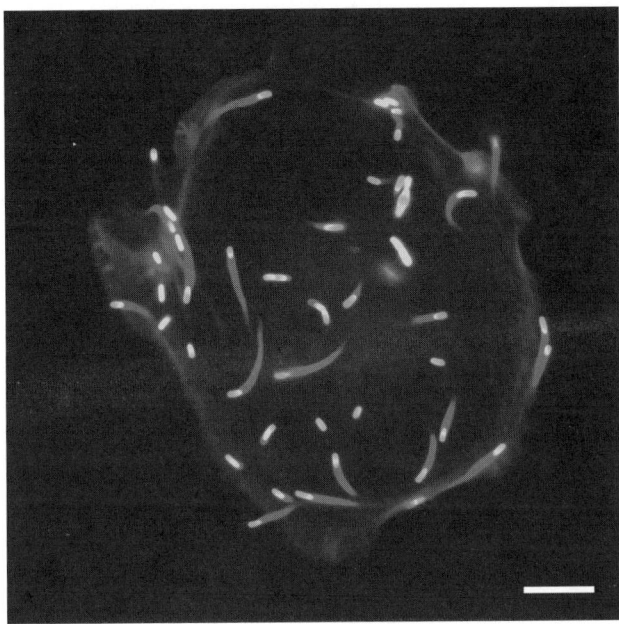

FIGURE 10.2. Image of actin comet tails formed by moving *L. monocytogenes* in infected tissue culture cells. Bacteria (green) were visualized by immunofluorescence using an anti-*L. monocytogenes* primary antibody and a FITC conjugated secondary antibody. Actin (red) was visualized using rhodamine-phalloidin. Image courtesy of Justin Skoble, Daniel Portnoy, and Matthew D. Welch. Scale bar 10 μm. (A color version of this figure appears between pages 196 and 197.)

apex of the tail. Bacteria located in membrane protrusions are also associated with actin comet tails that extend inward toward the cell body (Figure 10.1.). The ability to associate with actin is essential for spread as treatment of infected cells with cytochalasin D, an inhibitor of actin function, prevents bacterial movement to the plasma membrane and protrusion formation. Together these observations suggested a model for cell-to-cell spread in which the bacterium uses the host cell actin cytoskeleton to move through the cytoplasm of an infected cell and penetrate neighboring cells.

This basic model has been corroborated by more recent observations. Actin-based bacterial movement has been directly observed using video microscopy (Dabiri et al. 1990), and bacterial motility has been clocked at average speeds ranging from 1 to 36 μm/min, with the precise velocity differing between individual bacteria in a single cell and populations of bacteria in different cell types (Dabiri et al. 1990, Sanger et al. 1992, Theriot et al. 1992). Moreover, the coupling of actin-based movement and cell-to-cell spread has been directly observed in epithelial monolayers by time-lapse microscopy (Robbins et al. 1999). These live observations of spread show that moving bacteria often collide with the plasma membrane of the host cell, but collisions do not always result in the formation of protrusions that extend into neighboring cells. In many instances, bacteria ricochet off the membrane without forming a protrusion. Surprisingly, there is little correlation between the formation of productive protrusions and bacterial speed or angle of incidence. Instead, protrusion formation is correlated with the state of the host cell monolayer, suggesting that the nature or strength of cell–cell adhesions or the organization of the cortical cytoskeleton play critical roles. Nevertheless, if bacteria succeed in forming protrusions that extend into neighboring cells, the process of cell-to-cell transfer can occur.

It is now clear that this mechanism of cell-to-cell spread is not unique to *L. monocytogenes*. Numerous bacterial pathogens have been shown to use the host actin cytoskeleton to promote intracellular movement, including *Shigella flexneri* (Bernardini et al. 1989), *L. ivanovii* (Karunasagar et al. 1993), spotted fever group *Rickettsia* species (Heinzen et al. 1993), *Burkholderia pseudomallei* (Kespichayawattana et al. 2000), and *Mycobacterium marinum* (Stamm et al. 2003). Based on differences in the molecular mechanism by which they interact with actin (Gouin et al. 2005), it is likely that different pathogens have independently evolved the ability to manipulate the actin cytoskeleton of the host.

10.1.2. Properties of the Actin Cytoskeleton Relevant to L. monocytogenes *Motility*

Although it was initially surprising that *L. monocytogenes* and other pathogens use host actin for movement and spread, in hindsight, it makes perfect sense that pathogens evolved the ability to exploit actin for this purpose, given the key role that actin plays in eukaryotic cell motility. To appreciate how actin participates in bacterial movement, one needs to be familiar with the basic properties of actin, which have been reviewed elsewhere (Pollard et al. 2000) and are summarized briefly here.

Actin is an ATP-binding protein that exists in two forms in the cell: monomers (also called globular-actin or G-actin) and filaments (also called filamentous-actin or F-actin). Actin filament assembly results from the reversible polymerization of monomers onto the ends of filaments and is accompanied by ATP hydrolysis and release of inorganic phosphate (P_i). Actin filaments are polar and have two distinct ends: the fast-growing end (also called the barbed or plus (+) end) and the slow-growing end (also called the pointed or minus (−) end). Polymerization occurs primarily at barbed ends in cells and can be initiated by uncapping of existing filaments, severing of filaments, or nucleation of new filaments. Each of these processes is tightly controlled in the cell by many different actin-binding proteins.

In nonmuscle cells, actin filaments are primarily located at the cell cortex, where they are generally oriented with their fast-growing barbed ends facing the plasma membrane. Polymerization at barbed ends can generate the force that drives membrane protrusion at the leading edge during cell migration (Pollard and Borisy 2003). Moreover, polymerization can drive the intracellular movements of internal membrane-bound organelles such as endosomes (Taunton 2001). The ability of actin polymerization to be harnessed for force generation is exploited by *L. monocytogenes* to promote spread during infection.

10.2.   Intracellular Bacterial Movement

10.2.1.   *Actin Polymerization Provides the Driving Force for Propulsion*

Intracellular movement of *L. monocytogenes* is always accompanied by the formation of actin comet tails (Dabiri et al. 1990), suggesting that the assembly of actin into the tail is coupled to motility. The direct coupling of actin polymerization and movement was first demonstrated by marking segments of the actin network in comet tails using fluorescent actin derivatives and observing the behavior of the fluorescent marks over time (Sanger et al. 1992, Theriot et al. 1992). If newly polymerizing filaments are selectively marked, they are seen to assemble at the bacterial surface at the same rate as bacterial movement (Sanger et al. 1992). If older filaments away from the bacterium are marked, they remain stationary with respect to the cell as the bacterium moves forward (Theriot et al. 1992). The fact that actin polymerization occurs at the bacterial surface is also demonstrated by the observation that exogenously added actin assembles onto the surface of bacteria in permeabilized infected cells (Tilney et al. 1990). Together these experiments suggest that polymerization is directly linked to and may provide the driving force for motility.

Following polymerization at the bacterial surface, actin filaments in comet tails are depolymerized with a half-life of 30–60 s (Theriot et al. 1992, Nanavati et al. 1994), similar to actin in cortical regions of migrating cells (Theriot and Mitchison 1991, 1992). Depolymerization occurs stochastically and at a uniform rate throughout the bulk of the comet tail. The compartmentalization of polymerization at the bacterial surface, together with uniform depolymerization of filaments throughout the tail, leads to the characteristic comet shape where filament density decreases exponentially with distance away from the bacterium (Theriot et al. 1992). Depolymerization is critical for continued bacterial movement (Cramer 1999) as it regenerates a pool of actin monomer that is used to fuel new polymerization at the bacterial surface (Sanger et al. 1995).

In addition to filament dynamics, filament organization in *L. monocytogenes* tails plays a critical role in coupling polymerization to movement. Actin filaments within the tail are oriented with their barbed ends facing the bacterial surface (Tilney et al. 1992a,b), similar to actin orientation relative to the plasma membrane. This is consistent with the model that rapid barbed-end polymerization generates the force that drives motility. Filaments in comet tails are also linked together in stereotypical geometrical arrangements. In the outer shell of the tail, filaments are longer and are organized into parallel bundles that run along the axis of the tail (Zhukarev et al. 1995, Sechi et al. 1997). In the core of the tail and near the bacterial surface, filaments are shorter and are organized into a y-branched dendritic network (Tilney and Portnoy 1989, Zhukarev et al. 1995, Cameron et al. 2001). The highly cross-linked organization of filaments within the comet is thought to be critical for coupling polymerization to bacterial movement (Mogilner and Oster 2003b, Plastino and Sykes 2005), as is discussed in later sections.

10.2.2. Bacterial and Host Contributions to Actin Dynamics

Because actin dynamics plays a critical role in bacterial movement and spread, considerable attention has been devoted to understanding how actin is polymerized, depolymerized, and organized in *L. monocytogenes*' comet tails. We now have a relatively complete understanding of the molecular players involved in these processes and their mechanisms of action. However, before these players were identified, a simple question had to be answered: what contributions do the bacterium and the host each make to motility? Evidence that secreted bacterial proteins are required for actin polymerization initially came from the observation that inhibition of bacterial protein synthesis prevents actin assembly in infected host cells (Tilney et al. 1990). Complementary evidence that host proteins are also required came from the observation that purified actin cannot assemble on the surface of bacteria grown outside of host cells, but can assemble on the surface intracellular bacteria in permeabilized cells (Tilney et al. 1990, Tilney et al. 1992b). These observations provided a framework for the identification of the bacterial and host molecules and mechanisms that modulate actin dynamics, both in infected and uninfected host cells.

10.3. The ActA Surface Protein

10.3.1. ActA Mediates Actin Polymerization

A bacterial gene that is required for actin polymerization, called *actA*, was identified in a screen for transposon mutants that failed to spread between cells in a plaque assay (Kocks et al. 1992) and also by directed insertional mutagenesis (Domann et al. 1992). Mutations in *actA* cause a complete failure to assemble actin at the bacterial surface and to undergo intracellular motility. Entry and intracellular growth are not affected, but the motility defect results in the failure to spread from cell to cell. In frame deletion mutants in *actA* have also been generated and cause similar phenotypes (Brundage et al. 1993). Importantly, these *actA* deletion mutants have an LD_{50} (lethal dose, 50%) in mice of $\sim 10^7$, compared to $\sim 10^4$ for the wild-type strain, suggesting that the ability to polymerize actin is critical for virulence.

As expected for a protein that is required to manipulate the host actin cytoskeleton, ActA is present on the bacterial surface (Kocks et al.1992) (Figure 10.3.). ActA has a canonical signal sequence that mediates secretion and

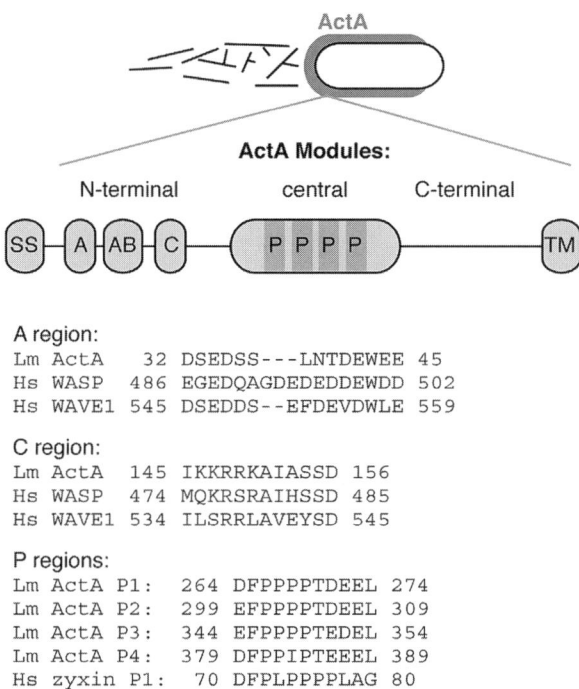

FIGURE 10.3. Cartoon diagram depicting ActA localization at the bacterial surface, ActA domain organization, and the sequences of different regions within ActA aligned with the sequences of homologous host proteins. ActA domains are indicated as follows: *SS* signal sequence, *A* acidic, *AB* actin-binding, *C* connector, *P* proline rich, *TM* transmembrane. Sequence alignments are for the listed proteins from the following species: *Lm, Listeria monocytogenes; Hs, Homo sapiens.*

a transmembrane domain that anchors it to the bacterial cytoplasmic membrane (Vazquez-Boland et al. 1992). It is observed only on the bacterial surface and is not found within actin comet tails (Kocks et al. 1993, Niebuhr et al. 1993), suggesting that surface-bound protein is the functional form. Interestingly, ActA exhibits a polarized distribution, with the highest concentration at the older pole and the lowest concentration at the newer pole or the septum (Kocks et al. 1993). The pole with the highest concentration corresponds to the location from which actin polymerizes and the comet tail emanates. Polarization of ActA is correlated with efficient comet tail formation and rapid motility (Rafelski and Theriot 2005), and therefore may play a key role in spread. How the polarized distribution of ActA is generated is not well understood, although polarization is known to occur after initial surface targeting (Rafelski and Theriot 2006).

In addition to being necessary for actin polymerization, ActA is also sufficient to promote actin polymerization in the absence of other bacterial proteins. This was first demonstrated by expressing ActA in uninfected eukaryotic cells (Pistor et al. 1994), where it targets to mitochondria and induces actin to polymerize around these organelles. However, mitochondria decorated with ActA do not undergo actin-based motility, perhaps because the surface distribution of ActA is not polarized. ActA can also be targeted to the plasma membrane rather than mitochondria by replacing its transmembrane domain with a CAAX box, which causes actin polymerization at the membrane and changes in cell shape (Friederich et al. 1995).

A direct demonstration of the sufficiency of ActA in actin-based motility was achieved by expressing or tethering ActA on the surface of bacteria that are not normally able to polymerize actin, such as *L. innocua* (Kocks et al. 1995) or *Streptococcus pneumoniae* (Smith et al. 1995), or by coating ActA on the surface of inert plastic beads (Cameron et al. 1999). The direct demonstration of sufficiency also relied on the ability to reconstitute *L. monocytogenes* actin-based motility in cell-free cytoplasmic extracts made from *Xenopus laevis* eggs (Theriot et al. 1994), which enables addition of ActA-coated bacteria or beads to cytoplasm without the complications of the infection process. When ActA-expressing *L. innocua* (Kocks et al. 1995), ActA-coated *S. pneumoniae* (Smith et al. 1995), or ActA-coated plastic beads (Cameron et al. 1999) are added to *Xenopus* egg extracts in place of *L. monocytogenes*, they undergo actin-based motility. Moreover, killed *L. monocytogenes* also undergo motility in cell extracts (Theriot et al. 1994). Thus, ActA is the only bacterial protein that is required for actin-based motility, and once ActA is displayed on their surface, bacteria can be passive participants while the host cell expends energy to move them within and between cells.

10.3.2. ActA Expression is Regulated During Infection

The expression of ActA is dramatically upregulated during infection (Bohne et al. 1994, Freitag and Jacobs 1999) to greater than 200 times the levels expressed during growth in broth (Moors et al. 1999b). Enhanced ActA expression requires that bacteria escape from the vacuole and enter the cytoplasm, suggesting that cytoplasmic conditions serve as a critical regulatory cue. Precise control over the levels of ActA expression is important, as reduced expression results in defects in actin polymerization and cell-to-cell spread in a plaque assay (Wong et al. 2004). As with many other *L. monocytogenes* virulence genes, *actA* expression is regulated by the transcription factor PrfA (Kreft and Vazquez-Boland 2001), a topic that is covered in more detail in other chapters within this book. The levels of surface-bound ActA may also be regulated at the level of protein stability (Moors et al. 1999a). ActA is a substrate for degradation in the host cell, and this is mediated in part by the N-end rule, as mutation of the N-terminal amino acid of ActA can result in decreased protein stability. Mutations that decrease ActA stability also decrease the efficiency of cell-to-cell spread in a plaque assay, suggesting that there is selective pressure for the presence of certain N-terminal residues to limit ActA degradation.

10.3.3. Modular Structure and Function of ActA

ActA is a modular scaffolding protein that contains multiple functional elements (Figure 10.3.). The sequences that control ActA secretion and surface localization are located at its termini. An N-terminal 29-amino acid signal sequence targets the protein for secretion and is then cleaved, and a C-terminal 22-amino acid transmembrane domain (with a 4-amino acid cytoplasmic tail) anchors the protein to the bacterial surface. A truncated protein that is missing the transmembrane domain is secreted from bacteria (Smith et al. 1995), a property that has been exploited to purify ActA from culture supernatants.

Apart from the extreme N- and C-termini, the remainder of ActA can be divided into three key functional modules. The bulk of the N- terminus consists of sequences that exhibit some similarity with iActA from *Listeria ivanovii* (Gouin et al. 1995, Kreft et al. 1995, Gerstel et al. 1996) and with host proteins in the Wiskott–Aldrich Syndrome protein (WASP) family (Bi and Zigmond 1999, Skoble et al. 2000, Boujemaa-Paterski et al. 2001, Zalevsky et al. 2001), which are key regulators of actin polymerization in cells (Welch and Mullins 2002). This region contains several key sequence elements (Figure 10.3.), including acidic (A) and connector (C) sequences that together mediate interaction with the host actin-nucleating Arp2/3 complex (Welch et al. 1998, Pistor et al. 2000, Skoble et al. 2000, Boujemaa-Paterski et al. 2001, Zalevsky et al. 2001), and actin-binding sequences (AB) that bind actin monomer (Lasa et al. 1997, Cicchetti et al. 1999, Skoble et al. 2001, Zalevsky et al. 2001). More detail about the interactions between ActA and these host factors is described in later sections. The N-terminal region of ActA also mediates dimerization (Mourrain et al. 1997) and phosphoinositide binding (Cicchetti et al. 1999, Steffen et al. 2000), although the importance of these interactions remains unclear.

The function of the N-terminal region of ActA in actin polymerization, motility, and spread has been addressed genetically. Deletion of this region causes a complete failure to polymerize actin when mutant ActA is targeted to mitochondria in uninfected cells (Pistor et al. 1995). More importantly, *L. monocytogenes* expressing N-terminal deletion mutants behave like *actA* null mutants—they fail to polymerize actin and undergo actin-based motility in *Xenopus* egg extracts (Lasa et al. 1995) and in infected cells (Skoble et al. 2000). Not surprisingly, these mutants also fail to spread in plaque assays (Skoble et al. 2000). Thus, the N-terminal region is absolutely essential for actin-based motility. It is also the only region that, when expressed along with the signal sequence and transmembrane domain, is sufficient to promote motility (Lasa et al. 1995, 1997).

The central module of ActA also contributes to actin-based motility. This region contains three or four proline-rich repeats (the number differs in natural isolates; Wiedmann et al. 1997), interspersed with longer repeat sequences (Smith et al. 1996). Each proline-rich repeat contains the core consensus motif (D/E)FPPPP(P/T)(D/E) (Niebuhr et al. 1997, Prehoda et al. 1999), which is also found in *L. ivanovii* iActA (Gerstel et al. 1996) as well as in host proteins involved in transmitting signals to the actin cytoskeleton, including the focal adhesion components zyxin and vinculin (Niebuhr et al. 1997), and the axon-guidance receptors Robo/SAX-3 (Kidd et al. 1998, Zallen et al. 1998). These core proline-rich sequences bind directly to host proteins in the enabled and vasodilator-stimulated phosphoprotein (Ena/VASP) family (Chakraborty et al. 1995, Pistor et al. 1995, Niebuhr et al. 1997, Prehoda et al. 1999). The role of Ena/VASP proteins in bacterial actin-based motility is described in greater detail in later sections.

Unlike the essential N-terminus, the central proline-rich repeat module is dispensable for actin-based motility. Deletion of this region reduces but does not abolish actin polymerization when a mutant variant of ActA is targeted to mitochondria (Pistor et al. 1995). More importantly, *L. monocytogenes* that express mutant ActA carrying deletions or point mutations of the proline-rich sequences move at one half to one third the rate of wild-type cells (Lasa et al. 1995, Smith et al. 1996, Niebuhr et al. 1997, Auerbuch et al. 2003), exhibit a reduction in the percentage of bacteria that initiate movement (Smith et al. 1996, Auerbuch et al. 2003), and display a change in the directional persistence of movement (Auerbuch et al. 2003). These mutants are also reduced in their ability to spread from cell to cell in a plaque assay (Smith et al. 1996, Niebuhr et al. 1997, Auerbuch et al. 2003) and are reduced in virulence in a mouse model of infection (Niebuhr et al. 1997, Auerbuch et al. 2001). Thus, the central region of ActA plays an important but nonessential role in pathogenesis.

Finally, the region of ActA between the central region and the C-terminal transmembrane domain appears to make no discernable contribution to actin-based motility (Lasa et al. 1995). Nevertheless, this region is thought to act as a spacer that spans the cell wall. If both the central and C-terminal modules are deleted, bacteria expressing the N-terminal module alone do not polymerize actin, suggesting that spacer sequences are required for the N terminus to span beyond the cell wall and interact with host proteins.

10.4.   Host Contributions to Actin-Based Motility

10.4.1.   *Reconstitution of Actin-Based Motility in Cell Extracts*

Although the ActA protein was identified rather rapidly following the initial discovery of *L. monocytogenes* actin-based motility, it took longer to identify the essential host factors that participate in actin polymerization. The delay was due to a lack of genetic tools in host organisms and a lack of biochemical tools for dissecting motility in vitro. A breakthrough that set the stage for the identification of host factors was the reconstitution of *L. monocytogenes* motility in vitro in concentrated cytoplasmic extracts made from *X. laevis* eggs (Theriot et al. 1994). Reconstitution was later achieved in human platelet cytoplasmic and

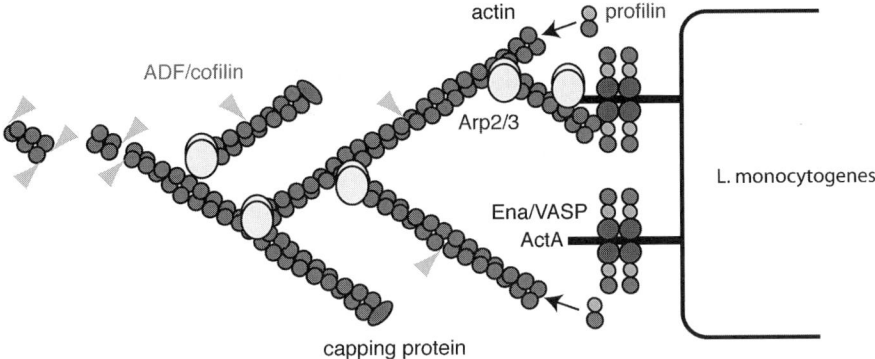

FIGURE 10.4. Cartoon diagram depicting the location and function of proteins known to participate in actin nucleation (Arp2/3 complex, yellow), elongation (VASP, dark blue; profilin, light blue), organization (Arp2/3 complex; capping protein, green), and depolymerization (ADF/cofilin, orange) during *L. monocytogenes* motility. (A color version of this figure appears between pages 196 and 197.)

cytoskeletal extracts, which are highly enriched in cytoskeletal proteins (Welch et al. 1997b). Actin-based motility in vitro requires the use of a bacterial strain (Leimeister-Wachter and Chakraborty 1989) that overexpresses ActA and other virulence factors, indicating that optimal levels of expression differ for motility in different environments. Nevertheless, the motility parameters in *Xenopus* egg and human platelet cytoplasmic extracts are remarkably similar to those in infected cells—bacteria move at an average speed of $\sim 6\,\mu\text{m}/\min$ (slightly slower than in infected cells) and assemble characteristic actin comet tails. The experimental advantage of these reconstitution systems is that the function of known host proteins can be assessed by addition or depletion, and the identification of unknown proteins can proceed by biochemical fractionation. Using this and other systems, a core set of host proteins that are involved in actin polymerization and motility has been identified (Figure 10.4.), as is described in the following sections.

10.4.2. The Arp2/3 Complex

The host factor that nucleates actin assembly with ActA was purified from human platelet extracts by fractionating and assaying for factors that could promote actin assembly at the *L. monocytogenes* surface (Welch et al. 1997b). This factor was identified as the Arp2/3 complex (Welch et al. 1997b), an evolutionarily conserved protein complex consisting of the actin-related proteins Arp2 and Arp3 and five additional subunits that was originally isolated from *Acanthamoeba castellanii* (Machesky et al. 1994). The Arp2/3 complex normally functions in host cells to nucleate actin assembly at the cortex during processes such as membrane protrusion, phagocytosis, and endocytosis (Welch and Mullins 2002). Additionally, Arp2/3 complex is necessary for bacterial actin-based motility as depletion of or interference with its activity in cell extracts (May et al. 1999, Yarar et al. 2002) and in infected cells (May et al. 1999) prevents actin polymerization and movement. In its purified form, Arp2/3 complex is sufficient to promote the formation of actin clouds surrounding *L. monocytogenes*, but it is not sufficient to promote actin-based motility, indicating that other factors are involved in this process (Welch et al. 1997b). Actin polymerization by purified Arp2/3 complex at the bacterial surface requires the presence of ActA (Welch et al. 1997b), consistent with the essential role of ActA in actin polymerization in cells.

The requirement for both ActA and Arp2/3 complex to assemble actin at the bacterial surface suggested that they function together in this process. This was confirmed by biochemical experiments, which indicate that ActA acts as a nucleation-promoting factor that stimulates two tightly coupled activities of the Arp2/3 complex: its capacity to nucleate actin assembly (Welch et al. 1998) and to cross-link actin filaments into y-branched structures, where the complex resides at the branch point (Skoble et al. 2001). Although the precise mechanism of Arp2/3 activation is not known, the prevailing model is that Arp2/3 binds to both ActA and to the side of an existing actin filament, and templates the formation of a new filament, which emanates from the side of the existing filament in a y-branched arrangement (Welch and Mullins 2002). The affinity of Arp2/3 for ActA is relatively low (Kd ~0.6 μM; Zalevsky et al. 2001), and hence the interaction is likely to be transient. Thus, ActA is thought to bind to and activate Arp2/3 complex and then release Arp2/3 complex at y-branches from the bacterial surface. This is consistent with the observed distribution of Arp2/3 complex throughout *L. monocytogenes* comet tails (Welch et al. 1997a,b) and the branched organization of filaments in the tails observed by electron microscopy (Cameron et al. 2001) (Figure 10.4.).

It is now clear that ActA evolved the capacity to mimic the activity of host nucleation-promoting factors (Welch and Mullins 2002). The N-terminal module of ActA, which is the minimal region that stimulates actin nucleation and y-branching with Arp2/3 complex (Welch et al. 1998, Skoble et al. 2000, Boujemaa-Paterski et al. 2001, Zalevsky et al. 2001), has sequences in common with the host WASP, N-WASP, and WAVE/Scar proteins (Figure 10.3.). The similarities are confined to the C and A regions that bind the Arp2/3 complex, and the AB region that binds actin monomers. Each of these sequences in ActA plays a role in promoting nucleation. In particular, the C region is essential for nucleation, as deletions or point mutations in this sequence cause a failure to promote actin assembly with Arp2/3 complex in vitro, undergo actin-based motility in cells, or spread from cell to cell (Pistor et al. 2000, Skoble et al. 2000, Boujemaa-Paterski et al. 2001, Lauer et al. 2001). Mutations in the A and AB regions only reduce the efficiency of actin nucleation, the rate of motility, and the efficiency of spread, indicating these regions are important but nonessential. Interestingly, mutations outside of the A, C, and AB regions, which do not directly affect actin nucleation with Arp2/3 complex in the context of purified proteins, do affect parameters of actin-based motility such as motility initiation, path curvature, and persistence of movement

(Lauer et al. 2001). Thus, the relationship between the biochemical activity of purified components in vitro and actin-based motility in cells is complex and is not yet fully understood.

10.4.3.   Ena/VASP and Profilin

Efficient *L. monocytogenes* actin polymerization and motility also require host proteins other than the Arp2/3 complex. These include the Ena/VASP proteins, which operate in conjunction with the actin monomer-binding protein profilin. In mammalian cells the Ena/VASP family is comprised of three members: VASP, Mena (mouse Ena), and EVL (Ena VASP like) (Krause et al. 2003). Each has a similar modular structure (Haffner et al. 1995) that includes an N-terminal Ena/VASP homology 1 (EVH1) domain that binds directly to the proline-rich motifs in ActA (Niebuhr et al. 1997), a central proline-rich region that binds to profilin (Reinhard et al. 1995, Kang et al. 1997), and a C-terminal Ena/VASP homology 2 region (EVH2) that binds to filamentous actin and also mediates multimerization (Bachmann et al. 1999). In uninfected host cells, Ena/VASP proteins and profilin are concentrated at the cortex and in cell–matrix adhesion complexes (Krause et al. 2003). During infection by *L. monocytogenes*, both are specifically recruited to and concentrated at the surface of intracellular bacteria (Theriot et al. 1994, Chakraborty et al. 1995) (Figure 10.4.), unlike the Arp2/3 complex, which is distributed throughout comet tails.

Several lines of evidence indicate that Ena/VASP proteins are functionally important, but not essential, for actin-based motility. Deletion or mutation of the proline-rich Ena/VASP binding sites in ActA prevents Ena/VASP surface recruitment (Smith et al. 1996, Niebuhr et al. 1997) and causes a reduction in the rate of movement (Auerbuch et al. 2003), the percentage of bacteria that initiate movement (Smith et al. 1996, Auerbuch et al. 2003), and the directional persistence of movement (Auerbuch et al. 2003). Moreover, bacteria exhibit reduced motility rates in MV^{D7} cells (Bear et al. 2000), which are deficient in the expression of all three Ena/VASP proteins (Geese et al. 2002). The similarity between the phenotypes caused by mutation of Ena/VASP binding sites in ActA and global depletion of these proteins from host cells suggests that Ena/VASP proteins function primarily via recruitment to the bacterial surface, and that the general cytoplasmic pool makes only a residual contribution to actin-based bacterial movement.

One key function of Ena/VASP proteins is to modulate the organization of actin filaments within comet tails. Ena/VASP proteins decrease the frequency of actin y-branching by the Arp2/3 complex in the context of purified proteins (Skoble et al. 2001), as well as in the comet tails formed by beads coated with ActA (Samarin et al. 2003) or host nucleation-promoting factors (Plastino et al. 2004). This is similar to the effect of Ena/VASP proteins on actin archi-tecture at the cortex of host cells (Bear et al. 2002). The ability to inhibit y-branching is correlated with an increase in the rates of actin-based motility (Samarin et al. 2003, Plastino et al. 2004). It has been proposed that this increased

bacterial speed is caused by an ability of Ena/VASP proteins to promote the dissociation of y-branched filaments from the bacterial surface, which would stimulate
the rate of polymerization and retard the drag force associated with filament–
surface attachments (Samarin et al. 2003). It is still unclear how Ena/VASP
exerts these activities. One hypothesis is that the effects on actin organization
and motility may be mediated by F-actin-binding activity. However, deletion of
the F-actin-binding region does not affect the rates of bacterial movement (Geese
et al. 2002, Auerbuch et al. 2003), suggesting that other activities must be more
relevant. The ability to bind F-actin does play a role in promoting straighter
movement trajectories, however, which contributes to the ability of the bacteria
to spread from cell to cell in a plaque assay (Auerbuch et al. 2003).

A second key activity of Ena/VASP proteins is their ability to recruit
profilin to the bacterial surface. Mutations in ActA that prevent Ena/VASP
binding also prevent profilin accumulation (Smith et al. 1996), highlighting the
functional connection between these molecules. The ability to recruit profilin
is critical for bacterial motility, as mutant Ena/VASP proteins lacking the
proline-rich profilin binding site are unable to promote rapid motility (Geese
et al. 2002, Auerbuch et al. 2003). The importance of profilin is also highlighted
by the observation that depletion of the protein from cell extracts (Theriot
et al. 1994, Marchand et al. 1995) and interference with its function in cells
(Southwick and Purich 1995, Grenklo et al. 2003) either halts motility or reduces
motility rates. Moreover, there is a correlation between the presence of profilin
at the bacterial surface and rapid bacterial movement—profilin is not recruited
to the surface of stationary bacteria, accumulates at the bacterial surface as
motility initiates, and disappears as motility slows (Geese et al. 2000). At
the biochemical level, profilin is thought to promote actin polymerization by
binding actin monomers, displacing them from the sequestering protein thymosin
$\beta 4$, and ushering them onto the barbed ends of actin filaments (Pantaloni and
Carlier 1993, Kang et al. 1999). Concentrating profilin at the bacterial surface
may in turn concentrate polymerization competent actin, enabling rapid actin
polymerization and bacterial movement.

10.4.4.   Capping Protein and ADF/Cofilin

The initiation of new actin polymerization must be spatially restricted to the
bacterial surface to ensure efficient coupling of polymerization to motility. This
function is fulfilled by capping protein (CapZ), a heterodimeric protein that
binds with high affinity to the barbed ends of actin filaments and prevents both
polymerization and depolymerization (Wear and Cooper 2004). Capping protein
is localized throughout actin comet tails (David et al. 1998) (Figure 10.4.).
However, its activity must be modulated such that new filaments remain
uncapped at the bacterial surface to enable force generation, but old filaments
become capped to prevent rampant polymerization and depletion of the monomer
pool. It is still not clear how this selectivity is achieved. One simple possibility
is that CapZ binds stochastically to barbed ends and dissociates slowly, so that

newly formed ends would tend to be uncapped, and older ends would tend to be capped. A second possibility is that *L. monocytogenes* actively inhibits capping by recruiting factors to the bacterial surface which antagonize the activity of capping protein (Marchand et al. 1995). It has been proposed that VASP may perform this function (Bear et al. 2002, Barzik et al. 2005), although this idea remains controversial (Boujemaa-Paterski et al. 2001, Samarin et al. 2003).

Continuous movement of *L. monocytogenes* in infected cells also requires disassembly of actin filaments and recycling of actin monomers, as treatment of cells with drugs that block actin depolymerization inhibits movement (Cramer 1999). Two proteins that sever actin filaments, ADF/cofilin and gelsolin, are localized to comet tails and have been implicated in actin disassembly (Rosenblatt et al. 1997, David et al. 1998, Laine et al. 1998). Depletion of ADF/cofilin from cell extracts does not alter the rate of actin polymerization or the speed of bacterial motility, but does dramatically reduce actin disassembly, which causes comet tails to become extremely long (Rosenblatt et al. 1997). Addition of excess ADF/cofilin has the opposite effect—it causes tail length to decrease (Carlier et al. 1997, Rosenblatt et al. 1997). Thus, ADF/cofilin clearly plays a major role in promoting actin disassembly during bacterial motility. Gelsolin may also play a role in actin disassembly in cells and in cell extracts, but its activity is dependent on Ca^{2+}. At physiological Ca^{2+} concentrations, gelsolin may augment the functions of ADF/cofilin, but does not appear to play a primary role in actin filament turnover (Larson et al. 2005).

10.4.5. Actin Bundling and Membrane-Binding Proteins

Tight cross-linking of actin filaments into a network in the comet tail is mechanically essential for enabling polymerization to generate the force that drives bacterial movement. As is discussed above, the organization of actin into y-branches by the Arp2/3 complex makes an important contribution to the architecture of the actin network in the comet tail. In addition, a long list of actin cross-linking and bundling proteins are present in *L. monocytogenes* comet tails, including α-actinin (Dabiri et al. 1990), fimbrin (Kocks and Cossart 1993), filamin (Van Kirk et al. 2000), and fascin (Brieher et al. 2004). Of these, only α-actinin has been shown to be essential for comet tail formation (Dold et al. 1994), and the function of the others remains to be evaluated. Bundling proteins may be of general importance, particularly in organizing actin into the parallel arrays that are observed in the outer sheath of the comet tail surrounding the y-branched core (Zhukarev et al. 1995, Sechi et al. 1997, Brieher et al. 2004). Interestingly, it was recently shown that *L. monocytogenes* motility initiated by Arp2/3 complex in vitro can continue in the absence of new nucleation by Arp2/3 complex if actin elongation persists and filaments are bundled by fascin (Brieher et al. 2004). Under these conditions, the actin comet tails lack the y-branched core and consist only of long parallel bundles resembling the outer sheath. Thus, actin elongation and bundling are sufficient to drive motility in the absence of nucleation, highlighting the functional importance of bundling to the mechanical integrity of the comet tail.

In addition to linkages between actin filaments, linkages of filaments and membranes are critical for *L. monocytogenes* cell-to-cell spread. Membrane–cytoskeletal linkages are mediated by the ezrin/radixin/moesin (ERM) family of proteins. In keeping with this function, ERM proteins are localized to the comet tails of bacteria in protrusions that are enveloped with membrane but not to comet tails of bacteria that are free within the cytoplasm (Temm-Grove et al. 1994, Sechi et al. 1997, Pust et al. 2005). Importantly, interference with ERM protein function by RNAi or expression of dominant negative protein fragments inhibits both protrusion formation and cell-to-cell spread, but does not affect the rates of intracellular motility (Pust et al. 2005). Thus, the ability to undergo actin-based motility may not in itself be sufficient to enable protrusion formation, and connections of actin filaments with membranes may be critical for maintaining the structure of the protrusion and for facilitating cell-to-cell bacterial transfer.

10.4.6. Reconstitution of Motility with Purified Proteins

Ten years of research aimed at understanding the biochemical mechanism of *L. monocytogenes* actin-based motility culminated in the reconstitution of this process with a system consisting only of purified proteins (Loisel et al. 1999). In a testament to the surprising biochemical simplicity of the system, the basic reconstitution mix consisted of only four factors: actin, Arp2/3 complex, capping protein, and ADF/cofilin. In the presence of these four core components, motility proceeds slowly ($0.5\,\mu m/min$). Addition of Ena/VASP and profilin to the core mix leads to an increase in the speed of movement (to $\sim 3\,\mu m/min$), consistent with the role of these factors in cells and cell extracts. In contrast to the apparently critical role of α-actinin in cells, it is not essential in vitro and its addition has no effect on movement velocity. Nevertheless, addition of α-actinin increases filament density in the comet tails and contributes to bacterium-filament attachments.

The ability to reconstitute actin-based bacterial motility unequivocally demonstrates that the process is driven by a core set of proteins that regulate actin assembly dynamics, and not by motor proteins. It also provides insights into the basic biochemical mechanisms that underlie host processes like leading edge protrusion during cell migration, which are also likely to be driven by the same set of factors. Despite the underlying biochemical similarities, the reconstitution approach highlights the fact that bacterial motility relies on a stripped-down set of core components, whereas the corresponding processes in host cells are more complex. For example, in host cells, the core actin assembly machinery is tightly regulated by signal transduction proteins that include Rho family GTPases and tyrosine kinases. By contrast, *L. monocytogenes* is remarkable in that its ActA protein has evolved the ability to bypasses activation of these host signaling pathways (Ebel et al. 1999, Frischknecht et al. 1999), allowing constitutive stimulation of Arp2/3 complex to promote continuous actin-based motility and cell-to-cell spread. Because it represents an efficient and stripped-down system, *L. monocytogenes* has served as a very useful model for cell biologists and biophysicists to study the basic mechanisms that control actin-based movement.

10.5.   Biophysics of Actin-Based Movement

Because the process of *L. monocytogenes* actin-based motility is relatively well understood at a biochemical level, it has captured the attention of biophysicists who are interested in understanding how actin polymerization generates motile force. One key question of interest centers on the magnitude of force generated to power *L. monocytogenes* motility. Recent measurements suggest a stall force ranging from 7 nN to upward of 150 nN, depending on the surface area over which polymerization occurs (Marcy et al. 2004, Parekh et al., 2005). This is considerably greater than the force required to deform the lipid bilayer of synthetic vesicles to generate long membrane tubules, which is in the 10 pN range (Hochmuth et al. 1996, Inaba et al. 2005). Thus, the force generated by actin polymerization is more than sufficient to promote formation of a protrusion, and the additional force may in fact be required to push the protrusion into the adjacent cell.

A second key question relates to how actin polymerization generates motile force. Several models have been proposed to explain this phenomenon (Figure 10.5.) (Plastino and Sykes 2005, Mogilner 2006). One genre of models is based on a classical "Brownian ratchet" type mechanism, where forward motion of the bacterium is allowed, but backward motion is prevented by the actin network in the comet tail. The most recent incarnation of these models is the "tethered elastic Brownian ratchet" model (Mogilner and Oster 2003a). This postulates the existence of two types of filaments at the bacterial surface: tethered and working. The tethered filaments are attached to the bacterial surface and hinder forward and backward motion. How filaments are attached is not understood, but this may involve interactions between ActA, Arp2/3 complex, and VASP. Attached filaments can detach over time to become working filaments, which are free to bend by thermal motion. When they bend away from the bacterial surface, a gap is created that allows polymerization and filament lengthening (Mogilner and Oster 1996). The energy derived from polymerization and the elastic energy stored in the bent filament are then converted into mechanical force when the lengthened filament contacts and pushes on the bacterial surface, which drives motility. In this model, the additional energy supplied from ATP hydrolysis by actin is used to recycle actin for multiple rounds of polymerization. The tethered elastic ratchet model has been tested in computer simulations that incorporate known biochemical characteristics of the core protein components needed to reconstitute motility, and it produces simulated bacterial motility that resembles the natural process (Alberts and Odell 2004).

Another proposed model is the "filament end tracking motor model" (Dickinson and Purich 2002, Dickinson et al. 2004), which postulates that individual filaments in the comet tail can be both tethered and working. In this model, an unknown end-tracking factor binds to ATP-containing subunits at the barbed end. As polymerization proceeds and ATP is hydrolyzed, the end-tracking protein dissociates, and then rebinds the terminal ATP-containing subunit, causing forward propulsion. Thus, in this case, the force-generating machine is directly

1. tethered elastic Brownian ratchet

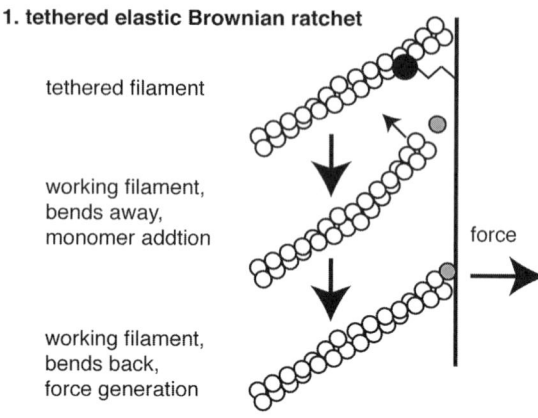

tethered filament

working filament,
bends away,
monomer addtion

working filament,
bends back,
force generation

force

2. filament end tracking motor

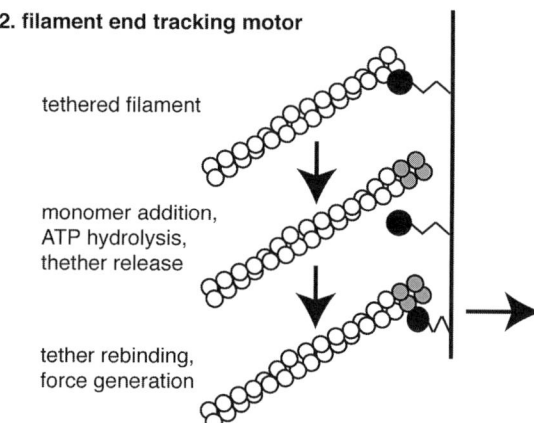

tethered filament

monomer addition,
ATP hydrolysis,
thether release

tether rebinding,
force generation

3. elastic network squeezing

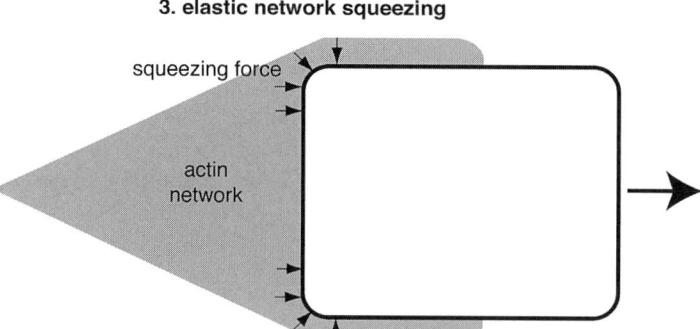

squeezing force

actin
network

FIGURE 10.5. Cartoon diagram depicting three models for how actin polymerization generates the force that drives bacterial propulsion. 1 The tethered elastic Brownian ratchet model. 2 The filament end-tracking motor model. 3 The elastic network squeezing model. In 1 and 2, the existing actin filament is represented by open circles, and the newly polymerizing monomers by shaded circles. The black circle represents the unknown filament tether.

coupled to the energy derived from ATP hydrolysis in the filament. One prediction derived from this model is that *L. monocytogenes* movement should proceed in steps that are proportional to the size of actin monomers adding on to the ends of helical filaments, a phenomenon that has been observed in infected cells (Kuo and McGrath 2000) and cell extracts (McGrath et al. 2003).

The third model for force generation differs from the first two in that it does not deal with molecular level phenomena, but instead addresses the elastic properties of the actin network as a whole (Gerbal et al. 2000, Bernheim-Groswasser et al. 2005). In this model, the polymerization of new actin filaments at the bacterial surface stores energy as elastic strain in the actin network of the comet tail. The strain then produces a squeezing force on the bacterial surface that pushes on the bacterium, generating propulsive force. In support of this model, it has been observed that ActA-coated liposomes induce actin-based motility cell extracts, and this is accompanied by squeezing of the liposome, which causes a normally spherical vesicle to be deformed into a teardrop shape (Giardini et al. 2003, Upadhyaya et al. 2003).

None of the models mentioned above are mutually exclusive, and hybrid mechanisms can be imagined. Other force-generating mechanisms may also operate. Further understanding of the mechanism of force generation will require a combination of biochemical, biophysical, and computational approaches.

10.6.   Conclusions

Nearly two decades after the discovery that *L. monocytogenes* exploits the host actin cytoskeleton to undergo intracellular movement, we have reached a reasonable understanding of how this process works at the molecular level. The key molecular components that are involved in motility have been identified, and a core set of proteins has been demonstrated to be sufficient for the reconstitution of the process in vitro. Nevertheless, much remains to be discovered. We do not yet know how actin polymerization is coupled to force generation to drive motility. Moreover, we do not understand the mechanism of cell-to-cell spread, or the cytoskeletal and membrane-associated factors which might be uniquely important for this process. Finally, we do not know whether the ability to associate with actin has yet undiscovered roles, for example in subverting the immune response. Answering these questions will be critical for attaining a full understanding of infection by *L. monocytogenes* and also by other pathogens that have evolved a similar capacity to undergo actin-based propulsion. The information gained from studying *L. monocytogenes* actin-based motility and cell-to-cell spread will also be of major importance in uncovering basic cell biological principles related to cytoskeletal function and regulation in the host.

Acknowledgments. I thank the Welch Lab and Dan Fletcher for stimulating discussions, and Dan Portnoy, Erin Goley, and Ken Campellone for their comments on this chapter. Work in the Welch Lab is supported by grant GM59609 from the NIH/NIGMS, an Established Investigator Award from the American Heart Association, and grant 2004-35607-14906 from the National Research Initiative of the USDA Cooperative State Research, Education and Extension Service (CSREES).

References

Alberts JB, Odell GM (2004). In silico reconstitution of *Listeria* propulsion exhibits nano-saltation. PLoS Biol 2:e412.

Auerbuch V, Lenz LL, Portnoy DA (2001). Development of a competitive index assay to evaluate the virulence of *Listeria monocytogenes actA* mutants during primary and secondary infection of mice. Infect Immun 69:5953–5957.

Auerbuch V, Loureiro JJ, Gertler FB, Theriot JA, Portnoy DA (2003). Ena/VASP proteins contribute to *Listeria monocytogenes* pathogenesis by controlling temporal and spatial persistence of bacterial actin-based motility. Mol Microbiol 49:1361–1375.

Bachmann C, Fischer L, Walter U, Reinhard M (1999). The EVH2 domain of the vasodilator-stimulated phosphoprotein mediates tetramerization, F-actin binding, and actin bundle formation. J Biol Chem 274:23549–23557.

Barzik M, Kotova TI, Higgs HN, Hazelwood L, Hanein D, Gertler FB, Schafer DA (2005). Ena/VASP proteins enhance actin polymerization in the presence of barbed end capping proteins. J Biol Chem 280:28653–28662.

Bear JE, Loureiro JJ, Libova I, Fassler R, Wehland J, Gertler FB (2000). Negative regulation of fibroblast motility by Ena/VASP proteins. Cell 101:717–728.

Bear JE, Svitkina TM, Krause M, Schafer DA, Loureiro JJ, Strasser GA, Maly IV, Chaga OY, Cooper JA, Borisy GG, Gertler FB (2002). Antagonism between Ena/VASP proteins and actin filament capping regulates fibroblast motility. Cell 109:509–521.

Bernardini ML, Mounier J, d'Hauteville H, Coquis-Rondon M, Sansonetti PJ (1989). Identification of *icsA*, a plasmid locus of *Shigella flexneri* that governs bacterial intra- and intercellular spread through interaction with F-actin. Proc Natl Acad Sci USA 86:3867–3871.

Bernheim-Groswasser A, Prost J, Sykes C (2005). Mechanism of actin-based motility: a dynamic state diagram. Biophys J 89:1411–1419.

Bi E, Zigmond SH (1999). Actin polymerization: where the WASP stings. Curr Biol 9:R161–R163.

Bohne J, Sokolovic Z, Goebel W (1994). Transcriptional regulation of *prfA* and PrfA-regulated virulence genes in *Listeria monocytogenes*. Mol Microbiol 11:1141–1150.

Boujemaa-Paterski R, Gouin E, Hansen G, Samarin S, Le Clainche C, Didry D, Dehoux P, Cossart P, Kocks C, Carlier MF, Pantaloni D (2001). *Listeria* protein ActA mimics WASp family proteins: it activates filament barbed end branching by Arp2/3 complex. Biochemistry 40:11390–11404..

Brieher WM, Coughlin M, Mitchison TJ (2004). Fascin-mediated propulsion of *Listeria monocytogenes* independent of frequent nucleation by the Arp2/3 complex. J Cell Biol 165:233–242.

Brundage RA, Smith GA, Camilli A, Theriot JA, Portnoy DA (1993). Expression and phosphorylation of the *Listeria monocytogenes* ActA protein in mammalian cells. Proc Natl Acad Sci USA 90:11890–11894.

Cameron LA, Footer MJ, van Oudenaarden A, Theriot JA (1999). Motility of ActA protein-coated microspheres driven by actin polymerization. Proc Nat Acad Sci USA 96:4906–4913.

Cameron LA, Svitkina TM, Vignjevic D, Theriot JA, Borisy GG (2001). Dendritic organization of actin comet tails. Curr Biol 11:130–135.

Carlier MF, Laurent V, Santolini J, Melki R, Didry D, Xia GX, Hong Y, Chua NH, Pantaloni D (1997). Actin depolymerizing factor (ADF/cofilin) enhances the rate of filament turnover: implication in actin-based motility. J Cell Biol 136: 1307–1322.

Chakraborty T, Ebel F, Domann E, Niebuhr K, Gerstel B, Pistor S, Temm-Grove CJ, Jockusch BM, Reinhard M, Walter U, Wehland J (1995). A focal adhesion factor directly linking intracellularly motile *Listeria monocytogenes* and *Listeria ivanovii* to the actin-based cytoskeleton of mammalian cells. EMBO J 14:1314–1321.

Cicchetti G, Maurer P, Wagener P, Kocks C (1999). Actin and phosphoinositide binding by the ActA protein of the bacterial pathogen *Listeria monocytogenes*. J Biol Chem 274:33616–33626.

Cramer LP (1999). Role of actin-filament disassembly in lamellipodium protrusion in motile cells revealed using the drug jasplakinolide. Curr Biol 9:1095–1105.

Dabiri GA, Sanger JM, Portnoy DA, Southwick FS (1990). *Listeria monocytogenes* moves rapidly through the host-cell cytoplasm by inducing directional actin assembly. Proc Natl Acad Sci USA 87:6068–6072.

David V, Gouin E, Troys MV, Grogan A, Segal AW, Ampe C, Cossart P (1998). Identification of cofilin, coronin, Rac and capZ in actin tails using a *Listeria* affinity approach. J Cell Sci 111:2877–2884.

Dickinson RB, Caro L, Purich DL (2004). Force generation by cytoskeletal filament end-tracking proteins. Biophys J 87:2838–2854.

Dickinson RB, Purich DL (2002). Clamped-filament elongation model for actin-based motors. Biophys J 82:605–617.

Dold FG, Sanger JM, Sanger JW (1994). Intact alpha-actinin molecules are needed for both the assembly of actin into the tails and the locomotion of *Listeria monocytogenes* inside infected cells. Cell Motil Cytoskeleton 28:97–107.

Domann E, Wehland J, Rohde M, Pistor S, Hartl M, Goebel W, Leimeister-Wachter M, Wuenscher M, Chakraborty T (1992). A novel bacterial virulence gene in *Listeria monocytogenes* required for host cell microfilament interaction with homology to the proline-rich region of vinculin. EMBO J 11:1981–1990.

Ebel F, Rohde M, von Eichel-Streiber C, Wehland J, Chakraborty T (1999). The actin-based motility of intracellular *Listeria monocytogenes* is not controlled by small GTP-binding proteins of the Rho- and Ras-subfamilies. FEMS Microbiol Lett 176:117–124.

Freitag NE, Jacobs KE (1999). Examination of *Listeria monocytogenes* intracellular gene expression by using the green fluorescent protein of *Aequorea victoria*. Infect Immun 67:1844–1852.

Friederich E, Gouin E, Hellio R, Kocks C, Cossart P, Louvard D (1995). Targeting of *Listeria monocytogenes* ActA protein to the plasma membrane as a tool to dissect both actin-based cell morphogenesis and ActA function. EMBO J 14:2731–2744.

Frischknecht F, Cudmore S, Moreau V, Reckmann I, Rottger S, Way M (1999). Tyrosine phosphorylation is required for actin-based motility of *vaccinia* but not *Listeria* or *Shigella*. Curr Biol 9:89–92.

Geese M, Loureiro JJ, Bear JE, Wehland J, Gertler FB, Sechi AS (2002). Contribution of Ena/VASP proteins to intracellular motility of *Listeria* requires phosphorylation and proline-rich core but not F-actin binding or multimerization. Mol Biol Cell 13:2383–2396.

Geese M, Schluter K, Rothkegel M, Jockusch BM, Wehland J, Sechi AS (2000). Accumulation of profilin II at the surface of *Listeria* is concomitant with the onset of motility and correlates with bacterial speed. J Cell Sci 113 (Pt 8):1415–1426.

Gerbal F, Chaikin P, Rabin Y, Prost J (2000). An elastic analysis of *Listeria monocytogenes* propulsion. Biophys J 79:2259–2275.

Gerstel B, Grobe L, Pistor S, Chakraborty T, Wehland J (1996). The ActA polypeptides of *Listeria ivanovii* and *Listeria monocytogenes* harbor related binding sites for host microfilament proteins. Infect Immun 64:1929–1936.

Giardini PA, Fletcher DA, Theriot JA (2003). Compression forces generated by actin comet tails on lipid vesicles. Proc Natl Acad Sci USA 100:6493–6498.

Gouin E, Dehoux P, Mengaud J, Kocks C, Cossart P (1995). *iactA* of *Listeria ivanovii*, although distantly related to *Listeria monocytogenes actA*, restores actin tail formation in an *L. monocytogenes actA* mutant. Infect Immun 63:2729–2737.

Gouin E, Welch MD, Cossart P (2005). Actin-based motility of intracellular pathogens. Curr Opin Microbiol 8:35–45.

Grenklo S, Geese M, Lindberg U, Wehland J, Karlsson R, Sechi AS (2003). A crucial role for profilin-actin in the intracellular motility of *Listeria monocytogenes*. EMBO Rep 4:523–529.

Haffner C, Jarchau T, Reinhard M, Hoppe J, Lohmann SM, Walter U (1995). Molecular cloning, structural analysis and functional expression of the proline-rich focal adhesion and microfilament-associated protein VASP. EMBO J 14:19–27.

Heinzen RA, Hayes SF, Peacock MG, Hackstadt T (1993). Directional actin polymerization associated with spotted fever group *Rickettsia* infection of vero cells. Infect Immun 61:1926–1935.

Hochmuth FM, Shao JY, Dai J, Sheetz MP (1996). Deformation and flow of membrane into tethers extracted from neuronal growth cones. Biophys J 70:358–369.

Inaba T, Ishijima A, Honda M, Nomura F, Takiguchi K, Hotani H (2005). Formation and maintenance of tubular membrane projections require mechanical force, but their elongation and shortening do not require additional force. J Mol Biol 348:325–333.

Kang F, Laine RO, Bubb MR, Southwick FS, Purich DL (1997). Profilin interacts with the Gly-Pro-Pro-Pro-Pro-Pro sequences of vasodilator-stimulated phosphoprotein (VASP): implications for actin-based *Listeria* motility. Biochemistry 36:8384–8392.

Kang F, Purich DL, Southwick FS (1999). Profilin promotes barbed-end actin filament assembly without lowering the critical concentration. J Biol Chem 274:36963–36972.

Karunasagar I, Krohne G, Goebel W (1993). *Listeria ivanovii* is capable of cell-to-cell spread involving actin polymerization. Infect Immun 61:162–169.

Kespichayawattana W, Rattanachetkul S, Wanun T, Utaisincharoen P, Sirisinha S (2000). *Burkholderia pseudomallei* induces cell fusion and actin-associated membrane protrusion: a possible mechanism for cell-to-cell spreading. Infect Immun 68:5377–5384.

Kidd T, Brose K, Mitchell KJ, Fetter RD, Tessier-Lavigne M, Goodman CS, Tear G (1998). Roundabout controls axon crossing of the CNS midline and defines a novel subfamily of evolutionarily conserved guidance receptors. Cell 92:205–215.

Kocks C, Cossart P (1993). Directional actin assembly by *Listeria monocytogenes* at the site of polar surface expression of the *actA* gene product involving the actin-bundling protein plastin (fimbrin). Infect Agents Dis 2:207–209.

Kocks C, Gouin E, Tabouret M, Berche P, Ohayon H, Cossart P (1992). *L. monocytogenes*-induced actin assembly requires the *actA* gene product, a surface protein. Cell 68:521–531.

Kocks C, Hellio R, Gounon P, Ohayon H, Cossart P (1993). Polarized distribution of *Listeria monocytogenes* surface protein ActA at the site of directional actin assembly. J Cell Sci 105:699–710.

Kocks C, Marchand JB, Gouin E, d'Hauteville H, Sansonetti PJ, Carlier MF, Cossart P (1995). The unrelated surface proteins ActA of *Listeria monocytogenes* and IcsA of *Shigella flexneri* are sufficient to confer actin-based motility on *Listeria innocua* and *Escherichia coli* respectively. Mol Microbiol 18:413–423.

Krause M, Dent EW, Bear JE, Loureiro JJ, Gertler FB (2003). Ena/VASP proteins: regulators of the actin cytoskeleton and cell migration. Annu Rev Cell Dev Biol 19:541–564.

Kreft J, Dumbsky M, Theiss S (1995). The actin-polymerization protein from *Listeria ivanovii* is a large repeat protein which shows only limited amino acid sequence homology to ActA from *Listeria monocytogenes*. FEMS Microbiol Lett 132:181–182.

Kreft J, Vazquez-Boland JA (2001). Regulation of virulence genes in *Listeria*. Int J Med Microbiol 291:145–157.

Kuo SC, McGrath JL (2000). Steps and fluctuations of *Listeria monocytogenes* during actin-based motility. Nature 407:1026–1029.

Laine RO, Phaneuf KL, Cunningham CC, Kwiatkowski D, Azuma T, Southwick FS (1998). Gelsolin, a protein that caps the barbed ends and severs actin filaments, enhances the actin-based motility of *Listeria monocytogenes* in host cells. Infect Immun 66:3775–3782.

Larson L, Arnaudeau S, Gibson B, Li W, Krause R, Hao B, Bamburg JR, Lew DP, Demaurex N, Southwick F (2005). Gelsolin mediates calcium-dependent disassembly of *Listeria* actin tails. Proc Natl Acad Sci USA 102:1921–1926.

Lasa I, Gouin E, Goethals M, Vancompernolle K, David V, Vandekerckhove J, Cossart P (1997). Identification of two regions in the N-terminal domain of ActA involved in the actin comet tail formation by *Listeria monocytogenes*. EMBO J 16:1531–1540.

Lasa I, Violaine D, Gouin E, Marchand J, Cossart P (1995). The amino-terminal part of ActA is critical for the actin-based motility of *Listeria monocytogenes*; the central proline-rich region acts as a stimulator. Mol Microbiol 18:425–436.

Lauer P, Theriot JA, Skoble J, Welch MD, Portnoy DA (2001). Systematic mutational analysis of the amino-terminal domain of the *Listeria monocytogenes* ActA protein reveals novel functions in actin-based motility. Mol Microbiol 42:1163–1177.

Leimeister-Wachter M, Chakraborty T (1989). Detection of listeriolysin, the thiol-dependent hemolysin in *Listeria monocytogenes*, *Listeria ivanovii*, and *Listeria seeligeri*. Infect Immun 57:2350–2357.

Loisel TP, Boujemaa R, Pantaloni D, Carlier MF (1999). Reconstitution of actin-based motility of *Listeria* and *Shigella* using pure proteins. Nature 401:613–616.

Machesky LM, Atkinson SJ, Ampe C, Vandekerckhove J, Pollard TD (1994). Purification of a cortical complex containing two unconventional actins from *Acanthamoeba* by affinity chromatography on profilin-agarose. J Cell Biol 127:107–115.

Marchand JB, Moreau P, Paoletti A, Cossart P, Carlier MF, Pantaloni D (1995). Actin-based movement of *Listeria monocytogenes*: actin assembly results from the local maintenance of uncapped filament barbed ends at the bacterium surface. J Cell Biol 130:331–343.

Marcy Y, Prost J, Carlier MF, Sykes C (2004). Forces generated during actin-based propulsion: a direct measurement by micromanipulation. Proc Natl Acad Sci USA 101:5992–5997.

May RC, Hall ME, Higgs HN, Pollard TD, Chakraborty T, Wehland J, Machesky LM, Sechi AS (1999). The Arp2/3 complex is essential for the actin-based motility of *Listeria monocytogenes*. Curr Biol 9:759–762.

McGrath JL, Eungdamrong NJ, Fisher CI, Peng F, Mahadevan L, Mitchison TJ, Kuo SC (2003). The force-velocity relationship for the actin-based motility of *Listeria monocytogenes*. Curr Biol 13:329–332.

Mogilner A (2006). On the edge: modeling protrusion. Curr Opin Cell Biol 18:32–39.

Mogilner A, Oster G (1996). Cell motility driven by actin polymerization. Biophys J 71:3030–3045.

Mogilner A, Oster G (2003a) Force generation by actin polymerization II: the elastic ratchet and tethered filaments. Biophys J 84:1591–1605.

Mogilner A, Oster G (2003b) Polymer motors: pushing out the front and pulling up the back. Curr Biol 13:R721–733.

Moors MA, Auerbuch V, Portnoy DA (1999a) Stability of the *Listeria monocytogenes* ActA protein in mammalian cells is regulated by the N-end rule pathway. Cell Microbiol 1:249–257.

Moors MA, Levitt B, Youngman P, Portnoy DA (1999b) Expression of listeriolysin O and ActA by intracellular and extracellular *Listeria monocytogenes*. Infect Immun 67:131–139.

Mounier J, Ryter A, Coquis-Rondon M, Sansonetti PJ (1990). Intracellular and cell-to-cell spread of *Listeria monocytogenes* involves interaction with F-actin in the enterocytelike cell line Caco-2. Infect Immun 58:1048–1058.

Mourrain P, Lasa I, Gautreau A, Gouin E, Pugsley A, Cossart P (1997). ActA is a dimer. Proc Natl Acad Sci USA 94:10034–10039.

Nanavati D, Ashton FT, Sanger JM, Sanger JW (1994). Dynamics of actin and alpha-actinin in the tails of *Listeria monocytogenes* in infected PtK2 cells. Cell Motil Cytoskeleton 28:346–358.

Niebuhr K, Chakraborty T, Rohde M, Gazlig T, Jansen B, Kollner P, Wehland J (1993). Localization of the ActA polypeptide of *Listeria monocytogenes* in infected tissue culture cell lines: ActA is not associated with actin "comets". Infect Immun 61:2793–2802.

Niebuhr K, Ebel F, Frank R, Reinhard M, Domann E, Carl UD, Walter U, Gertler FB, Wehland J, Chakraborty T (1997). A novel proline-rich motif present in ActA of *Listeria monocytogenes* and cytoskeletal proteins is the ligand for the EVH1 domain, a protein module present in the Ena/VASP family. EMBO J 16:5433–5444.

Pantaloni D, Carlier MF (1993). How profilin promotes actin filament assembly in the presence of thymosin beta 4. Cell 75:1007–1014.

Parekh SH, Chaudhuri O, Theriot JA, Fletcher DA (2005). Loading history determines the velocity of actin-network growth. Nat Cell Biol 7:1119–1123.

Pistor S, Chakraborty T, Niebuhr K, Domann E, Wehland J (1994). The ActA protein of *Listeria monocytogenes* acts as a nucleator inducing reorganization of the actin cytoskeleton. EMBO J 13:758–763.

Pistor S, Chakraborty T, Walter U, Wehland J (1995). The bacterial actin nucleator protein ActA of *Listeria monocytogenes* contains multiple binding sites for host microfilament proteins. Curr Biol 5:517–525.

Pistor S, Grobe L, Sechi AS, Domann E, Gerstel B, Machesky LM, Chakraborty T, Wehland J (2000). Mutations of arginine residues within the 146-KKRRK-150 motif of the ActA protein of *Listeria monocytogenes* abolish intracellular motility by interfering with the recruitment of the Arp2/3 complex. J Cell Sci 113:3277–3287.

Plastino J, Olivier S, Sykes C (2004). Actin filaments align into hollow comets for rapid VASP-mediated propulsion. Curr Biol 14:1766–1771.

Plastino J, Sykes C (2005). The actin slingshot. Curr Opin Cell Biol 17:62–66.

Pollard TD, Blanchoin L, Mullins RD (2000). Molecular mechanisms controlling actin filament dynamics in nonmuscle cells. Annu Rev Biophys Biomol Struct 29:545–576.

Pollard TD, Borisy GG (2003). Cellular motility driven by assembly and disassembly of actin filaments. Cell 112:453–465.

Prehoda KE, Lee DJ, Lim WA (1999). Structure of the enabled/VASP homology 1 domain-peptide complex: a key component in the spatial control of actin assembly. Cell 97:471–480.

Pust S, Morrison H, Wehland J, Sechi AS, Herrlich P (2005). *Listeria monocytogenes* exploits ERM protein functions to efficiently spread from cell to cell. EMBO J 24:1287–1300.

Racz P, Tenner K, Mero E (1972). Experimental *Listeria* enteritis. I. An electron microscopic study of the epithelial phase in experimental *Listeria* infection. Lab Invest 26:694–700.

Racz P, Tenner K, Szivessy K (1970). Electron microscopic studies in experimental keratoconjunctivitis listeriosa. I. Penetration of *Listeria monocytogenes* into corneal epithelial cells. Acta Microbiol Acad Sci Hung 17:221–236.

Rafelski SM, Theriot JA (2005). Bacterial shape and ActA distribution affect initiation of *Listeria monocytogenes* actin-based motility. Biophys J 89:2146–2158.

Rafelski SM, Theriot JA (2006). Mechanism of polarization of *Listeria monocytogenes* surface protein ActA. Mol Microbiol 59:1262–1279.

Reinhard M, Giehl K, Abel K, Haffner C, Jarchau T, Hoppe V, Jockusch BM, Walter U (1995). The proline-rich focal adhesion and microfilament protein VASP is a ligand for profilins. EMBO J 14:1583–1589.

Robbins JR, Barth AI, Marquis H, de Hostos EL, Nelson WJ, Theriot JA (1999). *Listeria monocytogenes* exploits normal host cell processes to spread from cell to cell. J Cell Biol 146:1333–1350.

Rosenblatt J, Agnew BJ, Abe H, Bamburg JR, Mitchison TJ (1997). *Xenopus* actin depolymerizing factor/cofilin (XAC) is responsible for the turnover of actin filaments in *Listeria monocytogenes* tails. J Cell Biol 136:1323–1332.

Samarin S, Romero S, Kocks C, Didry D, Pantaloni D, Carlier MF (2003). How VASP enhances actin-based motility. J Cell Biol 163:131–142.

Sanger JM, Mittal B, Southwick FS, Sanger JW (1995). *Listeria monocytogenes* intracellular migration: inhibition by profilin, vitamin D-binding protein and DNase I. Cell Motil Cytoskeleton 30:38–49.

Sanger JM, Sanger JW, Southwick FS (1992). Host cell actin assembly is necessary and likely to provide the propulsive force for intracellular movement of *Listeria monocytogenes*. Infect Immun 60:3609–3619.

Sechi AS, Wehland J, Small JV (1997). The isolated comet tail pseudopodium of *Listeria monocytogenes*: a tail of two actin filament populations, long and axial and short and random. J Cell Biol 137:155–167.

Skoble J, Auerbuch V, Goley ED, Welch MD, Portnoy DA (2001). Pivotal role of VASP in Arp2/3 complex-mediated actin nucleation, actin branch-formation, and *Listeria monocytogenes* motility. J Cell Biol 155:89–100.

Skoble J, Portnoy DA, Welch MD (2000). Three regions within ActA promote Arp2/3 complex-mediated actin nucleation and *Listeria monocytogenes* motility. J Cell Biol 150:527–538.

Smith GA, Portnoy DA, Theriot JA (1995). Asymetric distribution of the *Listeria monocytogenes* ActA protein is required and sufficient to direct actin-based motility. Mol Microbiol 17:945–951.

Smith GA, Theriot JA, Portnoy DA (1996). The tandem repeat domain in the *Listeria monocytogenes* ActA protein controls the rate of actin based motility, the percentage of moving bacteria, and the localization of vasodilator-stimulated phosphoprotein and profilin. J Cell Biol 135:647–660.

Southwick FS, Purich DL (1995). Inhibition of *Listeria* locomotion by mosquito oostatic factor, a natural oligoproline peptide uncoupler of profilin action. Infect Immun 63:182–190.

Stamm LM, Morisaki JH, Gao LY, Jeng RL, McDonald KL, Roth R, Takeshita S, Heuser J, Welch MD, Brown EJ (2003). *Mycobacterium marinum* escapes from phagosomes and is propelled by actin-based motility. J Exp Med 198:1361–1368.

Steffen P, Schafer DA, David V, Gouin E, Cooper JA, Cossart P (2000). *Listeria monocytogenes* ActA protein interacts with phosphatidylinositol 4,5-bisphosphate in vitro. Cell Motil Cytoskeleton 45:58–66.

Taunton J (2001). Actin filament nucleation by endosomes, lysosomes and secretory vesicles. Curr Opin Cell Biol 13:85–91.

Temm-Grove CJ, Jockusch BM, Rohde M, Niebuhr K, Chakraborty T, Wehland J (1994). Exploitation of microfilament proteins by *Listeria monocytogenes*: microvillus-like composition of the comet tails and vectorial spreading in polarized epithelial sheets. J Cell Sci 107:2951–2960.

Theriot JA, Mitchison TJ (1991). Actin microfilament dynamics in locomoting cells. Nature 352:126–131.

Theriot JA, Mitchison TJ (1992). Comparison of actin and cell surface dynamics in motile fibroblasts. J Cell Biol 118:367–377.

Theriot JA, Mitchison TJ, Tilney LG, Portnoy DA (1992). The rate of actin-based motility of intracellular *Listeria monocytogenes* equals the rate of actin polymerization. Nature 357:257–260.

Theriot JA, Rosenblatt J, Portnoy DA, Goldschmidt CP, Mitchison TJ (1994). Involvement of profilin in the actin-based motility of *L. monocytogenes* in cells and in cell-free extracts. Cell 76:505–517.

Tilney LG, Connelly PS, Portnoy DA (1990). Actin filament nucleation by the bacterial pathogen, *Listeria monocytogenes*. J Cell Biol 111:2979–2988.

Tilney LG, DeRosier DJ, Tilney MS (1992a) How *Listeria* exploits host cell actin to form its own cytoskeleton. I. Formation of a tail and how that tail might be involved in movement. J Cell Biol 118:71–81.

Tilney LG, DeRosier DJ, Weber A, Tilney MS (1992b) How *Listeria* exploits host cell actin to form its own cytoskeleton. II. Nucleation, actin filament polarity, filament assembly, and evidence for a pointed end capper. J Cell Biol 118:83–93.

Tilney LG, Portnoy DA (1989). Actin filaments and the growth, movement, and spread of the intracellular bacterial parasite, *Listeria monocytogenes*. J Cell Biol 109: 1597–1608.

Upadhyaya A, Chabot JR, Andreeva A, Samadani A, van Oudenaarden A (2003). Probing polymerization forces by using actin-propelled lipid vesicles. Proc Natl Acad Sci USA 100:4521–4526.

Van Kirk LS, Hayes SF, Heinzen RA (2000). Ultrastructure of *Rickettsia rickettsii* actin tails and localization of cytoskeletal proteins. Infect Immun 68:4706–4713.

Vazquez-Boland JA, Kocks C, Dramsi S, Ohayon H, Geoffroy C, Mengaud J, Cossart P (1992). Nucleotide sequence of the lecithinase operon of *Listeria monocytogenes* and possible role of lecithinase in cell-to-cell spread. Infect Immun 60:219–230.

Wear MA, Cooper JA (2004). Capping protein: new insights into mechanism and regulation. Trends Biochem Sci 29:418–428.

Welch MD, DePace AH, Verma S, Iwamatsu A, Mitchison TJ (1997a) The human Arp2/3 complex is composed of evolutionarily conserved subunits and is localized to cellular regions of dynamic actin filament assembly. J Cell Biol 138:375–384.

Welch MD, Iwamatsu A, Mitchison TJ (1997b) Actin polymerization is induced by the Arp2/3 protein complex at the surface of *Listeria monocytogenes*. Nature 385:265–269.

Welch MD, Mullins RD (2002). Cellular control of actin nucleation. Annu Rev Cell Dev Biol 18:247–288.

Welch MD, Rosenblatt J, Skoble J, Portnoy D, Mitchison TJ (1998). Interaction of human Arp2/3 complex and the *Listeria monocytogenes* ActA protein in actin filament nucleation. Science 281:105–108.

Wiedmann M, Bruce JL, Keating C, Johnson AE, McDonough PL, Batt CA (1997). Ribotypes and virulence gene polymorphisms suggest three distinct *Listeria monocytogenes* lineages with differences in pathogenic potential. Infect Immun 65:2707–2716.

Wong KK, Bouwer HG, Freitag NE (2004). Evidence implicating the 5' untranslated region of *Listeria monocytogenes* ActA in the regulation of bacterial actin-based motility. Cell Microbiol 6:155–166.

Yarar D, D'Alessio JA, Jeng RL, Welch MD (2002). Motility determinants in WASP family proteins. Mol Biol Cell 13:4045–4059.

Zalevsky J, Grigorova I, Mullins RD (2001). Activation of the Arp2/3 complex by the *Listeria* ActA protein. ActA binds two actin monomers and three subunits of the Arp2/3 complex. J Biol Chem 276:3468–3475.

Zallen JA, Yi BA, Bargmann CI (1998). The conserved immunoglobulin superfamily member SAX-3/Robo directs multiple aspects of axon guidance in *C. elegans*. Cell 92:217–227.

Zhukarev V, Ashton FT, Sanger JM, Sanger JW, Shuman H (1995). Steady state fluorescence polarization study of actin filament bundles in *Listeria*-infected cells. Cell Motil Cytoskeleton 30:229–246.

11
Adaptive Immunity to *Listeria monocytogenes*

Kelly A.N. Messingham and John T. Harty

Department of Microbiology, University of Iowa, Iowa City, IA 52242, USA

11.1. Murine Infection with *Listeria monocytogenes* as a Model Intracellular Pathogen

Serious complications resulting from human infection with *Listeria monocytogenes* are usually limited to pregnant women, the very young or very old, or otherwise immunocompromised individuals (Schlech 2000). However, infection of experimental animals with *L. monocytogenes* serves as an extremely useful immunological tool because the bacteria are well characterized, easily manipulated, and infect virtually all mammals (Sixl et al. 1978). In particular, murine listeriosis has been used for many decades to dissect the fundamental components of innate and adaptive immunity to intracellular pathogens (Mackaness 1962; North et al. 1997; Unanue 1997a,b,c; Finelli et al. 1999; Harty et al. 2000).

In a naturally occurring infection, *L. monocytogenes* is introduced into the gastrointestinal tract after consumption of contaminated food products where it binds to, and is taken up by, epithelial cells via interaction of bacterial internalin A and E-cadherin on the host cells (Gaillard et al. 1991). In comparison to humans, mice exhibit markedly reduced susceptibility to intestinal infection with *L. monocytogenes* due to a single amino-acid difference in mouse E-cadherin (Lecuit et al. 2001); therefore, intravenous (i.v.) or intraperitoneal (i.p.) infection of mice is used in most experimental systems. Regardless of the route of infection (i.p or i.v.), administration of one of the many laboratory *L. monocytogenes* strains available results in a highly reproducible infection that can be easily quantitated by assaying bacterial load (colony forming units; CFUs) in the spleen and liver at various days postinfection (White et al. 1999; Messingham et al. 2003). Mortality is dependent on the strain of bacteria used, with LD_{50} ranging from $\sim 1 \times 10^4$ virulent bacteria to 1×10^9 for some attenuated strains (Bouwer et al. 1999; Messingham et al. 2003; Badovinac et al. 2005). Upon infection, the bacteria are taken up by splenic and hepatic (primarily) phagocytes where the majority are killed within the phagosomes; however, a small percentage of bacteria are able to escape destruction and invade the cytosol where the race between bacterial replication and priming of the immune response begins.

Early after infection (hours to days), a cascade of innate immune events ensues that are critical for host survival; either the infection is limited or death results from an inability to control bacterial spread. Early reduction of bacterial numbers is mediated by a cytokine-dependent (primarily IFN-γ and TNF) inflammatory response that results in recruitment of additional activated macrophages and neutrophils primed for bacterial destruction (Nickol and Bonventre 1977; Bancroft et al. 1991). The presence of viable bacteria within a cell results in release of bacterial products into surrounding tissues and production of chemokines that facilitate recruitment of activated phagocytes and their subsequent release of bacteriocidal reactive oxygen species (North 1970; Rogers and Unanue 1993; Conlan and North 1994; Shiloh et al. 1999; Serbina et al. 2003) (see Chap. 12).

If innate immunity can adequately control the level of infection, the elaboration of the slower adaptive immune response results in *L. monocytogenes*-specific CD8$^+$ T-cell-dependent clearance of remaining infected cells (Mackaness 1962; McGregor et al. 1970; Kaufmann 1988; Kaufmann and Ladel 1994a,b; Ladel et al. 1994). Although the adaptive response to *L. monocytogenes* is comprised of a several of cell types (MHC Class I and II restricted CD8$^+$ and CD4$^+$ T cells, respectively), responding to a variety of bacterial antigens, bacterial clearance in an infected mouse is dependent on MHC Class Ia-restricted CD8$^+$ T cells; capable of antigen (Ag)-specific recognition of infected cells. It is the presence of these *L. monocytogenes* Ag-specific memory CD8$^+$ T cells that confer lifelong resistance to subsequent high dose rechallenge (Kaufmann 1988).

11.2. Adaptive Immunity to *L. monocytogenes*

Innate immunity to *L. monocytogenes* serves an essential role in the early control of bacterial numbers, thereby allowing time for the antigen (Ag)-specific adaptive immune response to achieve sterilizing immunity. In the absence of the adaptive response, innate immune mechanisms are unable to effect complete bacterial clearance. This was most clearly demonstrated by the inability of mouse strains that possess innate defenses but lack both T cell and humoral immunity (severe combined immunodeficient (SCID) mice, nude mice) to clear infection, which invariably results in death (Bancroft et al. 1991; Nickol and Bonventre 1977). Additionally, humoral immunity does not appear to play a significant role in the clearance of *L. monocytogenes*; antibody responses are very weak and serum transfer from immune mice does not improve outcome of infected naïve mice (Mackaness 1962; Miki and Mackaness 1964; Edelson and Unanue 2000). Thus, primary sterilizing immunity and long-term protective immunity to *L. monocytogenes* are entirely mediated by listerial-specific T cells. In this chapter, we will discuss the major elements involved in the initiation, execution, and regulation of the T-cell-mediated response to *L. monocytogenes* in the laboratory mouse.

11.3. Initiation of T-Cell-Mediated Immunity to *L. monocytogenes*

The Ag-specific immune response to *L. monocytogenes* is comprised of distinct populations of T cells responding to bacterial antigens presented in the context of MHC Class Ia, MHC Class Ib, or MHC Class II molecules. Regardless of MHC restriction, the Ag-specific response must be initiated through encounter of naïve T-cell clones bearing a TCR specific for bacterial peptide/MHC complexes on the surface of an antigen-presenting cell (APC). By virtue of their high level of expression of a variety of co-stimulatory molecules, dendritic cells (DCs) are the most potent activators of naïve CD4$^+$ and CD8$^+$ T cells (Heath and Carbone 2001; Muraille et al. 2005). In elegant studies using transgenic mice expressing the diphtheria toxin receptor (DTR) primarily on DCs, temporary depletion at the time of infection demonstrated that the listeria-specific CD8$^+$ T-cell response in vivo is dependent on antigen presentation by DCs (Jung et al. 2002).

Dendritic cells (or other APCs) could acquire *L. monocytogenes* antigens by being directly infected with the bacteria or by phagocytosing other infected (live or dead) cells and presenting the processed antigen, a phenomenon termed "cross presentation" or "cross priming" (Heath and Carbone 2001). The unique ability of DCs to present exogenous antigens on either MHC Class I or Class II molecules bypasses the requirement for the DCs themselves to be infected by *L. monocytogenes*. It has not been established how frequently DCs are actually infected by *L. monocytogenes*, although it is likely to occur in vivo. Thus, it is probable that both cross presentation of exogenous listerial antigens and direct presentation of intracellular bacterial antigens contribute to T-cell priming during a primary response to *L. monocytogenes*.

11.3.1. Specificity of MHC Class Ia-Restricted Responses

Early after infection,*L. monocytogenes* is taken up by activated phagocytes and is able to gain access to the cytoplasm through listeriolysin O (LLO)-mediated escape from the phagosome. CD8$^+$ T cells recognize listerial peptides, of typically 8–10 amino acids in length, presented by MHC Class I molecules on the surface of APCs or infected cells (Busch and Pamer 1998). Infection with *L. monocytogenes* results in efficient priming of MHC Class I restricted CD8$^+$ T cells due to the presence of bacterial antigens within the cytosol of the APC, where efficient processing by the endogenous MHC Class I presentation pathway can produce antigen peptides from binding to MHC Class I (Pamer and Cresswell 1998). Due to the intracytoplasmic location of this pathogen, it is not surprising that MHC Class I restricted CD8$^+$ T cells comprise the majority of T cells responding to *L. monocytogenes* infection and, therefore, have been studied extensively in this model.

To be accessible for MHC Class I presentation in an infected cell, a bacterial protein must be secreted into the cytoplasm for degradation by the proteasome, and resultant peptides must be transported to the golgi via the TAP transporter and loaded onto nascent MHC Class I molecules for expression on the cell surface (Germain 1994; Rock et al. 1994; Pamer and Cresswell 1998). In the case of *L. monocytogenes*, these proteins are virulence factors associated with phagosomal escape (LLO) (Kathariou et al. 1987; Portnoy et al. 1988; Pamer et al. 1991; Lety et al. 2001) or viability factors (p60) (Bubert et al. 1992; Wuenscher et al. 1993; Pamer 1994) that are essential for completion of the bacterial life cycle. Four major *L. monocytogenes* epitopes, presented by $H2 - K^d$ MHC Class Ia molecules, have been identified in infected BALB/c ($H-2^d$) mice (see Table 11.1. for a list of the major *L. monocytogenes* epitopes). Simultaneous recognition of these bacterial proteins results in a reproducible hierarchy of immunodominant and subdominant $CD8^+$ T-cell

TABLE 11.1. *Listeria monocytogenes* T cell epitopes.

Epitope	Antigen	MHC restriction	AA sequence	Reference
Class Ia				
LLO_{91-99}	LLO	K^d	GYKDGNEYI	Pamer et al. (1991) and Pamer (1994)
LLO_{88-99}	LLO	$H2-K^d$	PRKGYKDGNEY	Geginat et al. (2001)
$p60_{217-225}$	p60	$H2-K^d$	KYGVSVQDI	Pamer (1994)
$P60_{449-457}$	p60	$H2-K^d$	IYVGNGQMI	Sijts et al. (1996)
$P60_{476-484}$	p60	$H2-K^d$	KYLVGFGRV	Geginat et al. (2001) and Skoberne et al. (2001)
Mpl_{84-92}	Mpl	$H2-K^d$	GYLTDNDQI	Busch et al. (1997)
Class Ib				
f-MIGWII(A)	LemA	H2-M3	f-MIGWII(A)	Lenz et al. (1996) and Princiotta et al. (1998)
f-MIVTLF	AttM	H2-M3	f-MIVTLF	Princiotta et al. (1998)
f-MIVIL	unknown	H2-M3	f-MIVIL	Gulden et al. (1996), Pamer et al. (1992), and Princiotta et al. (1998)
Class II				
$LLO_{215-234}$	LLO	$I-E^K$ ($I-A^K$)	SQLIAKFGTAF KAVNNSLNV	Safley et al. (1991)
$LLO_{190-201}$	LLO	$I-A^b$	NEKYAQ AYPNVS	Geginat et al. (2001)
$LLO_{354-371}$	LLO	$I-E^K$ ($I-A^K$)	DEVQIIDGLNG DLRDILK	Safley et al. (1991)
$P60_{301-312}$	p60	$I-A^d$	EAAKPAPAPSTN	Geginat et al. (1998, 1999)
$3A1.1_{132-148}$	Bacterial surface proteins	$I-A^K$	IVDDTIDDRDNV VSIGF	Sanderson et al. (1995) and Campbell and Shastri (1998)
12A4.G7	Bacterial surface proteins	$I-A^K$	DDAVIYPISYDN AVLALDSR	Campbell and Shastri (1998)

For additional Class Ia and Class II epitopes (Geginat et al. 2001).

responses (Sercarz et al. 1993; Vijh and Pamer 1997; Busch et al. 1998a,b), represented as a greater or lesser frequency of responding cells that can be identified by peptide-stimulated intracellular cytokine production (IFN-γ, TNF-α) (Badovinac and Harty 2000), MHC Class I tetramer reagents (Altman et al. 1996; Busch et al. 1998a; Busch and Pamer 1999), or ELISPOT (Vijh and Pamer 1997; Skoberne et al. 2001).

The CD8$^+$ T-cell response to *L. monocytogenes* is multiclonal, comprised of T cells specific for multiple peptide epitopes, and the frequency of cells responding to each epitope can differ dramatically. At the peak ($\sim$day 7–9 postinfection; p.i.) of the primary response to sub-lethal *L. monocytogenes* infection, roughly 2–3% of the total CD8$^+$ T-cell population is specific for known listerial antigens (Vijh and Pamer 1997; Busch et al. 1998a,b; Mercado et al. 2000; Badovinac and Harty 2002; Messingham et al. 2003). In BALB/c mice the epitope stimulating the highest frequency of responding cells derives from the LLO protein (residues 91–99), accounting for approximately 1.5–2% of all CD8$^+$ cells in the spleen at the peak of the primary response. The response to p60 (residues 217–225) represents $\sim$0.5% of responding CD8$^+$ T cells. The LLO$_{91-99}$ and p60$_{217-225}$ epitopes comprise the immunondominant responses to *L. monocytogenes* in BALB/c mice. Other subdominant epitopes include p60$_{449-457}$ (Vijh and Pamer 1997) and the metalloprotease peptide, mpl$_{84-92}$ (Busch et al. 1997). These subdominant epitopes account for very few (0.05%) responding CD8$^+$ T cells. The evolution of the Ag-specific CD8$^+$ T-cell response is dependent on a variety of factors that influence bacterial peptide recognition, T-cell activation, and the magnitude of the response.

The factors that contribute to the relative immunodominance of one epitope over another are complex. The role of bacterial Ag secretion, rate of proteasomal degradation, efficiency of peptide loading onto MHC, and peptide/MHC stability have all been investigated (Pamer et al. 1997; Skoberne and Geginat 2002; Pamer 2004). It appears that each epitope, including those derived from the same peptide (p60), has unique properties (rate of initial peptide availability, peptide/MHC stability) that influence its ultimate availability for presentation on the surface of the APC (Sijts et al. 1996; Pamer et al. 1997). Kinetic analysis of Ag presentation by *L. monocytogenes*-infected cell lines or in vivo infected cells suggest that the presentation of each antigen is dynamic and the magnitude of the responding T-cell population is not dictated solely by the relative epitope abundance (Sijts et al. 1996; Pamer et al. 1997; Skoberne et al. 2001; Skoberne and Geginat 2002).

Independent of Ag presentation, the frequency of naive precursors displaying a TCR capable of binding a particular bacterial antigen/MHC Class I complex will influence the magnitude of the CD8$^+$ T-cell response. Although the number of available precursors of any given specificity is below our level of detection, estimates suggest that anywhere from 100's–1000 naïve precursors of each specificity exist in the spleen prior to infection (Bousso et al. 1998; Casrouge et al. 2000; Blattman et al. 2002). The overall magnitude of the *L. monocytogenes*-specific T-cell response is dictated by initial level of infection and is most likely a reflection of the number of available precursors recruited to undergo division (Shen et al. 1998; Kaech and Ahmed 2001).

It is important to note that nonsecreted listerial antigens are also capable of priming CD8[+] T-cell responses. This probably occurs via a cross-priming mechanism after DC uptake of digested bacteria within dead or dying neutrophils (Tvinnereim et al. 2004). However, CTL specific for nonsecreted antigens do not confer protective immunity due to the limited presentation of nonsecreted antigens within viable infected cells (Shen et al. 1998; Zenewicz et al. 2002). In this scenario, the majority of infected cells would escape detection during an acute infection because CD8[+] T cells specific for nonsecreted antigens would only encounter their cognate antigen through cross-presentation by an APC.

11.3.2. Specificity of the MHC Class Ib-Restricted Responses

The MHC Class Ib molecules share many structural similarities with MHC Class Ia but are much more highly conserved resulting in limited diversity even among different mouse strains. The most clearly defined nonclassical MHC molecule in mice, H2-M3, is capable of presenting peptides that contain N-formyl methionine (f-Met) at the amino terminus, a property exclusive to bacterial and mitochondrial proteins (Pamer and Cresswell 1998). Murine infection with *L. monocytogenes* results in the presentation of three known peptides by H2-M3. The presented peptides are relatively short and are referred to by amino acid sequence; f-MVIL, f-MIGWII(A), f-MIVTLF (Gulden et al. 1996; Lenz et al. 1996; Pamer et al. 1992; Princiotta et al. 1998; Tawab et al. 2002). It appears that individual responding clones are cross reactive so that a single bacterially derived N-formylated peptide is capable of activating H2-M3 restricted cells of multiple specificities (Ploss et al. 2003). While promiscuous antigen recognition is common to the innate response, this property is so far exclusive to the H2-M3 restricted adaptive response to bacterial antigens. There is limited additional evidence that presentation of *L. monocytogenes* antigens by Qa-1b MHC molecules also contributes to antilisterial immunity (Bouwer et al. 1997).

11.3.3. Specificity of the MHC Class II-Restricted Responses

In addition to the robust responses of MHC Class I restricted CD8[+] T cells, infection with *L. monocytogenes* also induces strong activation of MHC Class II restricted CD4[+] T cells. Several MHC Class II restricted listerial epitopes, derived primarily from LLO (Safley et al. 1991; Sanderson et al. 1995; Campbell and Shastri 1998) and p60 (Geginat et al., 1998, 1999, 2001), have been identified (see Table 11.1.). Typically, MHC Class II restricted antigens are acquired by APCs through phagocytosis of extracellular bacteria; a key component in the control of bacterial spread after intraveinous infection with *L. monocytogenes*.

However, the rapid LLO-mediated escape of bacteria into the cytosol would limit the accessibility of bacterial antigens to the MHC Class II presentation. Rather, it is likely that naïve CD4$^+$ T cells are stimulated by *L. monocytogenes* antigens that are cross-presented on the surface of DCs (Skoberne and Geginat 2002).

11.4.   Kinetics of the T-Cell Response to *L. monocytogenes*

11.4.1.   MHC Class Ia-Restricted Responses

Upon activation, massive clonal expansion of *L. monocytogenes* epitope-specific CD8$^+$ T cells results in amplification of virtually undetectable levels of naïve cells of a given specificity to levels that are readily detectable (1–2% cells in the spleen) (Busch et al. 1998b). To achieve this expansion, Ag-specific CD8$^+$ T cells exhibit doubling times of 6–8 h/division (Blattman et al. 2002). Although responses to *L. monocytogenes* are multiclonal, CD8$^+$ T-cell populations specific for independent antigens undergo expansion with coordinate kinetics. Within 7–9 days after infection, the rapidly expanding CD8$^+$ T cells of differing specificities (i.e., LLO$_{91-99}$, p60$_{217-225}$) reach their numerical peak in unison (Busch et al. 1998b). This point marks the onset of the death or contraction phase of the response where > 90% of cells specific for each epitope die within 3–5 days and the remaining cells comprise the Ag-specific memory cell pool (see Figure 11.1.). This initial memory cell pool is maintained in number and function for the life of the host (Ku et al. 2000); through homeostatic proliferation mechanisms independent of antigen (Lau et al. 1994; Murali-Krishna et al. 1999; Wong and Pamer 2001; Jabbari and Harty 2005), but dependent on the presence of cytokines, such as IL-15 (Ku et al. 2000).

After *L. monocytogenes* infection, the exact timing of the transition from the expansion to contraction phase of the CD8$^+$ T-cell response is dependent on the strain of bacteria used; the peak response to virulent *L. monocytogenes* is slightly delayed (8–9 days p.i.) compared to attenuated strains (day 7 p.i.) (Pope et al. 2001; Badovinac and Harty 2002; Wong and Pamer 2003; Porter and Harty 2006). Using highly sensitive methods of Ag detection ("Direct Ex vivo Antigen Display (DEAD)" and "functional Ag display" assays), it was demonstrated that infection with virulent *L. monocytogenes* results in delayed peaks in the bacterial load and resultant Ag presentation compared to infection with attenuated (*actA*-deficient) *L. monocytogenes* (Wong and Pamer 2003; Porter and Harty 2006). In either case, the peak of functional Ag display was followed ~ 5 days later by the transition from expansion to contraction of CD8$^+$ T-cell numbers. Although the reason for the 5-day interval between peak Ag levels and the onset of CD8$^+$ T-cell contraction is unknown, it is possible that the peak of functional Ag display stimulates the highest relative number of precursors programmed to undergo a set number of divisions resulting in a synchronized peak (expansion to contraction transition). Alternatively, it may be that continued

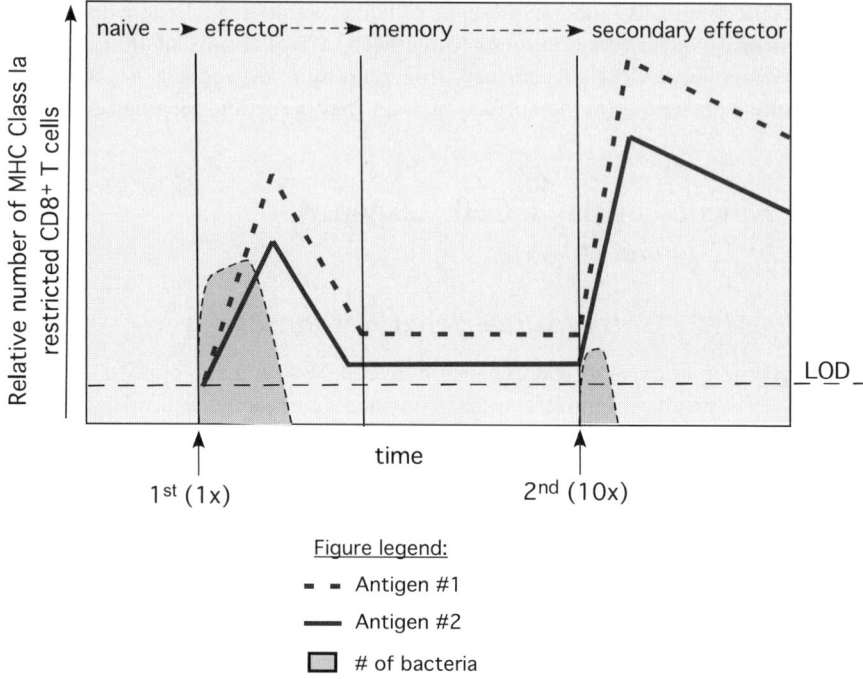

FIGURE 11.1. MHC Class Ia-restricted CD8$^+$ T cell response to *L. monocytogenes*. Total number of T cells specific for *L. monocytogenes*-derived antigens (antigens 1(*dashed*) and 2 (*solid*)). The coordinate expansion of Ag-specific CD8$^+$ T cells after infection (1×) is followed by a rapid and reproducible contraction phase leading to long-lived Ag-specific CD8$^+$ T cell memory. Reexposure to previously lethal doses of *L. monocytogenes* (10×) results in a more rapid and robust expansion resulting in Ag-specific CD8$^+$ T cell numbers in marked excess over the peak of the primary response. The responding Ag-specific CD8$^+$ T cells are capable of dramatically limiting bacterial burden. In comparison to primary CD8$^+$ T cell response, the contraction phase of the secondary response is delayed.

Ag interaction from the onset of infection to the peak of functional antigen display can influence the number of divisions achieved by responding CD8$^+$ T cells (Porter and Harty 2006). In this scenario, it is the loss or absence of Ag interaction shortly after peak Ag display that leads to conclusion of the "program" of expansion ~ 5 days later. Finally, both mechanisms may participate in the shaping of the CD8$^+$ T-cell response.

    Depending on the type of infection, a relatively diverse responding TCR repertoire will become focused so that cells of the highest affinity comprise the majority of the Ag-specifc cells (Malherbe et al. 2004). In the case of murine listeriosis, the responding T-cell populations have relatively diverse TCR repertoire utilization throughout the primary response into memory (Busch et al. 1998a; Opferman et al. 1999). As will be discussed later, some focusing of the TCR repertoire does occur during the secondary response to *L. monocytogenes* (Busch et al. 1998a).

Memory T cells are a phenotypically and functionally heterogeneous population and it is unknown when cells with memory characteristics develop during the immune response. In recent years, attempts have been made at identifying memory cell precursors shortly after Ag exposure by virtue of expression of a variety of surface markers found on memory cells. No single marker of "memory cell precursors" has been identified; however, these studies have identified two populations of memory cells defined largely by their tissue distribution (Sallusto et al. 1999; Wherry et al. 2003; Jabbari and Harty 2006). Effector (T_{EM}) and central (T_{CM}) memory populations are defined by their relative expression of low or high levels, respectively, of surface homing molecules, such as CD62L and CCR7, that permit entry into lymph nodes. It has been hypothesized that T_{EM} serve a primary surveillance role based on their increased ability for target-cell killing. T_{CM} may serve the complimentary role as a reservoir of Ag-specific cells poised for expansion after reencountering cognate Ag.

Expression of the IL-7 receptor (CD127) has been investigated as an early marker of memory. IL-7 is a critical cytokine for the survival of T cells, and it has been suggested that expression of its receptor in a small fraction of Ag-specific CD8$^+$ T cells at the peak of expansion may mark the cells that will survive contraction to become memory (Kaech et al. 2003). In support of this, increased CD127 expression is observed in experimental systems where contraction is limited or absent (Badovinac et al. 2004). On the other hand, recent studies utilizing a variety of vaccination models show that Ag-specific cells expressing high levels of CD127 contract normally (Badovinac et al. 2005; Lacombe et al. 2005). Taken together, these studies suggest that CD127 expression is not the defining factor in the identification of cells that will survive contraction to become memory T cells.

11.4.2. MHC Class Ib-Restricted Responses

During primary infection with *L. monocytogenes*, H2-M3 restricted CD8$^+$ T cells go through expansion and contraction and reach stable memory levels in a manner similar to classically restricted T cells (Kerksiek et al. 1999, 2001). During their expansion phase, H2-M3-restricted CD8$^+$ T cells reach peak frequencies within the spleen at 5–6 days p.i., preceding the peak of classically restricted CD8$^+$ T cells on days 7–9 p.i. Contraction of the H2-M3-restricted response also occurs more rapidly; memory levels are attained within 2–3 days after the peak (Kerksiek et al. 1999, 2001). Due to the specificity of H2-M3-restricted CD8$^+$ T cells for antigens common to all bacterial species, and their early appearance after primary infection, it is thought that the primary contribution is to aid (or prolong) the innate responses' early control of bacterial numbers (Kerksiek et al. 1999, 2001; Hamilton et al. 2004)).

11.4.3. MHC Class II-Restricted Responses

Upon interaction with their cognate antigen in the context of MHC Class II and costimulation, CD4$^+$ T cells progress through similar kinetics of expansion and contraction to memory levels as observed with MHC Class I-restricted CD8[+] T cells (Geginat et al. 2001; Skoberne and Geginat 2002; Corbin and Harty 2004). Likewise, the initiation of the CD4[+] T-cell response to *L. monocytogenes* appears to be regulated by initial antigen exposure (discussed in Programming of the T-cell response) (Corbin and Harty 2004). Although the precise role for CD4[+] T cells in the control of *L. monocytogenes* infection remains undefined, antigen recognition by these cells stimulates the production of copious quantities of Th1 cytokines that aid in bacterial clearance via activation of other cell types, including DCs and bacteriocidal macrophages (Hsieh et al. 1993; Bouwer et al. 1997).

Although there are many similarities in the CD4[+] and CD8[+] T-cell responses to *L. monocytogenes*, several differences also exist. In contrast to the stable number of memory CD8[+] T cells, *L. monocytogenes*-specific memory CD4[+] T-cell numbers decline over time (Schiemann et al. 2003). Additionally, secondary encounter with *L. monocytogenes* may result in selective expansion of high affinity CD4[+] T-cell clones resulting in repertoire focusing (Savage et al. 1999). Careful comparison of *L. monocytogenes*-specific CD4[+] and CD8[+] T-cell responses within the same host suggest that the elaboration of effector molecules is differentially regulated by Ag presence (Corbin and Harty 2005). It appears that while the production of cytokines (IFN-γ and TNF) by *L. monocytogenes*-specific CD8[+] T cells is rapidly down regulated in the absence of Ag, CD4[+] T cells will continue to proliferate and produce cytokines in response to persistent Ag exposure (Corbin and Harty 2005; Obst et al. 2005). An additional role for CD4[+] T cells in supporting the generation and maintenance of functional CD8[+] T-cell-mediated protective immunity will be discussed in the next section.

11.5.   Programming of the T-Cell Response

Upon infection with *L. monocytogenes*, the rapid response of the innate immune system functions to limit infection until the slower adaptive immune response can develop. This coordinated effort typically results in pathogen clearance within a week of infection (Harty and Bevan 1995; Badovinac and Harty 2000). The peak of the MHC Class Ia-restricted CD8[+] T-cell response is coincident with pathogen clearance followed by rapid contraction of Ag-specific cell numbers. The timing of these events led to the hypothesis that the CD8[+] T-cell response was dependent on prolonged antigen presentation, and thus the presence of infection. Recently, reports have emerged that support the concept that only a brief exposure to antigen is necessary to initiate all phases of the CD8[+] T-cell response (Mercado et al. 2000; Kaech and Ahmed 2001; Badovinac et al. 2002). These studies utilized antibiotic treatment of mice at various times after *L. monocytogenes* infection to kill all viable bacteria by day 3 p.i. and thus limit functional antigen display. Within the spleens of antibiotic-treated mice, Ag-specific CD8[+] T cells did not decrease their rate or peak of expansion, or the timing of contraction, when compared with mice that were not antibiotic treated (Mercado et al. 2000; Badovinac et al. 2002). However, antibiotic administration within 24 h of infection negatively influences the rate and magnitude of the anti-listerial CD8$^+$ T-cell response. Similar studies suggest that the kinetics and development of functional memory CD4$^+$ T cells responding to *L. monocytogenes* infection are also regulated by initial antigen exposure (Corbin and Harty 2004; Williams and Bevan 2004). These data suggest that events during the first few days of infection are critical for establishment of the maximal T-cell response to *L. monocytogenes*; once initiated, the kinetics and magnitude of the T-cell response are antigen independent.

In recent years, intense debate has focused on a role for CD4$^+$ T cells in the development of CD8$^+$ T-cell-mediated protective immunity. Previously, it was thought that because adoptive transfer of *L. monocytogenes*-specific CD8$^+$ T cells could protect naïve mice from high-dose bacterial challenge, CD4$^+$ T cells were dispensable in protective immunity. It was suggested that CD4$^+$ T cell help could contribute to protective immunity by stimulating a more robust expansion of Ag-specific CD8$^+$ T cells than could be realized by "unhelped" CD8$^+$ T cells (Bourgeois et al. 2002; Marzo et al. 2004). More detailed studies utilizing *L. monocytogenes* and other intracellular pathogens identify a possible role for CD4$^+$ T cells in the generation and maintenance of functional Ag-specific memory CD8$^+$ T cells. In these studies, the absence of CD4$^+$ T-cell stimulation, through either systemic depletion or absence of MHC Class II, resulted in a memory CD8$^+$ T-cell population that was not maintained long term (Shedlock et al. 2003; Sun and Bevan 2003). Surprisingly, this defect in the memory phase of the response occurred despite normal magnitude and kinetics of the primary response. Removal of CD4$^+$ T cells after CD8$^+$ T-cell priming did not effect the responding CD8$^+$ T-cell population. These findings suggest that CD4$^+$ T cell help plays a role in the initiation of the CD8$^+$ T-cell program, but are not required thereafter. In contrast, it has also been suggested that CD4$^+$ T cell help maintains functional CD8$^+$ T-cell memory and, therefore, is required during the memory phase of the response (Sun et al. 2004). The exact nature of the signal supplied by the CD4$^+$ T cells in these models is unclear. It may be that direct contact between CD4$^+$ and CD8$^+$ T cells (possibly through CD40/CD40L) (Bourgeois et al. 2002) is necessary for optimal CD8$^+$ T-cell memory; however, the requirement for CD40 ligation may depend on the type of infectious agent utilized (i.e., virus or intracellular bacteria) (Lee et al. 2003; Marzo et al. 2004).

11.6. Secondary Immunity to *L. monocytogenes*

11.6.1. MHC Class Ia-Restricted Responses

Protective immunity to *L. monocytogenes* is solely dependent on MHC Class Ia-restricted CD8$^+$ T cells. Re-exposure to *L. monocytogenes* results in a rapid mobilization of Ag-specific cells from memory CD8$^+$ T-cell pool, resulting in efficient clearance of challenge doses that are lethal for naïve mice (Lalvani et al. 1997). *L. monocytogenes*-specific memory CD8$^+$ T cells are increased in frequency and number over their naïve precursors, and possess increased capacity for proliferation and elaboration of effector functions in response to antigen levels that are reduced in comparison to a primary infection (Ahmed and Gray 1996). This increased ability to respond to secondary infection is also less dependent on the costimulaory molecules required for priming antigen-specific naïve T cells (Iezzi et al. 1998); however, it is likely that maximal responses are attained in the presence of robust costimulation supplied by DCs (Zammit et al. 2005). Together, unique qualities of memory CD8$^+$ T cells are responsible for very rapid recognition of, and response to, *L. monocytogenes* antigens resulting in the rapid and complete elimination of all infected cells.

After *L. monocytogenes* challenge, Ag-specific memory CD8$^+$ T cells reach their numerical peak approximately 5–6 days p.i. The elevated frequency of Ag-specific precursors in the memory pool, combined with their increased sensitivity for activation, results in secondary expansion to peak numbers in marked excess (10 to 50-fold; Badovinac et al. 2003, 2005) over primary memory numbers. The rapid expansion memory CD8$^+$ T cells exerts a negative influence on the expansion of naïve precursors so that the magnitude of the memory response is inversely proportional to the population of newly recruited cells (Badovinac et al. 2003). This is likely due to more rapid elimination of matured APCs, required for activation of naïve cells, by responding memory CD8$^+$ T cells (Wong and Pamer 2003; Yang et al. 2006). In comparison to the primary response to *L. monocytogenes*, the contraction of the secondary response is markedly protracted (see Figure 11.1.) (Badovinac et al. 2003).

During the secondary CD8$^+$ T-cell response to *L. monocytogenes*, a modest narrowing of the TCR repertoire occurs that is likely due to preferential activation of clones bearing TCRs with higher affinity for antigen (Busch et al. 1998a; Busch and Pamer 1999; Pamer 2004). Although the mechanism responsible for this preferential activation is not entirely known, it could result from a limited availability of antigen due to a very rapid clearance of bacteria after re-exposure.

Bacterial challenge in an *L. monocytogenes* immune mouse ultimately generates a new pool of secondary memory CD8$^+$ T cells. In comparison to primary memory cells, secondary memory CD8$^+$ T cells are more effective at reducing bacterial numbers and exhibit increased capacity for Ag-specific target cell killing and production of associated effector molecules (Jabbari and Harty 2006). Phenotypically, secondary memory CD8$^+$ T cells are slower to re-express hallmarks of memory, such as CD62L expression and IL-2 production, in response to Ag-stimulation. This difference in primary and secondary memory characteristics is associated with decreased expression of the receptor for IL-15, a cytokine known to be critical for the proliferation-driven maintenance of the memory cell pool. In these studies, increased proliferation of memory CD8$^+$ T cells resulted in increased expression of CD62L suggesting that memory cell phenotype and function are dynamic and specific to host environment.

11.6.2. MHC Class Ib-Restricted Responses

In contrast to the explosive expansion of MHC Class Ia-restricted CD8[+] T cells upon secondary infection with *L. monocytogenes*, the H2-M3-restricted memory CD8[+] T cells fail to undergo significant expansion, despite similar acquisition of the phenotypic markers of activation on both cell populations (Kerksiek et al. 2003). This finding led to the hypothesis that MHC Class Ib-restricted cells were functionally incapable of secondary expansion and, therefore, could not contribute to protective immunity (Urdahl et al. 2002; Kerksiek et al. 2003). In contrast, MHC Class Ib-restricted primary effector cells can contribute to protective immunity to *L. monocytogenes*, but only in the absence of an MHC Class Ia-restricted response (Kaufmann et al. 1988; Lukacs and Kurlander 1989; Seaman et al. 2000; D'Orazio et al. 2003). This is most simply demonstrated by the discovery that adoptive transfer of either effector or memory *L. monocytogenes*-specific CD8[+] T cells from MHC Class Ia-deficient hosts protect naïve mice from *L. monocytogenes* infection (Seaman et al. 2000). In WT mice undergoing a secondary response to *L. monocytogenes*, expansion of H2-M3-restricted memory CD8[+] T cells is suppressed by the memory Class Ia-restricted memory responses (Hamilton et al. 2004; Ploss et al. 2005). However, if primary responses are elicited using DC-peptide immunization for MHC Class Ib antigens, in the absence of MHC Class Ia responses, the H2-M3-restricted cells are capable of secondary expansion when challenged with LM, but do not contribute to protective immunity (Hamilton et al. 2004). Thus, MHC Class Ia-restricted memory responses limit expansion of an ineffective MHC Class Ib-restricted memory cell population. The precise mechanism of suppression remains unclear, but it appears that factors, possibly linked to the site of infection, such as antigen load, cytokine milieu, etc., may be involved.

11.6.3. MHC Class II-Restricted Responses

Because protective immunity to *L. monocytogenes* is dependent on MHC Class Ia-restriced CD 8[+] T cells, Ag-specific CD4[+] T cells have not been studied as extensively in this infection model. However, *L. monocytogenes*-specific CD4[+] T cells are capable of providing protective immunity to naïve mice independent of CD8[+] T cells, although to a lesser extent (Bishop and Hinrichs 1987; Czuprynski and Brown 1987; Lukacs and Kurlander 1989; Harty et al. 1992; Harty and Bevan 1996; Geginat et al. 1998). As mentioned previously, CD4[+] T-cell-mediated clearance of infected cells is dependent on Ag-specific cytokine-driven phagocyte recruitment and subsequent nonspecific target-cell killing.

11.7. CD8[+] T-Cell Effector Mechanisms

During expansion, activated T cells differentiate into effector populations, which leave the secondary lymphoid organs in search of *L. monocytogenes*-infected cells. Upon recognition of an infected cell, activated effector cells will produce cytokines known to recruit and/or activate microbiocidal effector cells, such as macrophages and neutrophils, and up-regulate molecules necessary for target cell lysis (Harty and Bevan 1999). Both expression of cytokine molecules and the execution of cytolytic effector mechansims by T cells are tightly regulated through TCR-dependent signals. Given that virtually every T-cell effector function is also employed by other cells (NK cells, macrophages, DC, etc.) of the immune system, understanding which mechanisms are important for cell-mediated resistance to infection is a substantial challenge. To this end, gene knockout mice, lacking specific T-cell effector molecules, have become important tools in addressing the biology of the adaptive response to *L. monocytogenes*. Although protective immunity to *L. monocytogenes* in WT mice is dependent on MHC Class Ia-restricted CD8$^+$ T cells, examination of gene knockout mice has not identified a single effector mechanism that is absolutely required for protective immunity.

11.7.1. IFN-γ

IFN-γ plays a central role in innate resistance to *L. monocytogenes* infection through the activation of phagocytic cells and resultant reduction in bacterial cell numbers. The absence of IFN-γ from the host mouse, through either antibody depletion or inactivation of the IFN-γ gene (GKO mice) or its receptor, results in a nearly 1000-fold decrease in the LD$_{50}$ of *L. monocytogenes* in naïve mice (Buchmeier and Schreiber 1985; Huang et al. 1993; Harty and Bevan 1995). These findings clearly demonstrate the requirement for IFN-γ in the innate response to *L. monocytogenes*. However, infection of GKO mice with an attenuated strain of *L. monocytogenes* (*actA*-deficient) that renders the bacteria deficient in cell-to-cell spread (Kocks et al. 1992) is tolerated similar to WT mice and results in efficient CD8$^+$ T-cell priming and long-lasting protective immunity to high-dose challenge. Once vaccinated, GKO demonstrate ~20,000-fold increase in resistance to virulent *L. monocytogenes* (Harty and Bevan 1995) and exhibit increased memory CD8$^+$ T-cell numbers due to protracted contraction (discussed below) in comparison to similarly vaccinated WT mice (Badovinac et al. 2000). These findings indicate that IFN-γ is not required for the development of functional protective immunity to *L. monocytogenes*. Upon infection, both CD8$^+$ and CD4$^+$ T cells produce IFN-γ in an antigen-specific manner (Hamilton et al. 2004; Corbin and Harty 2005). Whether the production of IFN-γ by *L. monocytogenes*-specific CD8$^+$ T cells contributes to, or improves, protective immunity is currently under investigation.

In addition to initial control of bacterial numbers, IFN-γ serves as a multi-potent regulator of Ag-specific CD8$^+$ T-cell homeostasis (Harty and White 1999; Harty and Badovinac 2002). It is well known that IFN-γ can enhance CD8$^+$ T-cell expansion by increasing the ability of APCs to process and present antigens to T cells (Fruh and Yang 1999). However, IFN-γ can also serve as a negative regulator of MHC Class I-restricted CD8$^+$ T-cell expansion in an epitope-specific fashion due to differential processing and presentation of the antigenic peptides (Skoberne and Geginat 2002). Infection of IFN-γ-deficient mice with attenuated *L, monocytogenes* results in CD8$^+$ T-cell response with an altered immunodominance hierarchy when compared to WT mice (Badovinac et al. 2000). Specifically, increased expansion of the CD8$^+$ T cells specific for the subdominant p60$_{217-225}$ epitope, relative to those specific for the LLO$_{91-99}$, was observed in IFN-γ-deficient mice. Thus, IFN-γ serves as a key regulator of immunodominance after infection.

The IFN-γ exerts its most potent effect on CD8$^+$ T-cell homeostasis through regulation of the contraction phase of the response to *L. monocytogenes* infection (Badovinac et al. 2000). After infection with attenuated *L. monocytogenes*, Ag-specific CD8$^+$ T cells from both WT and IFN-γ-deficient mice reach peak numbers on approximately day 7 p.i. followed by a rapid contraction, where the majority of responding cells are eliminated from the spleen of WT, but not GKO, mice by day 10–11 p.i. Elevated Ag-specific CD8$^+$ T-cell numbers persist in the spleen of IFN-γ deficient mice indefinitely, resulting in increased levels of *L. monocytogenes*-specific memory in GKO mice. Vaccinated IFN-γ deficient mice display a three- to –six-fold increase in the number of Ag-specific CD8$^+$ T cells in their spleens that is associated with increased protective immunity to high level challenge with virulent *L. monocytogenes* when compared to similarly vaccinated WT mice. A careful comparison of the per-cell protective capacity of Ag-specific memory CD8$^+$ T cells from WT and GKO mice will determine if the increased resistance to virulent *L. monocytogenes* results purely from the elevated memory levels in vaccinated GKO mice or if IFN-γ-deficient memory CD8$^+$ T cells are intrinsically different from WT. In addition, recent experiments also show that Ag-specific CD4$^+$ T cells do not contract in the absence of IFN-γ (Haring and Harty 2006).

11.7.2. TNF-α

Antigen-stimulated CD4$^+$ and CD8$^+$ T cells produce TNF, a cytokine that also plays an important role in the innate immune response to *L. monocytogenes* infection (Havell 1987; Nakane et al. 1988; Bancroft et al. 1989). Mice lacking the TNF-α gene (Rothe et al. 1993) or receptor (Endres et al. 1997) cannot survive primary infection with virulent *L. monocytogenes*. However, similar to IFN-γ, TNF is not required for the development of a protective CD8$^+$ T-cell response (White et al. 2000).

There are three possible mechanisms by which Ag-specific production of TNF could contribute to bacterial resistance; (1) through stimulation of bacteriocidal macrophages (Endres et al. 1997); (2) induction of apoptosis in target cells through receptor (TNFR1) ligation (Ashkenazi and Dixit 1998); and/or (3) stimulation/recruitment of phagocytes through increased adhesion molecule expression (Lukacs et al. 1994; Henninger et al. 1997; Kondo S and Sauder 1997). Adoptive transfer experiments utilizing *L. monocytogenes* memory CD8$^+$ T cells deficient in other modes (perforin/FAS) of cytolysis found that Ag-specific production of TNF can contribute to protective immunity (White and Harty 1998); however, the cellular source of TNF, and the nature of its contribution to anti-listerial immunity in a WT mouse, is unclear.

11.7.3. Cytolytic Effector Mechanisms: Granule Exocytosis and Receptor-Mediated Pathways

The presence of L. monocytogenes peptide–MHC complexes on the surface of infected cells serves as a beacon identifying targets for destruction by CD8$^+$ T cells. Ag-specific target cell lysis is a function largely limited to CD8$^+$ T cells. TCR recognition of an infected cell through binding cognate peptide–MHC complexes results in the directed exocytosis of cytotoxic granules containing perforin and granzymes that breach the target cell membrane (perforin) and activate the caspase cascade (granzymes) leading to apoptosis (Lieberman 2003). Regarding primary infection with L. monocytogenes, the contribution of perforin to control of bacterial numbers is marginal; perforin-deficient mice exhibit resistance (LD_{50}) similar to WT mice despite delayed bacterial clearance in the spleens, but not livers, of infected mice (Kagi et al. 1994; White et al. 1999).

In contrast to primary immunity, perforin-mediated cytotoxicity is critical for optimal protective immunity to virulent L. monocytogenes (Kagi et al. 1994; Messingham et al. 2003). Compared with WT mice, L. monocytogenes-vaccinated perforin-deficient (PKO) mice have elevated levels (two- to four-fold) of CD8$^+$ T-cell memory but exhibit reduced levels of protection against virulent L. monocytogenes. Although the existence of functional protective immunity to L. monocytogenes in vaccinated PKO mice underscores the existence of perforin-independent pathways for CD8$^+$ T-cell immunity to L. monocytogenes, perforin-deficient memory CD8$^+$ T cells display a fivefold reduction in the per-cell protective capacity when compared to WT cells five-fold (Messingham et al. 2003). Therefore, in some cases, increased levels of CD8$^+$ T-cell memory can compensate for the absence of an important effector molecule.

In recent years, a role for perforin in the regulation of the expansion phase of MHC Class I-restricted CD8$^+$ T-cell homeostasis has emerged (Harty and White 1999; Harty and Badovinac 2002). In an allogenic cell transfer model, where antigen load is identical perforin deficient CD8$^+$, but not CD4$^+$, T cells expand significantly more after antigen stimulation resulting in higher numbers at the peak of the response. This enhanced expansion of perforin-deficient cells appears to be the result of reduced CD8$^+$ T-cell death (AICD), rather than altered persistence of APC (Spaner et al. 1999; Ludewig et al. 2001). After L. monocytogenes infection of perforin-deficient mice, both MHC Class I (LLO_{91-99}, $p60_{217-225}$, $p60_{444-459}$) – and H2-M3 (f-MIGWII(A))-restricted CD8$^+$ T cells show three- to fourfold higher expansion despite similar rates of clearance of an attenuated strain of L. monocytogenes (Badovinac and Harty 2000; Messingham et al. 2003). Importantly, the contraction phase of the response after *L. monocytogenes* infection of perforin-deficient mice was normal (Badovinac and Harty 2000).

It also appears that the extent to which perforin influences $CD8^+$ T-cell expansion can be modulated by the host environment. For example, enhanced $CD8^+$ T-cell expansion is not observed in perforin-deficient mice ($H-2^d$) after primary infection with virulent *L. monocytogenes*, where deficient mice exhibit substantially delayed clearance in the spleen. However, during secondary infection, when clearance kinetics are virtually identical, $CD8^+$ T-cell expansion in perforin-deficient mice is increased (White et al. 1999). Ultimately, the increased expansion paired with a normal contraction phase has the net of effect of increasing the absolute number of functional Ag-specific memory cells and, in turn, increased resistance to *L. monocytogenes* rechallenge.

It should be noted that activated $CD8^+$ T cells also express CD95 ligand which can also activate the caspase cascade through ligation of its receptor (CD95/Fas) on the target cell. Normally, this pathway likely results in the elimination of self-reactive T cells that are repeatedly activated (AICD) (Van Parijs et al. 1998), and does not appear to make a significant contribution to the anti-listerial immunity in the presence or absence of perforin (White and Harty 1998). These findings also demonstrate functional T-cell-mediated immunity in the complete absence of the major pathways of cytolysis.

11.8. Conclusion

For many decades, murine listeriosis has been utilized as a highly reproducible model to safely study both innate and adaptive immune responses. Although T cells are not required for resistance to primary infection with *L. monocytogenes*, the identification of defined peptide epitopes derived from *L. monocytogenes* has allowed immunologists to utilize this infection model to dissect many aspects of the antigen-specific T-cell response to intracellular pathogens. The relative ease by which the bacterium can be manipulated to decrease virulence or express a plethora of exogenous T-cell epitopes has allowed this model to be used as a vehicle for the study of a wide variety of other pathogens. In this way, infection with *L. monocytogenes* also serves as a prototypical candidate for vaccine development for humans. In addition, human studies support the potential utility and relative safety of attenuated *L. monocytogenes* as a vaccine delivery vector (Angelakopoulos et al. 2002). As our knowledge base expands and the tools available improve, experimental infection with *L. monocytogenes* will no doubt remain a valuable system for the investigation of all aspects of immunity.

References

Ahmed R and Gray D (1996) Immunological memory and protective immunity: understanding their relation. Science 272(5258): 54–60

Altman JD, et al. (1996) Phenotypic analysis of antigen-specific T lymphocytes. Science 274(5284): 94–96

Angelakopoulos H, et al. (2002) Safety and shedding of an attenuated strain of *Listeria monocytogenes* with a deletion of actA/plcB in adult volunteers: a dose escalation study of oral inoculation. Infect Immun 70(7): 3592–3601

Ashkenazi A and Dixit VM (1998) Death receptors: signaling and modulation. Science 281(5381): 1305–1308

Badovinac VP and Harty JT (2000) Adaptive immunity and enhanced CD8+ T cell response to *Listeria monocytogenes* in the absence of perforin and IFN-gamma. J Immunol 164(12): 6444–6452

Badovinac VP and Harty JT (2000a) Intracellular staining for TNF and IFN-gamma detects different frequencies of antigen-specific CD8(+) T cells. J Immunol Methods 238(1–2): 107–117

Badovinac VP, et al. (2000) Regulation of antigen-specific CD8+ T cell homeostasis by perforin and interferon-gamma. Science 290(5495): 1354–1358

Badovinac VP and Harty JT (2002) CD8(+) T-cell homeostasis after infection: setting the 'curve'. Microbes Infect 4(4): 441–447

Badovinac VP, et al. (2002) Programmed contraction of CD8(+) T cells after infection. Nat Immunol 3(7): 619–626

Badovinac VP, et al. (2003) Regulation of CD8+ T cells undergoing primary and secondary responses to infection in the same host. J Immunol 170(10): 4933–4942

Badovinac VP, et al. (2004) CD8+ T cell contraction is controlled by early inflammation. Nat Immunol 5(8): 809–817

Badovinac VP, et al. (2005) Accelerated CD8+ T-cell memory and prime-boost response after dendritic-cell vaccination. Nat Med 11(7): 748–756

Bancroft GJ, et al. (1989) Tumor necrosis factor is involved in the T cell-independent pathway of macrophage activation in scid mice. J Immunol 143(1): 127–130

Bancroft GJ, et al. (1991) Natural immunity: a T-cell-independent pathway of macrophage activation, defined in the scid mouse. Immunol Rev 124: 5–24

Bishop DK and Hinrichs DJ (1987) Adoptive transfer of immunity to *Listeria monocytogenes*. The influence of in vitro stimulation on lymphocyte subset requirements. J Immunol 139(6): 2005–2009

Blattman JN, et al. (2002) Estimating the precursor frequency of naive antigen-specific CD8 T cells. J Exp Med 195(5): 657–664

Bourgeois C, et al. (2002) A role for CD40 expression on CD8+ T cells in the generation of CD8+ T cell memory. Science 297(5589): 2060–2063

Bourgeois C, et al. (2002) CD8 lethargy in the absence of CD4 help. Eur J Immunol 32(8): 2199–2207

Bousso P, et al. (1998) Individual variations in the murine T cell response to a specific peptide reflect variability in naive repertoires. Immunity 9(2): 169–178

Bouwer HG, et al. (1997) MHC class Ib-restricted cells contribute to antilisterial immunity: evidence for Qa-1b as a key restricting element for Listeria-specific CTLs. J Immunol 159(6): 2795–2801

Bouwer HG, et al. (1999) Existing antilisterial immunity does not inhibit the development of a *Listeria monocytogenes*-specific primary cytotoxic T-lymphocyte response. Infect Immun 67(1): 253–258

Bubert A, et al. (1992) Structural and functional properties of the p60 proteins from different Listeria species. J Bacteriol 174(24): 8166–8171

Buchmeier NA and Schreiber RD (1985) Requirement of endogenous interferon-gamma production for resolution of *Listeria monocytogenes* infection. Proc Natl Acad Sci USA 82(21): 7404–7408

Busch DH, et al. (1997) A nonamer peptide derived from *Listeria monocytogenes* metal-loprotease is presented to cytolytic T lymphocytes. Infect Immun 65(12): 5326–5329

Busch DH and Pamer EG (1998) MHC class I/peptide stability: implications for immuno-dominance, in vitro proliferation, and diversity of responding CTL. J Immunol 160(9): 4441–4448

Busch DH, et al. (1998a) Evolution of a complex T cell receptor repertoire during primary and recall bacterial infection. J Exp Med 188(1): 61–70

Busch DH, et al. (1998b) Coordinate regulation of complex T cell populations responding to bacterial infection. Immunity 8(3): 353–362

Busch DH and Pamer EG (1999) T lymphocyte dynamics during *Listeria monocytogenes* infection. Immunol Lett 65(1–2): 93–98

Busch DH and Pamer EG (1999) T cell affinity maturation by selective expansion during infection. J Exp Med 189(4): 701–710

Campbell DJ and Shastri N (1998) Bacterial surface proteins recognized by CD4+ T cells during murine infection with *Listeria monocytogenes*. J Immunol 161(5): 2339–2347

Casrouge A, et al. (2000) Size estimate of the alpha beta TCR repertoire of naive mouse splenocytes. J Immunol 164(11): 5782–5787

Conlan JW and North RJ (1994) Neutrophils are essential for early anti-Listeria defense in the liver, but not in the spleen or peritoneal cavity, as revealed by a granulocyte-depleting monoclonal antibody. J Exp Med 179(1): 259–268

Corbin GA and Harty JT (2004) Duration of infection and antigen display have minimal influence on the kinetics of the CD4+ T cell response to *Listeria monocytogenes* infection. J Immunol 173(9): 5679–5687

Corbin GA and Harty JT (2005) T cells undergo rapid ON/OFF but not ON/OFF/ON cycling of cytokine production in response to antigen. J Immunol 174(2): 718–726

Czuprynski CJ and Brown JF (1987) Dual regulation of anti-bacterial resistance and inflammatory neutrophil and macrophage accumulation by L3T4+ and Lyt 2+ Listeria-immune T cells. Immunology 60(2): 287–293

D'Orazio SE, et al. (2003) Class Ia MHC-deficient BALB/c mice generate CD8+ T cell-mediated protective immunity against *Listeria monocytogenes* infection. J Immunol 171(1): 291–298

Edelson BT and Unanue ER (2000) Immunity to Listeria infection. Curr Opin Immunol 12(4): 425–431

Endres R, et al. (1997) Listeriosis in p47(phox-/-) and TRp55-/- mice: protection despite absence of ROI and susceptibility despite presence of RNI. Immunity 7(3): 419–432

Finelli A, et al. (1999) MHC class I restricted T cell responses to *Listeria monocytogenes*, an intracellular bacterial pathogen. Immunol Res 19(2/3): 211–223

Fruh K and Yang Y (1999) Antigen presentation by MHC class I and its regulation by interferon gamma. Curr Opin Immunol 11(1): 76–81

Gaillard JL, et al. (1991) Entry of *L. monocytogenes* into cells is mediated by internalin, a repeat protein reminiscent of surface antigens from Gram-positive cocci. Cell 65: 1127–1141

Geginat G, et al. (1998) Th1 cells specific for a secreted protein of *Listeria monocytogenes* are protective in vivo. J Immunol 160(12): 6046–6055

Geginat G, et al. (1999) Enhancement of the *Listeria monocytogenes* p60-specific CD4 and CD8 T cell memory by nonpathogenic Listeria innocua. J Immunol 162(8): 4781–4789

Geginat G, et al. (2001) A novel approach of direct ex vivo epitope mapping identifies dominant and subdominant CD4 and CD8 T cell epitopes from *Listeria monocytogenes*. J Immunol 166(3): 1877–1884

Germain RN (1994) MHC-dependent antigen processing and peptide presentation: providing ligands for T lymphocyte activation. Cell 76(2): 287–299

Gulden PH, et al. (1996) A *Listeria monocytogenes* pentapeptide is presented to cytolytic T lymphocytes by the H2-M3 MHC class Ib molecule. Immunity 5(1): 73–79

Hamilton SE, et al. (2004) MHC class Ia-restricted memory T cells inhibit expansion of a nonprotective MHC class Ib (H2-M3)-restricted memory response. Nat Immunol 5(2): 159–168

Haring JS and Harty JT (2006) Aberrant Contraction of Ag-Specific CD4 T Cells in the Absence of IFN-g or its Receptor. Infect Immun 74(11): 6252–6263

Harty JT, et al. (1992) CD8 T cells can protect against an intracellular bacterium in an interferon gamma-independent fashion. Proc Natl Acad Sci U S A 89(23): 11612–11616

Harty JT and Bevan MJ (1995) Specific immunity to *Listeria monocytogenes* in the absence of IFN gamma. Immunity 3(1): 109–117

Harty JT and Bevan MJ (1996) CD8 T-cell recognition of macrophages and hepatocytes results in immunity to *Listeria monocytogenes*. Infect Immun 64(9): 3632–3640

Harty JT and Bevan MJ (1999) Responses of CD8(+) T cells to intracellular bacteria. Curr Opin Immunol 11(1): 89–93

Harty JT and White D (1999) A knockout approach to understanding CD8+ cell effector mechanisms in adaptive immunity to *Listeria monocytogenes*. Immunobiology 201(2): 196–204

Harty JT, et al. (2000) CD8+ T cell effector mechanisms in resistance to infection. Annu Rev Immunol 18: 275–308

Harty JT and Badovinac VP (2002) Influence of effector molecules on the CD8(+) T cell response to infection. Curr Opin Immunol 14(3): 360–365

Havell EA (1987) Production of tumor necrosis factor during murine listeriosis. J Immunol 139(12): 4225–4231

Heath WR and Carbone FR (2001) Cross-presentation, dendritic cells, tolerance and immunity. Annu Rev Immunol 19: 47–64

Henninger DD, et al. (1997) Cytokine-induced VCAM-1 and ICAM-1 expression in different organs of the mouse. J Immunol 158(4): 1825–1832

Hsieh CS, et al. (1993) Development of TH1 CD4+ T cells through IL-12 produced by Listeria-induced macrophages. Science 260(5107): 547–549

Huang S, et al. (1993) Immune response in mice that lack the interferon-gamma receptor. Science 259(5102): 1742–1745

Iezzi G, et al. (1998) The duration of antigenic stimulation determines the fate of naive and effector T cells. Immunity 8(1): 89–95

Jabbari A and Harty JT (2005) Cutting edge: differential self-peptide/MHC requirement for maintaining CD8 T cell function versus homeostatic proliferation. J Immunol 175(8): 4829–4833

Jabbari A and Harty JT (2006) Delayed acquisition of central memory phenotype by secondary memory CD8 T cells while maintaining heightened immunity. J Exp Med 203(4): 919–993

Jung S, et al. (2002) In vivo depletion of CD11c(+) dendritic cells abrogates priming of CD8(+) T cells by exogenous cell-associated antigens. Immunity 17(2): 211–220

Kaech SM and Ahmed R (2001) Memory CD8+ T cell differentiation: initial antigen encounter triggers a developmental program in naive cells. Nat Immunol 2(5): 415–422

Kaech SM, et al. (2003) Selective expression of the interleukin 7 receptor identifies effector CD8 T cells that give rise to long-lived memory cells. Nat Immunol 4(12): 1191–1198

Kagi D, et al. (1994) CD8+ T cell-mediated protection against an intracellular bacterium by perforin-dependent cytotoxicity. Eur J Immunol 24(12): 3068–3072

Kathariou S, et al. (1987) Tn916-induced mutations in the hemolysin determinant affecting virulence of *Listeria monocytogenes*. J Bacteriol 169(3): 1291–1297

Kaufmann SH (1988) Which T cells are relevant to resistance against *Listeria monocytogenes* infection? Adv Exp Med Biol 239: 135–150

Kaufmann SH, et al. (1988) Cloned *Listeria monocytogenes* specific non-MHC-restricted Lyt-2+ T cells with cytolytic and protective activity. J Immunol 140(9): 3173–3179

Kaufmann SH and Ladel CH (1994a) Application of knockout mice to the experimental analysis of infections with bacteria and protozoa. Trends Microbiol 2(7): 235–242

Kaufmann SH and Ladel CH (1994b) Role of T cell subsets in immunity against intracellular bacteria: experimental infections of knock-out mice with *Listeria monocytogenes* and Mycobacterium bovis BCG. Immunobiology 191(4–5): 509–519

Kerksiek KM, et al. (1999) H2-M3-restricted T cells in bacterial infection: rapid primary but diminished memory responses. J Exp Med 190(2): 195–204

Kerksiek KM, et al. (2001) Variable immunodominance hierarchies for H2-M3-restricted N-formyl peptides following bacterial infection. J Immunol 166(2): 1132–1140

Kerksiek KM, et al. (2003) H2-M3-restricted memory T cells: persistence and activation without expansion. J Immunol 170(4): 1862–1869

Kocks C, et al. (1992) *L. monocytogenes*-induced actin assembly requires the actA gene product, a surface protein. Cell 68(3): 521–531

Kondo S and Sauder DN (1997) Tumor necrosis factor (TNF) receptor type 1 (p55) is a main mediator for TNF-alpha-induced skin inflammation. Eur J Immunol 27(7): 1713–1718

Ku CC, et al. (2000) Control of homeostasis of CD8+ memory T cells by opposing cytokines. Science 288(5466): 675–678

Lacombe MH, et al. (2005) IL-7 receptor expression levels do not identify CD8+ memory T lymphocyte precursors following peptide immunization. J Immunol 175(7): 4400–4407

Ladel CH, et al. (1994) Studies with MHC-deficient knock-out mice reveal impact of both MHC I- and MHC II-dependent T cell responses on *Listeria monocytogenes* infection. J Immunol 153(7): 3116–3122

Lalvani A, et al. (1997) Rapid effector function in CD8+ memory T cells. J Exp Med 186(6): 859–865

Lau LL, et al. (1994) Cytotoxic T-cell memory without antigen. Nature 369(6482): 648–652

Lecuit M, et al. (2001) A transgenic model for listeriosis: role of internalin in crossing the intestinal barrier. Science 292(5522): 1722–1725

Lee BO, et al. (2003) CD40-deficient, influenza-specific CD8 memory T cells develop and function normally in a CD40-sufficient environment. J Exp Med 198(11): 1759–1764

Lenz LL, et al. (1996) Identification of an H2-M3-restricted Listeria epitope: implications for antigen presentation by M3. Immunity 5(1): 63–72

Lety MA, et al. (2001) Identification of a PEST-like motif in listeriolysin O required for phagosomal escape and for virulence in *Listeria monocytogenes*. Mol Microbiol 39(5): 1124–1139

Lieberman J (2003) The ABCs of granule-mediated cytotoxicity: new weapons in the arsenal. Nat Rev Immunol 3(5): 361–370

Ludewig B, et al. (2001) Perforin-independent regulation of dendritic cell homeostasis by CD8(+) T cells in vivo: implications for adaptive immunotherapy. Eur J Immunol 31(6): 1772–1779

Lukacs K and Kurlander RJ (1989) MHC-unrestricted transfer of antilisterial immunity by freshly isolated immune CD8 spleen cells. J Immunol 143(11): 3731–3736

Lukacs NW, et al. (1994) Intercellular adhesion molecule-1 mediates the expression of monocyte-derived MIP-1 alpha during monocyte-endothelial cell interactions. Blood 83(5): 1174–1178

Mackaness GB (1962) Cellular resistance to infection. J Exp Med 116: 381–406

Malherbe L, et al. (2004) Clonal selection of helper T cells is determined by an affinity threshold with no further skewing of TCR binding properties. Immunity 21(5): 669–679

Marzo AL, et al. (2004) Fully functional memory CD8 T cells in the absence of CD4 T cells. J Immunol 173(2): 969–975

McGregor DD, et al. (1970) The short lived small lymphocyte as a mediator of cellular immunity. Nature 228(5274): 855–856

Mercado R, et al. (2000) Early programming of T cell populations responding to bacterial infection. J Immunol 165(12): 6833–6839

Messingham KA, et al. (2003) Deficient anti-listerial immunity in the absence of perforin can be restored by increasing memory CD8+ T cell numbers. J Immunol 171(8): 4254–4262

Miki K and Mackaness GB (1964) The Passive Transfer Of Acquired Resistance To *Listeria monocytogenes*. J Exp Med 120: 93–103

Muraille E, et al. (2005) Distinct in vivo dendritic cell activation by live versus killed *Listeria monocytogenes*. Eur J Immunol 35(5): 1463–1471

Murali-Krishna K, et al. (1999) Persistence of memory CD8 T cells in MHC class I-deficient mice. Science 286(5443): 1377–1381

Nakane A, et al. (1988) Endogenous tumor necrosis factor (cachectin) is essential to host resistance against *Listeria monocytogenes* infection. Infect Immun 56(10): 2563–2569

Nickol AD and Bonventre PF (1977) Anomalous high native resistance to athymic mice to bacterial pathogens. Infect Immun 18(3): 636–645

North RJ (1970) The relative importance of blood monocytes and fixed macrophages to the expression of cell-mediated immunity to infection. J Exp Med 132(3): 521–534

North RJ, et al. (1997) Murine listeriosis as a model of antimicrobial defense. Immunol Rev 158: 27–36

Obst R, et al. (2005) Antigen persistence is required throughout the expansion phase of a CD4(+) T cell response. J Exp Med 201(10): 1555–1565

Opferman JT, et al. (1999) Linear differentiation of cytotoxic effectors into memory T lymphocytes. Science 283(5408): 1745–1748

Pamer E and Cresswell P (1998) Mechanisms of MHC class I-restricted antigen processing. Annu Rev Immunol 16: 323–358

Pamer EG, et al. (1991) Precise prediction of a dominant class I MHC-restricted epitope of *Listeria monocytogenes*. Nature 353(6347): 852–855

Pamer EG, et al. (1992) H-2M3 presents a *Listeria monocytogenes* peptide to cytotoxic T lymphocytes. Cell 70(2): 215–223

Pamer EG (1994) Direct sequence identification and kinetic analysis of an MHC class I-restricted *Listeria monocytogenes* CTL epitope. J Immunol 152(2): 686–694

Pamer EG, et al. (1997) MHC class I antigen processing of *Listeria monocytogenes* proteins: implications for dominant and subdominant CTL responses. Immunol Rev 158: 129–136

Pamer EG (2004) Immune responses to *Listeria monocytogenes*. Nat Rev Immunol 4(10): 812–823

Ploss A, et al. (2003) Promiscuity of MHC class Ib-restricted T cell responses. J Immunol 171(11): 5948–5955

Ploss A, et al. (2005) Distinct regulation of H2-M3-restricted memory T cell responses in lymph node and spleen. J Immunol 175(9): 5998–6005

Pope C, et al. (2001) Organ-specific regulation of the CD8 T cell response to *Listeria monocytogenes* infection. J Immunol 166(5): 3402–3409

Porter BB and Harty JT (2006) The onset of CD8+ T-cell contraction is influenced by the peak of *Listeria monocytogenes* infection and antigen display. Infect Immun 74(3): 1528–1536

Portnoy DA, et al. (1988) Role of hemolysin for the intracellular growth of *Listeria monocytogenes*. J Exp Med 167(4): 1459–1471

Princiotta MF, et al. (1998) H2-M3 restricted presentation of a Listeria-derived leader peptide. J Exp Med 187(10): 1711–1719

Rock KL, et al. (1994) Inhibitors of the proteasome block the degradation of most cell proteins and the generation of peptides presented on MHC class I molecules. Cell 78(5): 761–771

Rogers HW and Unanue ER (1993) Neutrophils are involved in acute, nonspecific resistance to *Listeria monocytogenes* in mice. Infect Immun 61(12): 5090–5096

Rothe J, et al. (1993) Mice lacking the tumour necrosis factor receptor 1 are resistant to TNF-mediated toxicity but highly susceptible to infection by *Listeria monocytogenes*. Nature 364(6440): 798–802

Safley SA, et al. (1991) Role of listeriolysin-O (LLO) in the T lymphocyte response to infection with *Listeria monocytogenes*. Identification of T cell epitopes of LLO. J Immunol 146(10): 3604–3616

Sallusto F, et al. (1999) Two subsets of memory T lymphocytes with distinct homing potentials and effector functions. Nature 401(6754): 708–712

Sanderson S, et al. (1995) Identification of a CD4+ T cell-stimulating antigen of pathogenic bacteria by expression cloning. J Exp Med 182(6): 1751–1757

Savage PA, et al. (1999) A kinetic basis for T cell receptor repertoire selection during an immune response. Immunity 10(4): 485–492

Schiemann M, et al. (2003) Differences in maintenance of CD8+ and CD4+ bacteria-specific effector-memory T cell populations. Eur J Immunol 33(10): 2875–2885

Schlech WF, 3rd (2000) Foodborne listeriosis. Clin Infect Dis 31(3): 770–775

Seaman MS, et al. (2000) MHC Class Ib-restricted CTL provide protection against primary and secondary *Listeria monocytogenes* infection. J Immunol 165(9): 5192–5201

Serbina NV, et al. (2003) Sequential MyD88-independent and -dependent activation of innate immune responses to intracellular bacterial infection. Immunity 19(6): 891–901

Sercarz EE, et al. (1993) Dominance and crypticity of T cell antigenic determinants. Annu Rev Immunol 11: 729–766

Shedlock DJ, et al. (2003) Role of CD4 T cell help and costimulation in CD8 T cell responses during *Listeria monocytogenes* infection. J Immunol 170(4): 2053–2063

Shen H, et al. (1998) Compartmentalization of bacterial antigens: differential effects on priming of CD8 T cells and protective immunity. Cell 92(4): 535–545

Shen H, et al. (1998) *Listeria monocytogenes* as a probe to study cell-mediated immunity. Curr Opin Immunol 10(4): 450–458

Shiloh MU, et al. (1999) Phenotype of mice and macrophages deficient in both phagocyte oxidase and inducible nitric oxide synthase. Immunity 10(1): 29–38

Sijts AJ, et al. (1996) Two *Listeria monocytogenes* CTL epitopes are processed from the same antigen with different efficiencies. J Immunol 156(2): 683–692

Sixl W, et al. (1978) Epidemiologic and serologic study of listeriosis in man and domestic and wild animals in Austria. J Hyg Epidemiol Microbiol Immunol 22(4): 460–469

Skoberne M, et al. (2001) Dynamic antigen presentation patterns of *Listeria monocytogenes*-derived CD8 T cell epitopes in vivo. J Immunol 167(4): 2209–2218

Skoberne M and Geginat G (2002) Efficient in vivo presentation of *Listeria monocytogenes*-derived CD4 and CD8 T cell epitopes in the absence of IFN-gamma. J Immunol 168(4): 1854–1860

Spaner D, et al. (1999) A role for perforin in activation-induced T cell death in vivo: increased expansion of allogeneic perforin-deficient T cells in SCID mice. J Immunol 162(2): 1192–1199

Sun JC and Bevan MJ (2003) Defective CD8 T cell memory following acute infection without CD4 T cell help. Science 300(5617): 339–342

Sun JC, et al. (2004) CD4+ T cells are required for the maintenance, not programming, of memory CD8+ T cells after acute infection. Nat Immunol 5(9): 927–933

Tawab A, et al. (2002) Recombinant lemA without adjuvant induces extensive expansion of H2-M3-restricted CD8 effectors, which can suppress primary listeriosis in mice. Int Immunol 14(2): 225–232

Tvinnereim AR, et al. (2004) Neutrophil involvement in cross-priming CD8+ T cell responses to bacterial antigens. J Immunol 173(3): 1994–2002

Unanue ER (1997a) Studies in listeriosis show the strong symbiosis between the innate cellular system and the T-cell response. Immunol Rev 158: 11–25

Unanue ER (1997b) Inter-relationship among macrophages, natural killer cells and neutrophils in early stages of Listeria resistance. Curr Opin Immunol 9(1): 35–43

Unanue ER (1997c) Why listeriosis? A perspective on cellular immunity to infection. Immunol Rev 158: 5–9

Urdahl KB, et al. (2002) Positive selection of MHC Class Ib-restricted CD8(+) T cells on hematopoietic cells. Nat Immunol 3(8): 772–779

Van Parijs L, et al. (1998) The Fas/Fas ligand pathway and Bcl-2 regulate T cell responses to model self and foreign antigens. Immunity 8(2): 265–274

Vijh S and Pamer EG (1997) Immunodominant and subdominant CTL responses to *Listeria monocytogenes* infection. J Immunol 158(7): 3366–3371

Wherry EJ, et al. (2003) Lineage relationship and protective immunity of memory CD8 T cell subsets. Nat Immunol 4(3): 225–234

White DW and Harty JT (1998) Perforin-deficient CD8+ T cells provide immunity to *Listeria monocytogenes* by a mechanism that is independent of CD95 and IFN-gamma but requires TNF-alpha. J Immunol 160(2): 898–905

White DW, et al. (1999) Perforin-deficient CD8+ T cells: in vivo priming and antigen-specific immunity against *Listeria monocytogenes*. J Immunol 162(2): 980–988

White DW, et al. (2000) Adaptive immunity against *Listeria monocytogenes* in the absence of type I tumor necrosis factor receptor p55. Infect Immun 68(8): 4470–4476

Williams MA and Bevan MJ (2004) Shortening the infectious period does not alter expansion of CD8 T cells but diminishes their capacity to differentiate into memory cells. J Immunol 173(11): 6694–6702

Wong P and Pamer EG (2001) Cutting edge: antigen-independent CD8 T cell proliferation. J Immunol 166(10): 5864–5868

Wong P and Pamer EG (2003) Feedback regulation of pathogen-specific T cell priming. Immunity 18(4): 499–511

Wuenscher MD, et al. (1993) The iap gene of *Listeria monocytogenes* is essential for cell viability, and its gene product, p60, has bacteriolytic activity. J Bacteriol 175(11): 3491–3501

Yang J, et al. (2006) Perforin-dependent elimination of dendritic cells regulates the expansion of antigen-specific CD8+ T cells in vivo. Proc Natl Acad Sci USA 103(1): 147–152

Zammit DJ, et al. (2005) Dendritic cells maximize the memory CD8 T cell response to infection. Immunity 22(5): 561–570

Zenewicz LA, et al. (2002) Nonsecreted bacterial proteins induce recall CD8 T cell responses but do not serve as protective antigens. J Immunol 169(10): 5805–5812

12
Immune Evasion and Modulation by *Listeria monocytogenes*

Lauren A. Zenewicz[1] and Hao Shen[2]

[1]*Section of Immunobiology, Yale University School of Medicine, New Haven, CT 06520, USA*
e-mail: lauren.zenewicz@yale.edu
[2]*Department of Microbiology, University of Pennsylvania School of Medicine, Philadelphia, PA 19104-6076, USA*

Abstract: *Listeria monocytogenes* has long served as a model pathogen for elucidating many functions of the immune response. Both the innate and adaptive immune systems are crucial to the recognition and elimination of this pathogen from the host. However, although *L. monocytogenes* infection induces robust immune responses, the bacterium has evolved mechanisms to evade and modulate the immune response. Its intracellular niche allows it to evade several aspects of the innate and adaptive immune systems. Expression of different virulence factors during infection leads to modulation of other arms of the immune response. In this chapter, we focus on the specific mechanisms *L. monocytogenes* has evolved to evade and modulate the host immune response.

12.1. Introduction: Evasion and Modulation of the Immune Response by Pathogens

As organisms evolved into multicellular life forms and devoted more time, resources, and energy to life, they needed to insure their investment with a system that could differentiate foreign invaders from self. The immune system allows for recognition of pathogens, in turn leading to an evaluation of the pathogen and then the appropriate response to control or eliminate the pathogen. However, pathogens have also been evolving alongside their hosts. In order to improve their odds for survival, pathogens often evade or modulate the immune response to create a more favorable host environment.

Avoiding immune recognition by evading the immune response is the most effective way for the pathogen to survive in a host. Examples of pathogen evasion include latency, antigenic variation, or sequestration in immune-privileged sites. Modulation of the immune response by pathogenic bacteria can be attributed to virulence factors, proteins, or other molecules that bacteria produce in order to complete a successful infection cycle (Hornef et al. 2002). Almost every immune response pathway has been shown to be modulated by one or more bacterial pathogens, including cytokine production, inflammation, macrophage apoptosis, T and B cell activation, and antigen presentation (Brennan and Cookson 2000, Dube et al. 2001, Sansonetti et al. 2000, Ullrich et al. 2000, Yao et al. 1999). Modulating one particular response can have resonating effects on other arms of the immune system. For example, downmodulation of antigen presentation by dendritic cells (DCs) can affect the CD4 T cell response, which may then influence antibody production and class-switching. Evasion and modulation of the immune response are important aspects of microbial pathogenesis that are not yet fully understood in many bacterial infections.

12.2.   Murine Listeriosis as a Model to Study Infection and Immunity

Listeria monocytogenes has been used as a model to study innate and acquired immunity since the early 1960s when Mackaness (1962) demonstrated that cellular immunity was critical for control of infection in mice. In this model, bacteria are intravenously injected into the bloodstream of mice. Within minutes, most bacteria can be found in the spleen and liver where they are quickly internalized by resident macrophages (Conlan 1996). In a sublethal infection, bacteria replicate until their numbers are controlled by activated macrophages. The development of a *Listeria*-specific T cell response is necessary to eliminate the bacteria (Bancroft et al. 1991, Bhardwaj et al. 1998).

Upon infection in the murine spleen, *L. monocytogenes* first localizes within macrophages in the marginal zone between the T cell-rich white pulp and the B cell-rich red pulp (Conlan 1996). These infected cells then migrate into the white pulp region and form the beginning of a focus of infection that expands as neighboring cells become infected by the intercellular spread of bacteria. Neutrophils are quickly recruited to these foci (Mandel and Cheers 1980), and these cells are important for the initial control of bacterial growth through their antimicrobial activities (Czuprynski et al. 1994, Rogers and Unanue 1993). Activated macrophages are then recruited to the site of infection and also aid in bacterial clearance by killing both extracellular and intracellular bacteria. Activated macrophages are also responsible for many of the cytokines induced by *L. monocytogenes* infection. High levels of TNF-α can be found in the serum and spleens of infected mice (Havell 1987, Kratz and Kurlander 1988). *L. monocytogenes* infection also induces high levels of the Th1-polarizing cytokine IL-12 (Hsieh et al. 1993, Tripp et al. 1993). These cytokines, as well as others, are important for the development of a proper immune response to control *L. monocytogenes* infection.

Although macrophages and neutrophils are important for initial control of *L. monocytogenes* infection, T cells are needed for final clearance of bacteria. CD4 and CD8 T cells are important for conferring sterilizing immunity since SCID mice develop a chronic infection (Bancroft et al. 1991, Bhardwaj et al. 1998). While both CD4 and CD8 T cells contribute to protective immunity, in vivo depletion and adoptive transfer studies have clearly demonstrated that memory CD8 T cells are the most effective T cell subset capable of mediating protection (Czuprynski and Brown 1990, Kaufmann et al. 1986).

CD8 T cells mediate antilisterial immunity through two synergistic mechanisms: (1) lysing infected target cells via perforin and granzymes to expose intracellular bacteria for killing by activated macrophages and (2) secreting IFN-γ to activate macrophages (Harty and Badovinac 2002). Complete protective immunity requires infection with live *L. monocytogenes* capable of escaping from phagosomes (Berche et al. 1987). This is in part due to the generation of the CD8 T cell response and a proper memory T cell population (Lauvau et al. 2001). The role of CD4 T cells in controlling *L. monocytogenes* infection is less well understood than that of CD8 T cells. Depletion of CD4 T cells during primary *L. monocytogenes* infection results in diminished granuloma formation (Mielke et al. 1988). *L. monocytogenes* induces a strong Th1 response, and like CD8 T cells, *L. monocytogenes*-specific CD4 T cells also secrete IFN-γ which may aid macrophage activation (Daugelat et al. 1994). (For a complete discussion on the role of adaptive immunity in *L. monocytogenes* infection, see Chap. 11.)

12.3. The Intracellular Niche of *L. monocytogenes* and Evasion of Immune Responses

Listeria monocytogenes can invade a wide range of cell types, including macrophages, hepatocytes, and enterocytes. Bacteria are phagocytosed by macrophages and can enter nonphagocytic cells by receptor-mediated endocytosis by bacterial encoded internalins, which are surface proteins that bind to receptors on host cells and mediate internalization of the bacterium (Cabanes et al. 2002). After entry into macrophages, most *Listeria* will be destroyed as the vacuole they reside within matures into a lysosome. However, about 10% of the bacteria escape the vacuole and enter the cytoplasm. Once in the cytosol, bacteria polymerize actin and can move throughout the cell (Domann et al. 1992, Kocks et al. 1992, Tilney and Portnoy 1989). Eventually, the bacteria propel themselves into neighboring cells through filopodia-like projections, escape from the newly formed double membrane vacuole, and continue this life cycle (Gedde et al. 2000, Smith et al. 1995). (See Chaps 9 and 10 for detailed descriptions of the intracellular life cycle of *L. monocytogenes*.)

The intracellular niche of *L. monocytogenes* allows it to evade some immune responses that otherwise are very effective against bacteria. As detailed below,

L. monocytogenes has evolved to avoid macrophage- and B cell-mediated killing, as well as reducing immune surveillance by residing within nonprofessional antigen presenting cell subsets. *L. monocytogenes* that cannot maintain their intracellular niche are unable to survive and replicate within the host.

12.3.1. Evasion of Macrophage-Mediated Killing

Many bacterial pathogens are rapidly killed after being engulfed by bactericidal macrophages. Once in the phagosome, the phagosome quickly undergoes maturation by fusing with lysosomal compartments. Due to production of reactive oxygen and nitrogen species, as well as pH changes, bacteria are rapidly killed and degraded. However, a select subset of bacterial pathogens has evolved to evade this destruction and survive intracellularly. This can be accomplished by two different means. The first mechanism is by bacterial interference with phagosome maturation so that bacteria can reside within a vacuole. For example, *Legionella* injects proteins into the host cell via a type IV secretion system that interfere with normal cell trafficking, leading to creation of a protective vacuole (Roy 2002). Instead of containing endocytic markers on its membrane, ER and Golgi-derived proteins are present. Within this vacuole, *Legionella* can replicate and survive within the cell until nutrients are limited, and the *Legionella* lyse the cell for transmission to the next host cell. The second mechanism for bacterial evasion of phagosome-mediated killing is for the bacteria to escape from the phagosome before maturation. This is the case for *L. monocytogenes* as well as other cytosolic intracellular pathogens such as *Shigella*. After engulfment by macrophages, approximately 10% of *L. monocytogenes* escape from the phagosome before their destruction (de Chastellier and Berche 1994). This is accomplished by the expression of the virulence factors listeriolysin O (LLO), phosphatidylinositol-specific phospholipase C (PI-PLC), and phosphatidylcholine-preferring phospholipase C (PC-PLC) (see Chap. 9 for a detailed description of how these virulence factors mediate *L. monocytogenes* escape from the primary and secondary vacuoles).

Macrophages are not the only cell type within which *L. monocytogenes* resides. *L. monocytogenes* invasion of nonprofessional antigen presenting cells, such as hepatocytes and enterocytes, also contributes to the bacterium's evasion of the immune response since these cells have limited antigen presentation capacities compared to macrophages. *L. monocytogenes* expresses on its surface several different ligands that, upon binding with receptors on host cells, initiate a signaling cascade that leads to internalization of the bacterium (see Chap. 8 for a detailed description of this process). The best studied ligand–receptor engagement is between internalin and E-cadherin, which is important for tight-junction formation between enterocytes (Cossart et al. 2003). Expression of internalin on the surface of *L. monocytogenes* is necessary for bacterial entry into enterocytes (Gaillard et al. 1991, Mengaud et al. 1996). This interaction is very specific, since human E-cadherin, but not mouse E-cadherin, can serve as a receptor (Lecuit et al. 1999). This mechanism to enter nonphagocytic cells has allowed *L. monocytogenes* to gain a niche where they are not subjected to the harsh antimicrobial activities or to professional antigen presentation of macrophages.

12.3.2. Evasion of Antibody-Mediated Killing

Antibodies, produced by B cells, are one of the two main arms of adaptive immunity. Antibodies contribute to the immune response against bacterial pathogens by (1) neutralization of bacteria and their toxins, (2) opsonization of bacteria which promotes uptake by phagocytic cells, and (3) complement activation which enhances opsonization. Humoral responses against many bacterial pathogens are sufficient for protection against disease. For example, vaccines to bacteria such as *Bordetella pertussis*, the causative agent of whopping cough, or *Clostridium tetani*, which is responsible for tetanus, induce high titers of antibodies and have high efficacies. Unlike these extracellular bacterial pathogens, the intracellular life cycle of *L. monocytogenes* allows it to evade this antibody-mediated protection.

Due to the intracellular nature of *L. monocytogenes* infection, little antibody response is induced during primary *L. monocytogenes* infection. Although some extracellular bacteria can be found during infection, B cells have little opportunity to encounter *L. monocytogenes* or its antigen. The low amounts of antibodies that are induced by infection are completely unable to confer protection during a rechallenge infection with *L. monocytogenes* (Mackaness 1962). Since the majority of the bacteria remain intracellular during infection and spread intercellularly without encountering the extracellular milieu, *L. monocytogenes*-specific antibodies are of limited to no use in controlling bacterial spread.

However, under certain experimental conditions antibodies can affect the course of infection. Although infection itself does not generate high titers of antibodies that are protective, a monoclonal antibody against the pore-forming virulence factor LLO can provide protection by acting intracellularly to neutralize LLO activity (Edelson et al. 1999, Edelson and Unanue 2001). This antibody treatment blocks bacterial escape from the phagosome in macrophages in vitro and in vivo that results in lower bacterial loads. Also, using B cell-deficient mice, natural antibodies in naive animals may play a role in reducing early dissemination of *L. monocytogenes* into vital organs (Ochsenbein et al. 1999). By trapping bacteria and their antigens in secondary lymphoid organs where specific immune responses are initiated, B cells and antibodies can facilitate the generation of the protective T cell response. Lastly, B cells have been shown to be important in the maintenance of memory CD8 T cells generated during *L. monocytogenes* infection (Shen et al. 2003). Thus, B cells and antibodies do play a minor, yet significant, role during *L. monocytogenes* infection by aiding other aspects of the immune response. However, due to the intracellular niche of *L. monocytogenes*, B cells have little impact on the control of infection.

12.4. *Listeria monocytogenes*-Induced Apoptosis Modulates the Immune Response

Apoptosis is a process of programmed cell death to eliminate damaged or unneeded cells. Apoptosis of T cells is a natural occurrence during T cell development and during contraction of a T cell response (Marsden and Strasser 2003). In contrast to necrosis, apoptosis is a process that minimizes inflammation. Macrophages express a phosphatidylserine receptor that allows them to recognize, engulf, and degrade apoptotic cells (Fadok et al. 1992). In the spleen, 48 h after *L. monocytogenes* infection, there are extensive lesions of apoptotic T cells (Merrick et al. 1997). *L. monocytogenes* has several different mechanisms for inducing apoptosis of immune cells thereby eliminating these cells from providing an anti-*Listeria* immune response.

Although intracellular bacteria have never been detected within T cells, during infection bacteria and T cells are often closely together in the foci of infection. There, T cells are potentially exposed to virulence factors secreted by the bacteria, and these factors may have direct effects on T cells. LLO serves functions in infection besides mediating bacterial escape from the phagosome. In vitro LLO treatment of T cells induces their apoptosis, and injection of LLO into the footpads of mice results in apoptotic lymphocytes in popliteal, but not inguinal, lymph nodes (Carrero et al. 2004). LLO may potentially disrupt the mitochondrial membrane, leading to cytochrome *c* release and initiation of apoptotic signaling pathways. In addition, studies from our laboratory have shown that LLO treatment of T cells leads to rapid upregulation of surface expression of FasL on these cells (Zenewicz et al. 2004). This suggests that LLO may not directly induce apoptosis but may mediate apoptosis through targeting other cells. Upregulation of FasL on T cells by *L. monocytogenes* may allow the pathogen to exploit Fas–FasL mediated apoptosis and associated cytokine production to modulate the host immune response and influence the course of infection.

In addition to LLO, PC-PLC also has a role in the induction of FasL upregulation. It has been proposed that cytolysins such as LLO form pores in the host cell membrane, allowing for entry of secreted bacterial proteins such as PC-PLC and PI-PLC into cells (Sibelius et al. 1996, Wadsworth and Goldfine 2002). PC-PLC and PI-PLC then further amplify the calcium signaling by inducing the release of intracellular calcium stores (Wadsworth and Goldfine 1999). Possibly, similar mechanisms induce calcium fluxes in T cells, resulting in degranulation and transport of intracellular FasL to the cell surface (Glass et al. 1996). Consistent with this model, purified LLO is sufficient for induction of FasL surface expression, while PC-PLC is able to induce FasL upregulation only in the presence of LLO (Zenewicz et al. 2004). The synergy between LLO and PC-PLC suggests that the induction of FasL upregulation represents a specific virulence mechanism involving the coordinated action of multiple virulence factors.

Listeria monocytogenes-induced apoptosis is likely to occur through several pathways since there are both caspase-3-dependent and independent roles in LLO-treated T cells (Carrero et al. 2004). In addition to the death receptor FasL, the death receptor TRAIL is also important during *L. monocytogenes* infection. Mice deficient in TRAIL have less cell death and have lower bacterial loads compared to wild-type littermates (Zheng et al. 2004).

Besides inducing apoptosis of T cells, *L. monocytogenes* also has toxic effects on innate immune cell subsets. *L. monocytogenes* infection of DCs induces apoptosis of these cells. This action may reduce this cell population which is very important for eliciting a *L. monocytogenes*-specific T cell response (Guzman et al. 1996, Jung et al. 2002). Also, *L. monocytogenes* infection of neutrophils induces more rapid apoptosis of the short-lived neutrophils compared to untreated cells (Kobayashi et al. 2003).

12.5.   Evasion of CD8 T Cell-Mediated Protective Immunity

The importance of CD8 T cells during *L. monocytogenes* infection is currently attributed to the cytosolic niche of the pathogen, allowing it to evade many aspects of immune surveillance. However, *L. monocytogenes* has also evolved mechanisms to allow it to avoid CD8 T cell-mediated immune responses. The ability of *L. monocytogenes* to spread intercellularly, from cell to cell, without encountering the extracellular milieu allows the bacteria to avoid perforin-mediated killing. The bacteria are also able to take advantage of the precision of the CD8 T cell immune response and avoid killing by nonspecific T cells.

In *L. monocytogenes*-immunized mice challenged with a second *L. monocytogenes* infection, memory CD8 T cells provide the majority of protection allowing for rapid clearance of bacteria from the mice. These cytolytic T cells have two main mechanisms for inducing death of target cells: (1) Fas–FasL interactions and (2) perforin-dependent cytolysis. In perforin-mediated killing, a CD8 T cell recognizes its target cell, resulting in the release of perforin and granzymes from intracellular stores, resulting in pore formation and apoptosis of the target cell (Harty et al. 2000). Mice deficient in perforin are not more susceptible to *L. monocytogenes* infection although they have slightly slower rates of clearance of bacteria from spleens but not livers (Kagi et al. 1994). However, perforin-deficient mice immunized with *L. monocytogenes* have a major defect in their ability to provide protective immunity upon re-challenge with the bacteria (Kagi et al. 1994). This protection defect can be partially overcome by increasing the number of perforin-deficient *L. monocytogenes*-specific T cells, suggesting that other effector mechanisms can moderately compensate for perforin-mediated cytolysis (Messingham et al. 2003).

Studies from our laboratory have shown that the ability of *L. monocytogenes* to evade perforin-mediated immunity is due to the capacity of the bacteria to spread intercellularly. Unlike wild-type *L. monocytogenes*, infection with bacteria deficient in ActA, which cannot spread cell to cell, is quickly controlled in immunized perforin-deficient mice (San Mateo et al. 2002). CD8 T cell cytolysis is critical for protective immunity to *L. monocytogenes* capable of cell-to-cell spread while protective immunity against spread-defective *L. monocytogenes* is largely independent of cytolysis. Thus, intercellular spread of *L. monocytogenes* allows for evasion of perforin-mediated cytolysis.

The intracellular niche of *L. monocytogenes* also allows the bacterium to avoid being killed by nonspecific CD8 T cell responses. Studies from our laboratory have shown that there is minimal killing of bystander bacteria by activated CD8 T cells in a recombinant *L. monocytogenes* system (Jiang et al. 2003). Mice were immunized with lymphocytic choriomeningitis virus (LCMV) to generate memory CD8 T cells and then challenged with a mixture of wild-type *L. monocytogenes* and recombinant *L. monocytogenes* expressing LCMV-derived antigen. Only the LCMV-antigen expressing strain was rapidly cleared by the memory CD8 T cells. The LCMV-immunized mice still had very high levels of wild-type *L. monocytogenes* in their spleens, comparable to levels found in infected unimmunized mice. The intracellular nature of *L. monocytogenes* may allow the bacteria to limit their exposure to the bystander killing by phagocytic cells that are activated by memory CD8 T cells. Thus, *L. monocytogenes* evades nonspecific CD8 T cell cytolysis, due in part to the inherent precision of the immune response.

12.6.   Induction of Type I Interferons by *L. monocytogenes* is Beneficial to the Bacteria

Recent studies have shown that *L. monocytogenes* infection induces type I interferons (McCaffrey et al. 2004, O'Riordan et al. 2002, Stockinger et al. 2004). Interferon-α and -β, which are usually associated with antiviral immune responses, are potent stimulators of antiviral genes, including proapoptotic and antigen presentation genes, as well as downregulating the cell machinery that viruses hijack in order to replicate (Decker et al. 2005). Induction of type I interferons is critical for the immune system to combat many viruses. On the contrary, induction of type I interferons by *L. monocytogenes* appears to be beneficial to the bacteria.

Induction of type I interferons by *L. monocytogenes* requires escape of intracellular bacteria into the cytosol since heat-killed or LLO-deficient bacteria are unable to induce this response (McCaffrey et al. 2004, O'Riordan et al. 2002). In vitro experiments with gene microarrays have shown that *L. monocytogenes* infection of macrophages triggers two distinct, temporally separate waves of gene induction (McCaffrey et al. 2004). The first wave induces genes dependent on NF-κB, which likely occurs through Toll-like receptors (TLRs), and is not dependent on the invasion of cells by live bacteria. However, the later wave of gene induction is dependent on *L. monocytogenes* escaping from the phagosome. These genes include type I interferons, as well as genes associated with interferon-signaling, including Stats and Jaks, and other interferon-dependent genes. The exact mechanism for interferon induction is not known; however, it is independent of TLR2, TLR4, or receptor interacting protein-2 (RIP-2) signaling, but dependent on the transcription factor IFN regulatory factor 3 (IRF3) and the serine–threonine kinase TNFR-associated NF-κB kinase (TANK)-binding kinase 1 (TBK1) (O'Connell et al. 2005, Stockinger et al. 2004). Recent data also indicate that members of the NOD and NALP family play a role in intracellular recognition of *L. monocytogenes* (Kobayashi et al. 2005, Mariathasan et al. 2006).

Type I interferons appear to be important for *L. monocytogenes* infection. Mice that have elevated type I interferon levels as a result of poly I:C treatment prior to *L. monocytogenes* infection have increased bacterial loads compared to untreated mice (O'Connell et al. 2004). Also, in the absence of type I interferon signaling, in mice deficient in the type I interferon receptor (IFNAR1) or IRF3, *L. monocytogenes* cannot reach as high titers in mice (Auerbuch et al. 2004, Carrero et al. 2004, O'Connell et al. 2004). Type I interferons appear to mediate the apoptosis of T cells seen early in infection since mice deficient in type I interferon signaling lack this apoptotic event (Carrero et al. 2004, O'Connell et al. 2004). Early apoptosis of T cells appears to be beneficial for infection due to the induction of expression of antiinflammatory cytokines, such as IL-10, by the phagocytic cells that clear the postapoptotic T cells (Carrero et al. 2006). Thus, induction of type I interferons by *L. monocytogenes* may allow it to modulate the immune response in its favor.

12.7.   Conclusions

Although *L. monocytogenes* is an excellent model pathogen for studying the immune response due to the strong innate and adaptive immune responses it induces, this bacterium has many mechanisms for evading and modulating the immune response. Due to its intracellular niche, *L. monocytogenes* has evolved to limit immune system recognition and responses to infection. Every pathogen is unique. However, some of what has been elucidated about evasion and modulation of the immune response by *L. monocytogenes* can be applied to other pathogens.

A resonating theme from these studies is that there is a balance of power during infection. Most nonpersistent bacterial pathogens, such as *L. monocytogenes*, are not attempting to kill the host but only want to propagate and be transmitted to the next host. By evading and/or modulating the immune response they are in effect exerting control over the host. On the other hand, the immune response needs to control infection so that the pathogen does not cause damage, reducing their fitness. In host–pathogen interactions, the organism with the most control has the power. Understanding this balance of power and how pathogens evade and modulate the immune response to shift the balance in their favor is important for a better appreciation of infection and immunity.

References

Auerbuch V, Brockstedt DG, Meyer-Morse N, O'Riordan M and Portnoy DA (2004). Mice Lacking the Type I Interferon Receptor Are Resistant to *Listeria monocytogenes*. J Exp Med 200:527–533.

Bancroft GJ, Schreiber RD and Unanue ER (1991). Natural immunity: a T-cell-independent pathway of macrophage activation, defined in the SCID mouse. Immunol Rev 124:5–24.

Berche P, Gaillard JL and Sansonetti PJ (1987). Intracellular growth of *Listeria monocytogenes* as a prerequisite for in vivo induction of T cell-mediated immunity. J Immunol 138:2266–2271.

Bhardwaj V, Kanagawa O, Swanson PE and Unanue ER (1998). Chronic *Listeria* infection in SCID mice: requirements for the carrier state and the dual role of T cells in transferring protection or suppression. J Immunol 160:376–384.

Brennan MA and Cookson BT (2000). *Salmonella* induces macrophage death by caspase-1-dependent necrosis. Mol Microbiol 38:31–40.

Cabanes D, Dehoux P, Dussurget O, Frangeul L and Cossart P (2002). Surface proteins and the pathogenic potential of *Listeria monocytogenes*. Trends Microbiol 10:238–245

Carrero JA, Calderon B and Unanue ER (2004). Listeriolysin O from *Listeria monocytogenes* is a lymphocyte apoptogenic molecule. J Immunol 172:4866–4874

Carrero JA, Calderon B and Unanue ER (2004). Type I interferon sensitizes lymphocytes to apoptosis and reduces resistance to *Listeria* infection. J Exp Med 200:535–540

Carrero JA, Calderon B and Unanue ER (2006). Lymphocytes are detrimental during the early innate immune response against *Listeria monocytogenes*. J Exp Med 203:933–940

Conlan JW (1996). Early pathogenesis of *Listeria monocytogenes* infection in the mouse spleen. J Med Microbiol 44:295–302

Cossart P, Pizarro-Cerda J and Lecuit M (2003). Invasion of mammalian cells by *Listeria monocytogenes*: functional mimicry to subvert cellular functions. Trends Cell Biol 13:23–31

Czuprynski CJ and Brown JF (1990). Effects of purified anti-Lyt-2 mAb treatment on murine listeriosis: comparative roles of Lyt-2+ and L3T4+ cells in resistance to primary and secondary infection, delayed-type hypersensitivity and adoptive transfer of resistance. Immunology 71:107–112

Czuprynski CJ, Brown JF, Wagner RD and Steinberg H (1994). Administration of antigranulocyte monoclonal antibody RB6-8C5 prevents expression of acquired resistance to *Listeria monocytogenes* infection in previously immunized mice. Infect Immun 62:5161–5163

Daugelat S, Ladel CH, Schoel B and Kaufmann SH (1994). Antigen-specific T-cell responses during primary and secondary *Listeria monocytogenes* infection. Infect Immun 62:1881–1888

de Chastellier C and Berche P (1994). Fate of *Listeria monocytogenes* in murine macrophages: evidence for simultaneous killing and survival of intracellular bacteria. Infect Immun 62:543–553

Decker T, Muller M and Stockinger S (2005). The yin and yang of type I interferon activity in bacterial infection. Nat Rev Immunol 5:675–687

Domann E, Wehland J, Rohde M, Pistor S, Hartl M, Goebel W, Leimeister-Wachter M, Wuenscher M and Chakraborty T (1992). A novel bacterial virulence gene in *Listeria monocytogenes* required for host cell microfilament interaction with homology to the proline-rich region of vinculin. Embo J 11:1981–1990

Dube PH, Revell PA, Chaplin DD, Lorenz RG and Miller VL (2001). A role for IL-1 alpha in inducing pathologic inflammation during bacterial infection. Proc Natl Acad Sci U S A 98:10880–10885

Edelson BT, Cossart P and Unanue ER (1999). Cutting edge: paradigm revisited: antibody provides resistance to *Listeria* infection. J Immunol 163:4087–4090

Edelson BT and Unanue ER (2001). Intracellular antibody neutralizes *Listeria* growth. Immunity 14:503–512

Fadok VA, Voelker DR, Campbell PA, Cohen JJ, Bratton DL and Henson PM (1992). Exposure of phosphatidylserine on the surface of apoptotic lymphocytes triggers specific recognition and removal by macrophages. J Immunol 148:2207–2216

Gaillard JL, Berche P, Frehel C, Gouin E and Cossart P (1991). Entry of *L. monocytogenes* into cells is mediated by internalin, a repeat protein reminiscent of surface antigens from gram-positive cocci. Cell 65:1127–1141

Gedde MM, Higgins DE, Tilney LG and Portnoy DA (2000). Role of listeriolysin O in cell-to-cell spread of *Listeria monocytogenes*. Infect Immun 68:999–1003.

Glass A, Walsh CM, Lynch DH and Clark WR (1996). Regulation of the Fas lytic pathway in cloned CTL. J Immunol 156:3638–3644

Guzman CA, Domann E, Rohde M, Bruder D, Darji A, Weiss S, Wehland J, Chakraborty T and Timmis KN (1996). Apoptosis of mouse dendritic cells is triggered by listeriolysin, the major virulence determinant of *Listeria monocytogenes*. Mol Microbiol 20:119–126

Harty JT and Badovinac VP (2002). Influence of effector molecules on the CD8(+) T cell response to infection. Curr Opin Immunol 14:360–365

Harty JT, Tvinnereim AR and White DW (2000). CD8+ T cell effector mechanisms in resistance to infection. Annu Rev Immunol 18:275–308

Havell EA (1987). Production of tumor necrosis factor during murine listeriosis. J Immunol 139:4225–4231

Hornef MW, Wick MJ, Rhen M and Normark S (2002). Bacterial strategies for overcoming host innate and adaptive immune responses. Nat Immunol 3:1033–1040

Hsieh CS, Macatonia SE, Tripp CS, Wolf SF, O'Garra A and Murphy KM (1993). Development of Th1 CD4+ T cells through IL-12 produced by *Listeria*-induced macrophages. Science 260:547–549

Jiang J, Zenewicz LA, San Mateo LR, Lau LL and Shen H (2003). Activation of antigen-specific CD8 T cells results in minimal killing of bystander bacteria. J Immunol 171:6032–6038

Jung S, Unutmaz D, Wong P, Sano G, De los Santos K, Sparwasser T, Wu S, Vuthoori S, Ko K, Zavala F, Pamer EG, Littman DR and Lang RA (2002). In vivo depletion of CD11c(+) dendritic cells abrogates priming of CD8(+) T cells by exogenous cell-associated antigens. Immunity 17:211–220

Kagi D, Ledermann B, Burki K, Hengartner H and Zinkernagel RM (1994). CD8+ T cell-mediated protection against an intracellular bacterium by perforin-dependent cytotoxicity. Eur J Immunol 24:3068–3072

Kaufmann SH, Hug E and De Libero G (1986). *Listeria monocytogenes*-reactive T lymphocyte clones with cytolytic activity against infected target cells. J Exp Med 164:363–368

Kobayashi KS, Chamaillard M, Ogura Y, Henegariu O, Inohara N, Nunez G and Flavell RA (2005). Nod2-dependent regulation of innate and adaptive immunity in the intestinal tract. Science 307:731–734

Kobayashi SD, Braughton KR, Whitney AR, Voyich JM, Schwan TG, Musser JM and DeLeo FR (2003). Bacterial pathogens modulate an apoptosis differentiation program in human neutrophils. Proc Natl Acad Sci U S A 100:10948–10953

Kocks C, Gouin E, Tabouret M, Berche P, Ohayon H and Cossart P (1992). *L. monocytogenes*-induced actin assembly requires the *actA* gene product, a surface protein. Cell 68:521–531

Kratz SS and Kurlander RJ (1988). Characterization of the pattern of inflammatory cell influx and cytokine production during the murine host response to *Listeria monocytogenes*. J Immunol 141:598–606

Lauvau G, Vijh S, Kong P, Horng T, Kerksiek K, Serbina N, Tuma RA and Pamer EG (2001). Priming of memory but not effector CD8 T cells by a killed bacterial vaccine. Science 294:1735–1739

Lecuit M, Dramsi S, Gottardi C, Fedor-Chaiken M, Gumbiner B and Cossart P (1999). A single amino acid in E-cadherin responsible for host specificity towards the human pathogen *Listeria monocytogenes*. Embo J 18:3956–3963

Mackaness GB (1962). Cellular resistance to infection. J Exp Med 116:381–406

Mandel TE and Cheers C (1980). Resistance and susceptibility of mice to bacterial infection: histopathology of listeriosis in resistant and susceptible strains. Infect Immun 30:851–861

Mariathasan S, Weiss DS, Newton K, McBride J, O'Rourke K, Roose-Girma M, Lee WP, Weinrauch Y, Monack DM and Dixit VM (2006). Cryopyrin activates the inflammasome in response to toxins and ATP. Nature 440:228–232

Marsden VS and Strasser A (2003). Control of apoptosis in the immune system: Bcl-2, BH3-only proteins and more. Annu Rev Immunol 21:71–105

McCaffrey RL, Fawcett P, O'Riordan M, Lee KD, Havell EA, Brown PO and Portnoy DA (2004). A specific gene expression program triggered by Gram-positive bacteria in the cytosol. Proc Natl Acad Sci U S A 101:11386–11391

Mengaud J, Ohayon H, Gounon P, Mege RM and Cossart P (1996). E-cadherin is the receptor for internalin, a surface protein required for entry of *L. monocytogenes* into epithelial cells. Cell 84:923–932

Merrick JC, Edelson BT, Bhardwaj V, Swanson PE and Unanue ER (1997). Lymphocyte apoptosis during early phase of *Listeria* infection in mice. Am J Pathol 151:785–792

Messingham KA, Badovinac VP and Harty JT (2003). Deficient anti-listerial immunity in the absence of perforin can be restored by increasing memory CD8+ T cell numbers. J Immunol 171:4254–4262

Mielke ME, Ehlers S and Hahn H (1988). T cell subsets in DTH, protection and granuloma formation in primary and secondary *Listeria* infection in mice: superior role of Lyt-2+ cells in acquired immunity. Immunol Lett 19:211–215

O'Connell RM, Saha SK, Vaidya SA, Bruhn KW, Miranda GA, Zarnegar B, Perry AK, Nguyen BO, Lane TF, Taniguchi T, Miller JF and Cheng G (2004). Type I Interferon Production Enhances Susceptibility to *Listeria monocytogenes* Infection. J Exp Med 200:437–445

O'Connell RM, Vaidya SA, Perry AK, Saha SK, Dempsey PW and Cheng G (2005). Immune activation of type I IFNs by *Listeria monocytogenes* occurs independently of TLR4, TLR2, and receptor interacting protein 2 but involves TNFR-associated NF kappa B kinase-binding kinase 1. J Immunol 174:1602–1607

O'Riordan M, Yi CH, Gonzales R, Lee KD and Portnoy DA (2002). Innate recognition of bacteria by a macrophage cytosolic surveillance pathway. Proc Natl Acad Sci U S A 99:13861–13866

Ochsenbein AF, Fehr T, Lutz C, Suter M, Brombacher F, Hengartner H and Zinkernagel RM (1999). Control of early viral and bacterial distribution and disease by natural antibodies. Science 286:2156–2159

Rogers HW and Unanue ER (1993). Neutrophils are involved in acute, nonspecific resistance to *Listeria monocytogenes* in mice. Infect Immun 61:5090–5096.

Roy CR (2002). Exploitation of the endoplasmic reticulum by bacterial pathogens. Trends Microbiol 10:418–424

San Mateo LR, Chua MM, Weiss SR and Shen H (2002). Perforin-mediated CTL cytolysis counteracts direct cell-cell spread of *Listeria monocytogenes*. J Immunol 169:5202–5208

Sansonetti PJ, Phalipon A, Arondel J, Thirumalai K, Banerjee S, Akira S, Takeda K and Zychlinsky A (2000). Caspase-1 activation of IL-1beta and IL-18 are essential for *Shigella flexneri*-induced inflammation. Immunity 12:581–590

Shen H, Whitmire JK, Fan X, Shedlock DJ, Kaech SM and Ahmed R (2003). A specific role for B cells in the generation of CD8 T cell memory by recombinant *Listeria monocytogenes*. J Immunol 170:1443–1451

Sibelius U, Chakraborty T, Krogel B, Wolf J, Rose F, Schmidt R, Wehland J, Seeger W and Grimminger F (1996). The listerial exotoxins listeriolysin and phosphatidylinositol-specific phospholipase C synergize to elicit endothelial cell phosphoinositide metabolism. J Immunol 157:4055–4060

Smith GA, Marquis H, Jones S, Johnston NC, Portnoy DA and Goldfine H (1995). The two distinct phospholipases C of *Listeria monocytogenes* have overlapping roles in escape from a vacuole and cell-to-cell spread. Infect Immun 63:4231–4237

Stockinger S, Reutterer B, Schaljo B, Schellack C, Brunner S, Materna T, Yamamoto M, Akira S, Taniguchi T, Murray PJ, Muller M and Decker T (2004). IFN regulatory factor 3-dependent induction of type I IFNs by intracellular bacteria is mediated by a TLR- and Nod2-independent mechanism. J Immunol 173:7416–7425

Tilney LG and Portnoy DA (1989). Actin filaments and the growth, movement, and spread of the intracellular bacterial parasite, *Listeria monocytogenes*. J Cell Biol 109:1597–1608

Tripp CS, Wolf SF and Unanue ER (1993). Interleukin 12 and tumor necrosis factor alpha are costimulators of interferon gamma production by natural killer cells in severe combined immunodeficiency mice with listeriosis, and interleukin 10 is a physiologic antagonist. Proc Natl Acad Sci U S A 90:3725–3729

Ullrich HJ, Beatty WL and Russell DG (2000). Interaction of *Mycobacterium avium*-containing phagosomes with the antigen presentation pathway. J Immunol 165:6073–6080

Wadsworth SJ and Goldfine H (1999). *Listeria monocytogenes* phospholipase C-dependent calcium signaling modulates bacterial entry into J774 macrophage-like cells. Infect Immun 67:1770–1778

Wadsworth SJ and Goldfine H (2002). Mobilization of protein kinase C in macrophages induced by *Listeria monocytogenes* affects its internalization and escape from the phagosome. Infect Immun 70:4650–4660

Yao T, Mecsas J, Healy JI, Falkow S and Chien Y (1999). Suppression of T and B lymphocyte activation by a *Yersinia pseudotuberculosis* virulence factor, yopH. J Exp Med 190:1343–1350

Zenewicz LA, Skinner JA, Goldfine H and Shen H (2004). *Listeria monocytogenes* virulence proteins induce surface expression of Fas ligand on T lymphocytes. Mol Microbiol 51:1483–1492

Zheng SJ, Jiang J, Shen H and Chen YH (2004). Reduced apoptosis and ameliorated listeriosis in TRAIL-null mice. J Immunol 173:5652–5658

13
Bacteriophages of *Listeria*

Steven Hagens and Martin J. Loessner

Institute of Food Science and Nutrition, ETH Zurich, Switzerland
e-mail: martin.loessner@ilw.agrl.ethz.ch

Abstract: Bacteriophages have been shown to influence the evolution of their host and, in several cases, have a major effect on pathogenicity and/or virulence of bacterial pathogens. Several mechanisms allow phages to change the biology and associated phenotypes of their host. This chapter aims at explaining these mechanisms in the context of *Listeria* evolution and pathogenesis, using examples from other pathogens. Our current knowledge on the biology and applications of a few selected *Listeria* phages is reviewed and discussed. The lack of evidence for their influence on the phenotype of lysogenized host bacteria likely reflects our fragmentary knowledge about *Listeria* phages, especially on the molecular level. Clearly, much more research is required to understand the full impact of phages on their hosts, both in an ecological and evolutionary context.

13.1. Bacteriophages

13.1.1. Introduction

Bacterial viruses—known as bacteriophages or phages—are the most abundant self-replicating genetic elements, with estimates of total numbers ranging as high as 10^{31}. These numbers are based on electron microscopy images taken in attempts to enumerate phages in environmental samples. The first such study done on aquatic phages revealed an average of 10^7 virus-like particles (VLPs) per milliliter with typical head–tail morphology (Bergh et al. 1989), and later studies indicated even substantially higher numbers (Wommack and Colwell 2000). With an estimated number of 10^{24} infections per second in order to maintain the total population calculated above, phages are bound to have an enormous impact on bacterial communities both in evolutionary terms and on a global ecological scale (Wilhelm and Suttle 1999).

Phages are highly varied, both in shape and size of their enveloping capsids and in the composition and complexity of their genetic material. They have in common an absolute dependency on their host for replication, featuring no metabolism of their own. In general, phages exhibit very narrow host ranges, mostly infecting only one particular genus, species, or, in some cases, even specific strains. Recognition of the host by phage receptor-binding proteins is largely but not solely responsible for the observed specificity. *Escherichia coli* phage λ, for example, recognizes the outer-membrane protein LamB of its host, and when this protein was heterologously expressed in various related and unrelated species, successful infection by λ of many, albeit not all, of the modified bacteria could be demonstrated (de Vries et al. 1984).

Phages have been classified based on morphological traits, and the most abundant (96%) and perhaps best-studied of these viruses belong to the order *Caudovirales* (the tailed phages), comprised of large, tailed phages with an isometric capsid containing a dsDNA chromosome. This group is further subdivided into *Myoviridae, Siphoviridae*, and *Podoviridae* characterized by long contractile tails, long flexible tails, or short noncontractile tails, respectively. For more information on bacteriophage classification and different types of phages, readers are referred to the website of the International Committee on Taxonomy of Viruses (ICTV): www.ncbi.nlm.nih.gov/ICTVdb/.

In general, bacteriophages have adopted two different life styles. Infection by a strictly virulent phage invariably results in the production of phage proteins, replication of phage DNA, and ends in lysis of the host and release of progeny phage. In contrast, a complex and not fully understood series of events, involving both environmental factors and the physiological state of the host may result in the fate of the temperate phages after infection. The lytic pathway described above for virulent phages is one option, but the infection process may alternatively result in integration of the phage genome into the host bacterial chromosome. This state, known as lysogeny, where the host is called lysogenic and the phage becomes a prophage and is replicated along with the host chromosome, can be maintained over many generations (Little et al. 1999). Subsequent excision of the phage DNA and entry into the lytic cycle again is a very complex and poorly understood process depending on many factors. It has been demonstrated in several cases that phage induction is mediated by the host cell SOS-response to certain stress stimuli. In these cases, it was shown that the more severe the DNA damage is—and thus the less likely host survival becomes—the proportion of prophage excision and lysis increases (Little and Mount 1982). The recent increases in the number of bacterial genome sequences demonstrated that many (if not all) bacterial genomes also contain cryptic phages, which are prophages that have lost essential functions and are no longer able to excise, replicate, and form infectious particles. Nevertheless, the viral DNA is present in the bacterial genomes and may directly on indirectly affect the phenotype of the host cell.

13.1.2.   Phages and Pathogenicity

The tremendous impact of phages on the pathogenicity of their host has only recently become the focus of detailed research.

Phages can introduce new bacterial DNA into their host after infection. Although regulated, the normal packaging of phage DNA into capsids can in certain cases result in incorporation of host DNA into phage particles. This process is called transduction and occurs in two different forms. In *specialized transduction*, improper excision of phage DNA results in packaging of host genes immediately adjacent to the phage genome into infectious phage particles (Matsushiro 1963). This process is unlikely to have a large impact on the pathogenicity of bacteria which are infected by these particles, since only very small portions of host DNA are transduced, and these portions are always those adjacent to the phage integration sites. In *generalized transduction*, however, random host DNA roughly equal in size to a normal phage genome may be packaged into the empty phage heads. The resulting particles are identical to normal phage except for their information content, and they can perform the first two steps of the infectious process, attachment and injection of DNA (Ikeda and Tomizawa, 1965). A schematic overview of generalized transduction is shown in Figure 13.1. It is conceivable that such a more or less randomly selected stretch of host DNA may also contain virulence factors which can, upon recombination with the chromosome of the infected bacteria, influence the phenotype and pathogenicity of the recipient.

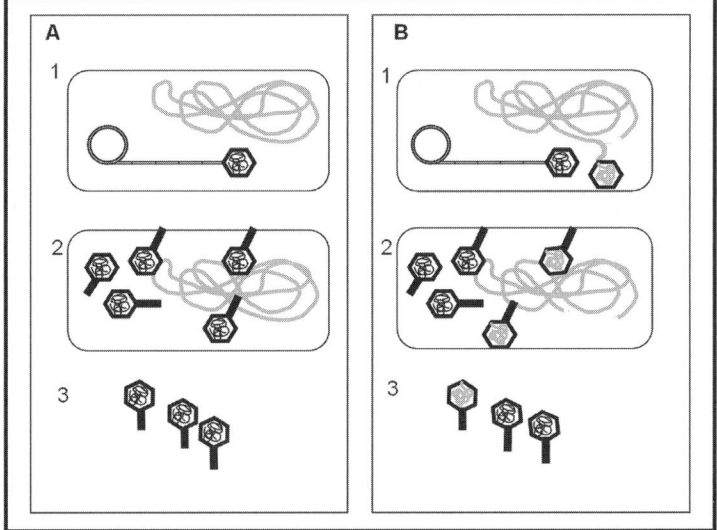

FIGURE 13.1. The principle of generalized transduction is shown here. In *panel A*, normal phage development is depicted. After rolling-circle replication of the phage DNA into multigenome concatemers, the genomes are individually packaged into empty heads (**1**). Phage particles are assembled (**2**), and progeny virions are released to infect new host cell (**3**). In *panel B*, host DNA is packaged into phage head particles (**1**). Particles containing bacterial DNA are normally assembled (**2**), and phage progeny is released. These pseudoinfective particles containing host DNA can inject the mistakenly packaged genetic material into susceptible host bacteria (**3**).

Evidence for this mechanism has been put forward, implicating phages in the distribution of a virulence-associated region in the genome of the animal pathogen *Dichelobacter nodosus* and various other bacteria (Cheetham and Katz 1995).

A phenomenon called lysogenic conversion is often involved in the modulation of host pathogenesis by phage. After incorporation of a temperate phage genome into the host chromosome, most prophage genes are silenced, especially those involved in virus morphogenesis and host cell lysis. In contrast, the genes needed to maintain the lysogenic state are normally expressed during lysogeny. However, bioinformatic analyses have demonstrated that many phage-encoded genes have unknown functions, and it is generally assumed that temperate phages can serve as vectors to introduce novel genetic information into their host that may enhance their fitness in certain environments. These coding sequences may themselves directly specify new properties or act by influencing the expression of existing genes. If this new environment happens to be the human body, the results can be dramatic. Table 13.1. shows several examples of pathogens, their prophages, and the toxins or virulence factors that are encoded by the phages. Such lysogenic conversion has also been reported for *Mycoplasma, Staphylococcus,* and *Streptococcus* (for a comprehensive overview of the state of research concerning these matters, the interested reader is referred to recent overviews on phage-related virulence of pathogens; Waldor et al. 2005).

Interestingly, the transcription promoter for the CTX cholera toxin from *Vibrio cholerae* (Table 13.1.), encoded by genes *ctxA* and *ctxB*, is not phage-regulated but controlled by the master *V. cholerae* virulence regulator, transcription factor ToxR (Skorupski and Taylor 1997). This is in contrast to another well-known example of lysogenic conversion: that of shiga-like toxin (*stx*) converting *E. coli* phages (Table 13.1.), where the transcription of *stx* genes is largely controlled by

TABLE 13.1. Examples of temperate phages encoding pathogenicity and/or virulence factors required for bacterial pathogenesis.

Host species	Phages	Genes	Virulence factor	Reference
Vibrio cholerae	CTXΦ	*ctxA/ctxB*	Cholera toxin CTX	Waldor and Mekalanos (1996)
Escherichia coli (STEC, EHEC)	H19-B 933W	$stx_{1A}/stx_{1B}stx_{2A}/stx_{2B}$	Shiga-like toxins STX1 and STX2	Smith et al. (1983) and O'Brien et al. (1984)
Salmonella enterica	Fels-1 SopEΦ	*nanH* *sopE*	Neuraminidase Type III-translocated G nucleotide exchange factor	Hardt and Galan (1997) and Figueroa-Bossi et al. (2001)
Clostridium botulinum	1D	*botD*	Neurotoxin BoNT	Eklund et al. (1971)
Corynebacterium diphtheriae	β	*tox*	DT	Freeman (1951)

late phage promoters, and the highest levels of STX transcription are observed during lysis of part of the bacterial population (Yee et al. 1993). Virulence modulation after phage induction has recently also been reported for *Staphylococcus aureus* (Goerke et al. 2006). The observations indicate that not only the phage per se, but also the highly specific interaction of both virus and host cell is responsible for full expression of the phenotype.

13.2. *Listeria* Phages

13.2.1. Introduction

The study of phages has provided valuable insights into many genetic principles, such as restriction–modification, the workings of promoters, and the concept of the operon, but at the same time has also contributed to a better understanding of their hosts. Approximately 400 *Listeria* phages have been isolated and described to date. Many of these are specific for *L. monocytogenes*, but phages able to infect *L. ivanovii, L. innocua, L. seeligeri* and *L. welshimeri* have also been described. However, still very little is known about *Listeria* phage biology, especially on the molecular level. (Sword and Pickett 1961; Jasinska 1964; Hamon and Peron 1966; Audurier et al. 1977; Chiron et al. 1977; Ortel 1981; Rocourt et al. 1982; Ortel and Ackermann 1985; Rocourt et al. 1985; Rocourt 1986; Loessner 1991; Gerner-Smidt et al. 1993; Loessner et al. 1994; Hodgson 2000).

All *Listeria* phages isolated thus far belong to the order *Caudovirales* (tailed phages), and of these most belong to the family of *Siphoviridae* (long, flexible noncontractile tail). The few exceptions belong to the family of *Myoviridae* (long, inflexible contractile tail). To date, no *Podoviridae* (short or missing tail) for *Listeria* have been isolated. Many *Listeria* phages have been characterized and, with three exceptions (see below), all of these are presumably temperate. Most isolates were derived from lysogenic strains either after UV or chemical induction or after spontaneous lysis of their hosts. The few environmental isolates mostly stem from silage or sewage plants.

The temperate *Listeria* phages are extremely host-specific, infecting only particular serovar groups. As mentioned above, this appears to be mainly due to the ability to recognize and attach to specific cell-wall ligands (phage receptors), which in the case of *Listeria* are serovar-specific sugar substituents on the polyribitol phosphate teichoic acids (Wendlinger et al. 1996; Tran et al. 1999).

Currently, only three complete nucleotide sequences of *Listeria* phages are available (A118, PSA, P100) (Loessner et al. 2000; Zimmer et al. 2003; Carlton et al. 2005).

13.2.2. The Temperate Phages

The A118 (Figure 13.2.A) and PSA are temperate phages and belong to the *Siphoviridae* family, characterized by dsDNA-containing isometric capsids and long, flexible tails.

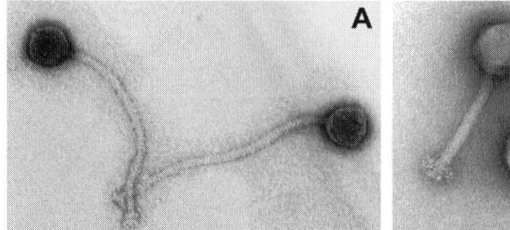

FIGURE 13.2. Electron micrographs of bacteriophages infecting *Listeria*. *Panel A* shows the temperate Siphovirus A118, and *panel B* the broad host range virulent Myovirus A511.

The A118 genome consists of 40,834 bp encoding 72 open reading frames (ORFs). The DNA packaged into phage heads is larger and consists of approximately 43.3 kb, which indicated about 6% redundancy (Loessner et al. 2000). Circular permutation along with terminal redundancy is common among phages and requires the circularization of the phage genome after entry into the host cell. Evidence indicated that after rolling circle replication of A118 phage DNA, sequential packaging starts at a random point of the concatamer, and genomes exceeding the one unit genome size by approximately 2.5 kb of DNA are incorporated into the heads.

Based on sequence similarities, functions could be assigned to 26 of the A118 ORFs. Clustering of genes reflecting different life cycles is common in phages, and especially pronounced in temperate phages. These gene clusters can be divided into those involved in DNA replication, phage morphogenesis, and lysis and those genes necessary for the establishment and maintenance of lysogeny which (as in A118 and PSA) are often oriented in the opposite direction of transcription. Many temperate phages with the same hosts have significant homology over parts of their genomes. Superinfection of lysogenized hosts is common and results in a close encounter of different virus genomes. The theory introducing modular evolution of phage genomes has been postulated more than 25 years ago (Susskind and Botstein 1978) and has since been supported by detailed study of many phages. However, except for a similar arrangement of gene clusters, surprisingly little overall similarity exists between the genomes of A118 (serovar 1/2 host strains) and PSA (Zimmer et al. 2003), the latter of which only infects *L. monocytogenes* serovar 4 strains. This suggests that these phage lineages may have diverged at an earlier stage. Consequently, because phages can only evolve together with their host bacteria, one must assume that the corresponding host strain lineages also diverged relatively early.

While A118 integrates into a region homologous to the *Bacillus subtilis comK* gene which is then disrupted, the integration site of PSA is a t-RNA$_{Arg}$ gene, where the *attB* is functionally complemented by prophage nucleotides. These sites appear to be frequently used by different *Listeria* phages to integrate, and possible ramifications on host pathogenesis will be discussed at a later stage. t-RNA genes are generally known to harbor prophages, in many different bacteria (Campbell 2003).

The 37,618-bp PSA genome is not circularly permuted nor is it terminally redundant, but features 3'overhanging ends of 10 bp. Such cohesive (cos) ends are the alternative to circular permutation and allow direct circularization of the genome after entry into the host, without the requirement for a recombinase. Of the 57 ORFs found in the PSA genome, functional assignments could be made for 33 putative and confirmed gene products based on sequence homologies to known genes. Gene products from all life cycle-specific regions were identified, and, although there is little homology, they appear to serve similar functions as in A118. Two genes differ radically between the two phages. Because of their different substrates, the phage integrases are completely unrelated. Whereas the A118 integrase is a serine recombinase similar to Tn10 resolvase and *Salmonella* Hin invertase, the PSA integrase is homologous to the *E. coli* XerD protein. XerD is involved in resolving multimeric plasmids containing a *xer* site (Alen et al. 1997; Loessner and Calendar 2006). The endolysins responsible for host cell-wall degradation in lysis encoded by the two phages also differ. Endolysins have a two-domain organization, dividing the enzyme into two functional parts. The C-terminal part is responsible for substrate recognition, and binding to the cell wall is serovar-specific. The cell-wall binding domain (CBD) of phage PSA recognizes ligands on cell walls of serovars 4, 5, and 6, and the A118 CBD binds to serovar 1/2 and 3 cell walls. Both CBDs lack known motifs involved in the recognition of cell-wall anchors. The N-terminal, enzymatically active domains (EAD) also differ significantly from each other. The PSA EAD is an *N*-acetylmuramoyl-L-alanine amidase (Zimmer et al. 2003; Korndoerfer et al. 2006), whereas the A118 EAD is L-alanine-D-glutamate endopeptidase (Loessner et al. 1995). Interestingly, the endolysin of A500, another serovar 4b-specific phage, has a CBD which is almost identical to that of PSA, but its EAD is a peptidase related to Ply118. This supports the theory of modular evolution, with genetic exchange between phages. Phage endolysins have been employed to design attenuated suicide *Listeria monocytogenes* strains for delivery of antigen-encoding eukaryotic expression vectors (Dietrich et al. 1998). Phage endolysins and their applications with a focus on *Listeria* phage endolysins have recently been reviewed (Loessner 2005).

13.2.3. *Virulent Phages*

Phage A511 (Figure 13.2.B) and P100 are polyvalent, virulent bacteriophages able to infect most strains of different serovars within the genus *Listeria*; they can multiply on about 95% of *L. monocytogenes* strains of serovars 1/2 and 4. They belong to the family of *Myoviridae* with dsDNA-containing isometric capsid and a contractile tail. Sequencing of the complete P100 genome (131,384 bp) was recently completed, and an in-depth analysis of the predicted ORFs was performed (Carlton et al. 2005). A total of 174 putative gene products and

18 tRNAs were predicted. Twenty-five putative assignments could be made comprising phage structural proteins as well as genes responsible for DNA replication, transcription, and lysis. Homology searches did not reveal significant hits for the remaining ORFs. An in-depth analysis did not reveal any homology to genes, proteins, or other factors known or suspected to be involved in pathogenesis or virulence in microorganisms. This result is not surprising since infection invariably leads to host lysis, and the phage would gain nothing from carrying such genes. Phage P100 is highly similar to phage A511, with a large overall homology and some genes being identical on the nucleotide level. The host range, although overlapping for a large part, is not identical. The A511 genome has recently been sequenced completely (Dorscht et al. submitted), and comparison of the two genomes may help to reveal the molecular basis for the differences in specificity.

P35 is a somewhat unusual virulent phage. The virus is a member of the *Siphoviridae* family and can infect approximately 75% of serovar 1/2 strains (Hodgson 2000; Loessner, unpublished information). Its genome sequence was recently completed (Dorscht et al. submitted) and showed P35 life cycle-specific genes to be organized in the clustering typical for temperate *Listeria* phages. However, no lysogeny control region could be identified. It is possible that the module was initially present, but somehow lost as a result of an illegitimate recombination. However, as a consequence, the resulting phage featured a broad host range, because lack of the immunity region also eliminates repression of phage infection upon infection of host cells carrying prophages with homologous repressor proteins. Clearly, being able to lysogenize hosts has both advantages and disadvantages. In order to prevent extinction, at least one single progeny virulent phage has to find a new host and replicate. In environments with low cell density, this is a disadvantage, and the prophage state appears more attractive. Being so intimately linked to your host cells genetic information may also lead to a very narrow host range, however. Perhaps the scarcity of truly virulent *Listeria* phages and the abundance of temperate phages reflect the ecology of the host bacterium. In fact, although listeriae are ubiquitous, they are rarely found in large numbers.

13.2.4.   Phage-Based Tools for the Study of Listeria

Within the context of *Listeria* pathogenicity, the most important role phages have played to date is their use as a typing tool in epidemiological studies. Phage typing has been instrumental in establishing food as the primary contamination source in humans (Fleming et al. 1985). Several phage typing sets have been established, and the low cost together with the relative ease of use makes it a useful tool to this day (Audurier et al. 1979, 1984; Rocourt et al. 1985; Loessner and Busse 1990; Loessner 1991; Estela and Sofos 1993; Gerner-Smidt et al. 1993; McLauchlin et al. 1996; van der Mee-Marquet et al. 1997). However, typability of *Listeria* isolates is variable. Strains of serovar 4 exhibit the highest degree of phage sensitivity, followed by strains of serovar 1/2. Strains of serovar

3 are largely resistant to phage infection, only few strains are susceptible to broad-range virulent phages such as A511 and P100. This variability in typing limits the use of phage typing as a universal discriminatory method, and additional phages for typing would be desirable (van der Mee-Marquet et al. 1997; Capita et al. 2002).

Genetically modified *Listeria* reporter bacteriophages can be used for confirming the presence of live *Listeria* cells in a sample, especially contaminated food (Loessner et al. 1996, 1997). Transducing phages will later be considered for their possible role in pathogenesis, but within a research context, they have been employed for studying transposon insertion and resulting mutant phenotypes and in strain construction (Freitag 2000). Another phage-based tool which can readily be used for the study of pathogenesis are two integrative *E. coli/Listeria* shuttle vectors, pPL1 and pPL2. A plasmid vector able to replicate in *E. coli* was equipped with genes encoding two different phage integrases which lead to plasmid integration in the chromosome at the respective phage integration sites after introduction into *Listeria*. The researchers were able to demonstrate full reestablishment of pathogenicity in complementation studies with knock-out mutants of *hly* and *actA*, which had not been possible with nonintegrative plasmids. At the same time, integration of the plasmids had no effect on virulence in wild-type strains (Lauer et al. 2002). These plasmids have also been used to study the contribution to virulence of a second *secA* gene *secA*$_2$ found in *L. monocytogenes* (Lenz and Portnoy 2002; Lenz et al. 2003), and further plasmid derivatives were used in studying the contribution to virulence of MogR, a transcriptional repressor required for virulence (Grundling et al. 2004).

13.2.5. *Phage Therapy of* Listeria

Because of the unique intracellular life style of *L. monocytogenes* during infection of the mammalian host, it is impossible to employ a classical phage therapy concept in the treatment of the disease. The same barrier that prevents our immune system from attacking and antibiotics from reaching the bacteria would very likely also keep any phages from interacting with and infecting their bacterial hosts.

Within this context, the possibility of using *Listeria* phage as a preventive measure as food additive in high-risk foods should be mentioned. In the first study for biocontrol of *Listeria* with bacteriophage, the researchers were able to demonstrate a significant reduction of bacterial growth in melon and fruit (Leverentz et al. 2003). A recent study has shown that such a phage therapy approach for food using the virulent phage P100 can eradicate or significantly reduce the growth of *Listeria* in soft cheese, depending on the phage-dose, and may thus be able to reduce the risk of contracting the disease (Carlton et al. 2005). Bioinformatic analyses indicated that the P100 genome does not specify any known pathogenicity or virulence factors, and the encoded polypeptides are not likely to act as food allergens. Moreover, an experimental repeated oral-dose toxicity study in rats showed that P100 had no effects on the health of the treated animals.

Recently, a comprehensive study of the effect of A511 and P100 on *L. monocytogenes* in various foodstuffs treated under different conditions has been completed (Günther et al. submitted). Both phages were highly effective in reducing or eradicating *Listeria* contaminations from almost all food types tested. These promising results appear to be mostly due to the phages used; A511 and P100 are broad host range, strictly virulent phages, and infection invariably leads to host cell lysis.

A phage-based decontamination regimen for surfaces and machines at risk of contamination in food processing may also be considered.

13.2.6. Listeria *Phages and Pathogenesis*

Although most strains, including clinical isolates of *Listeria*, contain prophages—indeed many are polylysogens (Rocourt 1986) containing multiple prophages or cryptic phages—no evident phenotype has yet been associated with the presence of prophages under laboratory conditions nor could a (pro)phage be linked to epidemic strains or outbreaks. However, this situation may reflect our limited understanding of *Listeria* phage genetics and host cell interactions. What seems clear, however, is that the presence of prophages does not attenuate the pathogenic potential or virulence of strains from outbreaks. Phage PSA has been isolated by UV-induction from the genome of the epidemic *L. monocytogenes* strain Scott A (Fleming et al. 1985; Loessner et al. 1994), and the strains isolated from the large Vacherin soft cheese outbreak in Switzerland (Bille 1988) also contain inducible prophages (unpublished information).

Phage A118 integrates itself into a homologue of *B. subtilis comK* gene. In the latter organism, the corresponding gene product represents a global regulator for competence, the complex series of events resulting in the uptake of DNA from the environment. ComK also regulates noncompetence-related genes (van Sinderen et al. 1995). Although both classes of genes are also present in the *Listeria* genomes, they do not seem to be naturally transformable. Whatever the actual function of the gene may be, strains with phage-disrupted *comK* do not show any immediately apparent distinct phenotype under standard laboratory conditions and show no reduced virulence in model systems (Lauer et al. 2002). Again, this might reflect our lack of understanding regarding *Listeria* and biology of its phages. Another possible explanation for the lack of apparent effects of *comK* gene disruption might be the functional replacement by phage-encoded factors.

Within the A118 genome, the putative gene products encoded by ORFs 61 and 66 show a significant similarity to LmaD and LmaC of *L. monocytogenes* (Loessner et al. 2000). Although the function of the four genes in the *lma*DCBA operon of *L. monocytogenes* is not clear, their presence appears to be restricted to the pathogenic strains (Schaferkordt and Chakraborty 1997). Bioinformatic analyses and sequence alignments of A118 gp61 and gp66 also suggested a possible role of these proteins in the regulation of gene expression, which would correlate well with the localization of the genes within the A118 "early genes" cluster, driving DNA recombination and replication during host cell infection.

The possible role of generalized transduction in the development of pathogenesis has already been mentioned. In one study, the ability of some broad host range phages (including A511 and P35) and more than 50 serovar-specific temperate phages (including A118 and PSA) to transduce bacterial DNA from one *L. monocytogenes* host to another was investigated. Donor cells contained antibiotic resistance markers on transposable elements, and recipients were tested for antibiotic resistance after infection with phages propagated on the donors. As expected, none of the broad host range virulent phages were capable of transduction. However, most of the serovar-specific phages transduced marker DNA in various frequencies. The proportion of transducing particles to normal virions ranged from approximately 5×10^{-2} to $10^{-4,}$ depending on the strain and phages tested. Both A118 and P35 proved capable of transduction. Frequencies and amount of DNA transduced were higher in A118, but the broader host range of P35 enables transduction into a wider variety of strains. Interestingly, PSA was not capable of transduction. At this time, the unit genome structure was not known, but the later finding that PSA features cohesive ends and therefore packages its DNA dependent on a terminase recognition site (Zimmer et al. 2003) explained the differences reported by Hodgson (2000) and established a link between genome structure and the ability to mistakenly package host bacterial DNA into empty virus particles.

Close inspection of the LIPI-2 pathogenicity island of *L. ivanovii* revealed that it was localized near the same t-RNA$_{Arg}$ gene that temperate phages may use to integrate, and the authors speculate that its initial introduction into *L. ivanovii* may have been mediated by a temperate phage (Dominguez-Bernal et al. 2006). It has previously been postulated that the organization of the virulence gene cluster *prfA-plcA-hly-mpl-actA-plcB* may have been due to phage transduction (Chakraborty et al. 2000). However, there exists no proof or at least preliminary experimental evidence for any of these events.

It may be concluded that the biological basis for the introduction of new genes from related strains is provided by the transducing bacteriophages and that genes acquired through other means may subsequently be distributed in various populations by phages. While transduction is certainly playing a major role in the intraspecies genetic exchange, its role in the evolution of traits is unclear. However, given that the temperate phages isolated from *L. innocua* are generally able to infect *L. ivanovii* and can then reinfect the nonpathogenic host, the interspecies transfer of virulence genes from this animal pathogen to *L. innocua* is at least theoretically possible.

Phages can shape the genetic composition of bacterial strains in any given population. If prophages manage to enhance the fitness or virulence of a particular pathogenic strain, they may eventually contribute to the occurrence of the disease. Therefore, even with our limited understanding of the processes underlying the host evolution driven by phages, it seems obvious that bacterial pathogenesis can be influenced by bacterial viruses in some way, and further research should be directed to elucidate the precise nature of this influence.

Acknowledgments. We thank Rudi Lurz (Berlin) for providing the electron micrograph of A511, and Julia Dorscht (Munich) and Susanne Guenther (Zurich) for providing unpublished information.

References

Alen C, Sherratt DJ, Colloms SD (1997). Direct interaction of aminopeptidase A with recombination site DNA in Xer site-specific recombination. EMBO J 16:5188–5197.

Audurier A, Rocourt J, Courtieu AL (1977). [Isolation and characterization of "*Listeria monocytogenes*" bacteriophages (author's translation)]. Ann Microbiol (Paris) 128:185–198.

Audurier A, Chatelain R, Chalons F, Piechaud M (1979). [Bacteriophage typing of 823 "*Listeria monocytogenes*" strains isolated in France from 1958 to 1978]. Ann Microbiol (Paris) 130B:179–189.

Audurier A, Taylor AG, Carbonnelle B, McLauchlin J (1984). A phage typing system for *Listeria monocytogenes* and its use in epidemiological studies. Clin Invest Med 7:229–232.

Bergh O, Borsheim KY, Bratbak G, Heldal M (1989). High abundance of viruses found in aquatic environments. Nature 340:467–468.

Bille J (1988). Listeriosis. Schweiz Rundsch Med Prax 77:173–175.

Campbell A (2003). Prophage insertion sites. Res Microbiol 154:277–282.

Capita R, Alonso-Calleja C, Mereghetti L, Moreno B, del Camino Garcia-Fernandez M (2002). Evaluation of the international phage typing set and some experimental phages for typing of *Listeria monocytogenes* from poultry in Spain. J Appl Microbiol 92:90–96.

Carlton RM, Noordman WH, Biswas B, de Meester ED, Loessner MJ (2005). Bacteriophage P100 for control of *Listeria monocytogenes* in foods: genome sequence, bioinformatic analyses, oral toxicity study, and application. Regul Toxicol Pharmacol 43:301–312.

Chakraborty T, Hain T, Domann E (2000). Genome organization and the evolution of the virulence gene locus in *Listeria* species. Int J Med Microbiol 290:167–174.

Cheetham BF, Katz ME (1995). A role for bacteriophages in the evolution and transfer of bacterial virulence determinants. Mol Microbiol 18:201–208.

Chiron JP, Maupas P, Denis F (1977). Ultrastructure of *Listeria monocytogenes* bacteriophages. C R Seances Soc Biol Fil 171:488–491.

de Vries GE, Raymond CK, Ludwig RA (1984). Extension of bacteriophage lambda host range: selection, cloning, and characterization of a constitutive lambda receptor gene. Proc Natl Acad Sci USA 81:6080–6084.

Dietrich G, Bubert A, Gentschev A, Sokolovic Z, Simm A, Catic A, Kaufmann SH, Hess J, Szalay AA, Goebel W (1998). Delivery of antigen-encoding plasmid DNA into the cytosol of macrophages by attenuated suicide *Listeria monocytogenes*. Nat Biotechnol 16:181–185.

Dominguez-Bernal G, Muller-Altrock S, Gonzalez-Zorn B, Scortti M, Herrmann P, Monzo HJ, Lacharme L, Kreft J, Vazquez-Boland JA (2006). A spontaneous genomic deletion in *Listeria ivanovii* identifies LIPI-2, a species-specific pathogenicity island encoding sphingomyelinase and numerous internalins. Mol Microbiol 59:415–432.

Eklund MW, Poysky FT, Reed SM, Smith CA (1971). Bacteriophage and the toxigenicity of *Clostridium botulinum* type C. Science 172:480–482.

Estela LA, Sofos JN (1993). Comparison of conventional and reversed phage typing procedures for identification of *Listeria* spp. Appl Environ Microbiol 59:617–619.

Figueroa-Bossi N, Uzzau S, Maloriol D, Bossi L (2001). Variable assortment of prophages provides a transferable repertoire of pathogenic determinants in *Salmonella*. Mol Microbiol 39:260–271.

Fleming DW, Cochi SL, MacDonald KL, Brondum J, Hayes PS, Plikaytis BD, Holmes MB, Audurier A, Broome CV, Reingold AL (1985). Pasteurized milk as a vehicle of infection in an outbreak of listeriosis. N Engl J Med 312:404–407.

Freeman VJ (1951). Studies on the virulence of bacteriophage-infected strains of *Corynebacterium diphtheriae*. J Bacteriol 61:675–688.

Freitag N (2000). Genetic Tools for Use with *Listeria monocytogenes*, In: Fischetti VA (ed), Gram-Positive Pathogens, vol. 1. ASM Press, Washington, DC, pp. 488–498.

Gerner-Smidt P, Rosdahl VT, Frederiksen W (1993). A new Danish *Listeria monocytogenes* phage typing system. Apmis 101:160–167.

Goerke C, Koller J, Wolz C (2006). Ciprofloxacin and trimethoprim cause phage induction and virulence modulation in *Staphylococcus aureus*. Antimicrob Agents Chemother 50:171–177.

Grundling A, Burrack LS, Bouwer HG, Higgins DE (2004). *Listeria monocytogenes* regulates flagellar motility gene expression through MogR, a transcriptional repressor required for virulence. Proc Natl Acad Sci USA 101:12318–12323.

Hamon Y, Peron Y (1966). On the nature of bacteriocins produced by *Listeria monocytogenes*. C R Acad Sci Hebd Seances Acad Sci D 263:198–200.

Hardt WD, Galan JE (1997). A secreted Salmonella protein with homology to an avirulence determinant of plant pathogenic bacteria. Proc Natl Acad Sci USA 94:9887–9892.

Hodgson DA (2000). Generalized transduction of serotype 1/2 and serotype 4b strains of *Listeria monocytogenes*. Mol Microbiol 35:312–323.

Ikeda H, Tomizawa JI (1965). Transducing fragments in generalized transduction by phage P1. II. Association of DNA and protein in the fragments. J Mol Biol 14:110–119.

Jasinska S (1964). Bacteriophages of lysogenic strains of *Listeria monocytogenes*. Acta Microbiol Pol 13:29–43.

Korndoerfer, IP, Danzer, J, Schmelcher, M, Zimmer, M, Skerra, A, Loessner, MJ(2006). The crystal structure of the bacteriophage PSA endolysin reveals a unique fold responsible for specific recognition of *Listeria* cell walls. J MOl Biol 364:678–689.

Lauer P, Chow MY, Loessner MJ, Portnoy DA, Calendar R (2002). Construction, characterization, and use of two *Listeria monocytogenes* site-specific phage integration vectors. J Bacteriol 184:4177–4186.

Lenz LL, Portnoy DA (2002). Identification of a second *Listeria* secA gene associated with protein secretion and the rough phenotype. Mol Microbiol 45:1043–1056.

Lenz, LL, Mohammadi S, Geissler A, Portnoy DA (2003). SecA2-dependent secretion of autolytic enzymes promotes *Listeria monocytogenes* pathogenesis. Proc Natl Acad Sci USA 100:12432–12437.

Leverentz B, Conway WS, Camp MJ, Janisiewicz WJ, Abuladze T, Yang M, Saftner R, Sulakvelidze A (2003). Biocontrol of *Listeria monocytogenes* on fresh-cut produce by treatment with lytic bacteriophages and a bacteriocin. Appl Environ Microbiol 69:4519–4526.

Little JW, Mount DW (1982). The SOS regulatory system of *Escherichia coli*. Cell 29:11–22.

Little JW, Shepley DP, Wert DW (1999). Robustness of a gene regulatory circuit. Embo J 18:4299–4307.

Loessner MJ (1991). Improved procedure for bacteriophage typing of *Listeria* strains and evaluation of new phages. Appl Environ Microbiol 57:882–884.

Loessner MJ (2005). Bacteriophage endolysins—current state of research and applications. Curr Opin Microbiol 8:480–487.

Loessner MJ, Busse M (1990). Bacteriophage typing of *Listeria* species. Appl Environ Microbiol 56:1912–1918.

Loessner MJ Calendar R (2006). The *Listeria* Bacteriophages. In: Calendar R (ed), The Bacteriophages, vol. 1. Oxford University Press, New York, pp. 593–601.

Loessner MJ, Estela LA, Zink R, Scherer S (1994). Taxonomical classification of 20 newly isolated *Listeria* bacteriophages by electron microscopy and protein analysis. Intervirology 37:31–35.

Loessner MJ, Wendlinger G, Scherer S (1995). Heterogeneous endolysins in *Listeria monocytogenes* bacteriophages: a new class of enzymes and evidence for conserved holin genes within the siphoviral lysis cassettes. Mol Microbiol 16:1231–1241.

Loessner MJ, Rees CE, Stewart GS, Scherer S (1996). Construction of luciferase reporter bacteriophage A511::luxAB for rapid and sensitive detection of viable *Listeria* cells. Appl Environ Microbiol 62:1133–1140.

Loessner MJ, Rudolf M, Scherer S (1997). Evaluation of luciferase reporter bacteriophage A511::luxAB for detection of *Listeria monocytogenes* in contaminated foods. Appl Environ Microbiol 63:2961–2965.

Loessner MJ, Inman IB, Lauer P, Calendar R (2000). Complete nucleotide sequence, molecular analysis and genome structure of bacteriophage A118 of *Listeria monocytogenes*: implications for phage evolution. Mol Microbiol 35:324–340.

Matsushiro A (1963). Specialized transduction of tryptophan markers in *Escherichia coli* K12 by bacteriophage phi-80. Virology 19:475–482.

McLauchlin J, Audurier A, Frommelt A, Gerner-Smidt P, Jacquet C, Loessner MJ, van der Mee-Marquet N, Rocourt J, Shah S, Wilhelms D (1996). WHO study on subtyping *Listeria monocytogenes*: results of phage-typing. Int J Food Microbiol 32:289–299.

O'Brien AD, Newland JW, Miller SF, Holmes RK, Smith HW, Formal SB (1984). Shiga-like toxin-converting phages from *Escherichia coli* strains that cause hemorrhagic colitis or infantile diarrhea. Science 226:694–696.

Ortel S (1981). Lysotyping of *Listeria monocytogenes*. Z Gesamte Hyg 27:837–840.

Ortel S, Ackermann HW (1985). Morphology of new *Listeria* phages. Zentralbl Bakteriol Mikrobiol Hyg [A] 260:423–437.

Rocourt J (1986). Bacteriophages and bacteriocins of the genus *Listeria*. Zentralbl Bakteriol Mikrobiol Hyg [A] 261:12–28.

Rocourt J, Schrettenbrunner A, Seeliger HP (1982). Isolation of bacteriophages from *Listeria monocytogenes* Serovar 5 and *Listeria innocua*. Zentralbl Bakteriol Mikrobiol Hyg [A] 251:505–511.

Rocourt J, Catimel B, Schrettenbrunner A (1985). Isolation of *Listeria seeligeri* and *L. welshimeri* bacteriophages. Lysotyping of *L. monocytogenes*, *L. ivanovii*, *L. innocua*, *L. seeligeri* and *L. welshimeri*. Zentralbl Bakteriol Mikrobiol Hyg [A] 259: 341–350.

Rocourt J, Audurier A, Courtieu AL, Durst J, Ortel S, Schrettenbrunner A, Taylor AG (1985). A multi-centre study on the phage typing of *Listeria monocytogenes*. Zentralbl Bakteriol Mikrobiol Hyg [A] 259:489–497.

Schaferkordt S, Chakraborty T (1997). Identification, cloning, and characterization of the lma operon, whose gene products are unique to *Listeria monocytogenes*. J Bacteriol 179:2707–2716.

Skorupski K, Taylor RK (1997). Control of the ToxR virulence regulon in *Vibrio cholerae* by environmental stimuli. Mol Microbiol 25:1003–1009.

Smith HW, Green P, Parsell Z (1983). Vero cell toxins in *Escherichia coli* and related bacteria: transfer by phage and conjugation and toxic action in laboratory animals, chickens and pigs. J Gen Microbiol 129:3121–3137.

Susskind MM, Botstein D (1978). Molecular genetics of bacteriophage P22. Microbiol Rev 42:385–413.

Sword CP, Pickett MJ (1961). The isolation and characterization of bacteriophages from *Listeria monocytogenes*. J Gen Microbiol 25:241–248.

Tran HL, Fiedler F, Hodgson DA, Kathariou S (1999). Transposon-induced mutations in two loci of *Listeria monocytogenes* serotype 1/2a result in phage resistance and lack of N-acetylglucosamine in the teichoic acid of the cell wall. Appl Environ Microbiol 65:4793–4798.

van der Mee-Marquet N, Loessner MJ, Audurier A (1997). Evaluation of seven experimental phages for inclusion in the international phage set for the epidemiological typing of *Listeria monocytogenes*. Appl Environ Microbiol 63:3374–3377.

van Sinderen D, Luttinger A, Kong L, Dubnau D, Venema G, Hamoen L (1995). comK encodes the competence transcription factor, the key regulatory protein for competence development in *Bacillus subtilis*. Mol Microbiol 15:455–462.

Waldor MK, Mekalanos JJ (1996). Lysogenic conversion by a filamentous phage encoding cholera toxin. Science 272:1910–1914.

Waldor MK, Friedmann DI, Adhya SL (2005). Phages: Their Role in Bacterial Pathogenesis and Biotechnology, ASM Press, Washington, DC.

Wendlinger G, Loessner MJ, Scherer S (1996). Bacteriophage receptors on *Listeria monocytogenes* cells are the N-acetylglucosamine and rhamnose substituents of teichoic acids or the peptidoglycan itself. Microbiology 142 (Pt 4):985–992.

Wilhelm SW, Suttle CA (1999). Viruses and nutrient cycles in the sea. Bioscience 49:781–788.

Wommack KE, Colwell RR (2000). Virioplankton: viruses in aquatic ecosystems. Microbiol Mol Biol Rev 64:69–114.

Yee AJ, De Grandis S, Gyles CL (1993). Mitomycin-induced synthesis of a Shiga-like toxin from enteropathogenic *Escherichia coli* H.I.8. Infect Immun 61:4510–4513.

Zimmer M, Sattelberger E, Inman RB, Calendar R, Loessner MJ (2003). Genome and proteome of *Listeria monocytogenes* phage PSA: an unusual case for programmed +1 translational frameshifting in structural protein synthesis. Mol Microbiol 50:303–317.

Zink R, Loessner MJ, Scherer S (1995). Characterization of cryptic prophages (monocins) in *Listeria* and sequence analysis of a holin/endolysin gene. Microbiology 141 (Pt 10):2577–2584.

Index